Lecture Notes in Physics

Springer
Berlin
Heidelberg
New York
Hong Kong
London
Milan
Paris
Tokyo

Physics and Astronomy ONLINE LIBRARY

http://www.springer.de/phys/

The Editorial Policy for Edited Volumes

The series *Lecture Notes in Physics* (LNP), founded in 1969, reports new developments in physics research and teaching - quickly, informally but with a high degree of quality. Manuscripts to be considered for publication are topical volumes consisting of a limited number of contributions, carefully edited and closely related to each other. Each contribution should contain at least partly original and previously unpublished material, be written in a clear, pedagogical style and aimed at a broader readership, especially graduate students and nonspecialist researchers wishing to familiarize themselves with the topic concerned. For this reason, traditional proceedings cannot be considered for this series though volumes to appear in this series are often based on material presented at conferences, workshops and schools.

Acceptance

A project can only be accepted tentatively for publication, by both the editorial board and the publisher, following thorough examination of the material submitted. The book proposal sent to the publisher should consist at least of a preliminary table of contents outlining the structure of the book together with abstracts of all contributions to be included. Final acceptance is issued by the series editor in charge, in consultation with the publisher, only after receiving the complete manuscript. Final acceptance, possibly requiring minor corrections, usually follows the tentative acceptance unless the final manuscript differs significantly from expectations (project outline). In particular, the series editors are entitled to reject individual contributions if they do not meet the high quality standards of this series. The final manuscript must be ready to print, and should include both an informative introduction and a sufficiently detailed subject index.

Contractual Aspects

Publication in LNP is free of charge. There is no formal contract, no royalties are paid, and no bulk orders are required, although special discounts are offered in this case. The volume editors receive jointly 30 free copies for their personal use and are entitled, as are the contributing authors, to purchase Springer books at a reduced rate. The publisher secures the copyright for each volume. As a rule, no reprints of individual contributions can be supplied.

Manuscript Submission

The manuscript in its final and approved version must be submitted in ready to print form. The corresponding electronic source files are also required for the production process, in particular the online version. Technical assistance in compiling the final manuscript can be provided by the publisher's production editor(s), especially with regard to the publisher's own LaTeX macro package which has been specially designed for this series.

LNP Homepage (springerlink.com)

On the LNP homepage you will find:

–The LNP online archive. It contains the full texts (PDF) of all volumes published since 2000. Abstracts, table of contents and prefaces are accessible free of charge to everyone. Information about the availability of printed volumes can be obtained.

–The subscription information. The online archive is free of charge to all subscribers of the printed volumes.

–The editorial contacts, with respect to both scientific and technical matters.

–The author's / editor's instructions.

M. Falcke D. Malchow (Eds.)

Understanding Calcium Dynamics

Experiments and Theory

Springer

Editors

Martin Falcke
Hahn-Meitner-Institut
Glienicker Str.100
14109 Berlin, Germany

Dieter Malchow
University of Konstanz
Fachbereich Biologie
78457 Konstanz, Germany

Cataloging-in-Publication Data applied for

A catalog record for this book is available from the Library of Congress.

Bibliographic information published by Die Deutsche Bibliothek
Die Deutsche Bibliothek lists this publication in the Deutsche Nationalbibliografie; detailed bibliographic data is available in the Internet at http://dnb.ddb.de

ISSN 0075-8450
ISBN 3-540-40236-5 Springer-Verlag Berlin Heidelberg New York

Springer-Verlag Berlin Heidelberg New York
a member of BertelsmannSpringer Science+Business Media GmbH

springer.de

Printed in Germany

Typesetting: Camera-ready by the authors/editor
Cover design: *design & production*, Heidelberg

Printed on acid-free paper
54/3141/du - 5 4 3 2 1 0

Preface

Intracellular Ca^{2+} is an important second messenger in living cells. Ca^{2+} is required for the heartbeat, muscle contraction, stabilizing of the bones and brain functionality. In fact, Ca^{2+} is essential for eucaryotic life, not only in simple organisms like yeast and cellular slime moulds, but also in higher plants, animals and human. Ca^{2+} activates eggs at fertilization, effects differentiation and development and is essential in central functions like mitosis, ATP synthesis, oxidative phosphorylation, motility and cellular signalling. Another important aspect is secretion of hormones and neurotransmitter release. On the other hand, if Ca^{2+} concentrations rise too high, cell death occurs. This may be a desired outcome as in apoptosis (controlled cell death) or deleteriours if uncontrolled, neuronal cell death impairs brain function.

Ca^{2+} dynamics is the exchange of Ca^{2+} ions between intracellular stores and the cytosol, the interior and exterior of a cell or between cells, as well as the binding of Ca^{2+} to Ca^{2+} binding proteins. The active elements of the exchange processes are channels and pumps in the membranes bounding the intracellular stores and the cell as a whole. Channels have an open and a closed state. They allow for flux of Ca^{2+} down its electro-chemical gradient, when they are open. Pumps transport Ca^{2+} against its electro-chemical gradient incurring consumption of energy. The most important Ca^{2+} binding proteins for the dynamics of Ca^{2+} itself are buffers that are located in Ca^{2+} stores. They bind most of the Ca^{2+} in a concentration dependent manner and thus affect time scales of Ca^{2+} transients. The cytosol contains many different Ca^{2+}-binding proteins. These are activated or inhibited by increasing Ca^{2+} concentrations. A central mediator is calmodulin that regulates many enzymes and channels.

The concentration dependent binding of Ca^{2+} to buffers serves as an indicator of the concentration of free Ca^{2+} concentration in intracellular measurements. Certain buffers change their fluorescence wavelength upon binding of a calcium ion. The intensity of fluorescence on that wavelength provides information on the concentration of buffer bound Ca^{2+} and hence on the concentration of free Ca^{2+} as well.

This book addresses students as well as established scientists, biologists, physicists and mathematicians. It provides an introduction to experimental and theoretical results and methods on selected topics of Ca^{2+} regulation. At the same time, it reports the ideas currently discussed, since the authors are experts in their field and are actively involved in ongoing research.

This volume of "Lecture Notes in Physics" appears in a physics series. That indicates the growing interest of physicists in the biological application of their discipline. However, there are reasons special to Ca^{2+} dynamics, why physicists should pay attention this field. The complex temporal and spatial structures created by the concentration patterns stimulated the interest of theorists especially in the last two decades. In mathematical and physical terms, intracellular Ca^{2+} dynamics is a nonlinear system operating far from thermodynamic equilibrium. The local event is excitable, oscillatory or bistable. Excitable dynamics responds to small perturbations of its stationary state with a relaxation to that state but amplifies perturbations larger than a critical threshold. Oscillatory dynamics exhibits sustained oscillations of the concentration and bistability means that there are two stable stationary states with different concentration levels. As is typical of these three dynamic regimes, propagating concentration changes of intracellular calcium – called waves – are observed experimentally. The theory of intracellular Ca^{2+} dynamics is today one of the most developed branches of theoretical biology and areas of contribution to research in physics on complex oscillations and chaotic behaviour. The theoretical physics of spatio-temporal structure formation profited from results on wave instabilities (spiral breakup) and bifurcations (dispersion gap, T-point), waves in discrete media or with concentration dependent diffusion.

One of the fascinating aspects of intracellular Ca^{2+} regulation is the ability to observe how global events build up from elemental local events such as puffs and sparks. Small localized concentration changes cooperate to form oscillations and waves. Global and elemental events can be observed in the same experiment. Moreover, the elemental events are stochastic and occur at spatially discrete locations. Altogether, intracellular Ca^{2+} dynamics may turn out to be a random medium showing spatio-temporal structure formation. Since the most important parameters are experimentally accessible, it provides a unique opportunity to study global pattern formation starting at the elemental event.

Mathematical descriptions of intracellular Ca^{2+} dynamics must start from the biological ideas on morphology, function and control of channels, pumps and buffers and link it to the diffusion processes in the cell. Hence, we present in the first chapters of the book the results of biological experimental research and subsequently a mathematical description and analysis.

A ubiquitous channel on the membrane of the endoplasmic reticulum is the Inositol 1,4,5-trisphosphate receptor channel (IP_3R). It controls release of Ca^{2+} from the endoplasmic reticulum, which is one of the intracellular Ca^{2+} stores. Its function and control are introduced in Chap. 1. The endoplasmic reticulum is the subject of Chap. 2. Besides its role in Ca^{2+} storage and release, the endoplasmic reticulum is the central site of synthesis and folding of membrane and secretory proteins. Chapter 3 considers intracellular Ca^{2+} fluxes including the uptake and release of mitochondria, which can accumulate massive amounts of Ca^{2+}. Mitochondrial Ca^{2+} stimulates respiration and ATP synthesis. Raised intramitochondrial Ca^{2+} thus supports energy-requiring events. The regulation of development is one of the multiple functions of intracellular Ca^{2+} transients as demonstrated by the example of neuronal differentiation in Chap. 4. Chapter 5 is about mobilization of Ca^{2+} from intracellular stores by a mechanism distinct from those regulated by Inositol 1,4,5-trisphosphate. Muscle cells store Ca^{2+} - among other compartments - in the sarcoplasmic reticulum. The ryanodine receptor channel (RyR) is mostly found on the membrane of this storage compartment. It is introduced in Chap. 6. Calcium waves not only occur inside cells, but coordinate the response of groups of cells on the tissue level, too. Chapter 7 deals with these intercellular waves.

The theoretical part of this book consists of Chaps. 8-13. The experimental findings addressed by the theory of a chapter are summarized again in each case.

Small cells usually show a globally synchronized behaviour of the intracellular Ca^{2+} concentration. Hence, spatial aspects of the dynamics can be neglected and the cell can be modelled by equations considering temporal changes only (ordinary differential equations). Chapter 8 gives a summary of the mathematical description and analysis of these Ca^{2+} oscillations and their regulation. Subsequently, Chap. 9 depicts how the signal encoded in oscillations is read by the addressee (target) in the cell. Decoding Ca^{2+} signalling is a major question in cell biology and in neurobiology and the answer to this question is only partly known.

Waves of high cytosolic Ca^{2+} concentration travelling through the cell are observed in many cells but as a whole in cells larger than the wavelength only. Intercellular communication can be mediated by waves travelling from cell to cell. The modelling of both wave types is the subject of Chap. 10. The necessity for application of advanced theoretical tools in the analysis of fluorescence measurements is the topic of Chap. 11. Chapters 12 and 13 deal with stochastic aspects of intracellular Ca^{2+} regulation. Stochastic effects are to be expected, if only a few elements with random behaviour form a functional unit of a system. Such functional units are small groups of channels on the membrane of the endoplasmic reticulum (clusters). These channel groups can generate random local release events the modelling of which is demonstrated in Chap. 12. Spatial coupling by Ca^{2+} diffusion is to weak in a wide range of parameters to provide for spatial averaging as necessary to observed deterministic dynamics. Therefore, global events like waves and oscillations may still be caused by random local events. Chapter 13 illustrates that by focusing on the theoretical description of the transition from local to global events.

The Ca^{2+} field will still expand. The reason is that so many players take part in Ca^{2+} regulation. We are only beginning to know some Ca^{2+} entry channels as well as their regulation. The gene of the store-operated Ca^{2+} channel has not yet been identified. It is unknown, how the filling state of a Ca^{2+} store is transmitted to regulate Ca^{2+} entry. Different kinases and phosphatases may modify Ca^{2+} release channels. The role of the many different small Ca^{2+} binding proteins is obscure as is how they might come into play. The function of fatty acids in Ca^{2+} regulation is only beginning to be elucidated. Moreover, it seems that each cell type is handling Ca^{2+} somewhat differently. Nevertheless, the principal players in Ca^{2+} dynamics are known and therefore meaningful theories and models can be formulated. We hope that the reader will be inspired to do so for his own field of work.

Martin Falcke
Dieter Malchow

Contents

1. Regulation of IP_3 Receptors by IP_3 and Ca^{2+}

Colin W. Taylor and Jane E. Swatton

Department of Pharmacology, University of Cambridge, Tennis Court Road, Cambridge, CB2 1PD U.K.

Abstract. Inositol 1,4,5-trisphosphate (IP_3) receptors are intracellular Ca^{2+} channels that mediate release of Ca^{2+} from intracellular stores. The channels are oligomeric assemblies of four subunits, each of which has an N-terminal IP_3-binding domain and each of which contributes to formation of the Ca^{2+} channel. In mammals, three different genes encode IP_3 receptors subunits and the type 1 receptor (and perhaps the type 2 receptor) is also expressed as splice variants. Further diversity arises from assembly of the receptor in hetero- and homo-tetrameric channels. The subtypes differ in their expression and regulation, but they share the key property of being regulated by both IP_3 and cytosolic Ca^{2+}. All three mammalian IP_3 subtypes, and probably also the IP_3 receptors expressed in invertebrates, are biphasically regulated by cytosolic Ca^{2+}, although the underlying mechanisms appear to differ between subtypes. The interactions between IP_3 and Ca^{2+} in controlling IP_3 receptor gating, and the physiological significance of such regulation will be reviewed.

1.1. An Introduction to IP_3 Receptors

Inositol 1,4,5-trisphosphate (IP_3) receptors are intracellular Ca^{2+} channels that mediate the release of Ca^{2+} from intracellular stores in response to the many physiological stimuli that stimulate IP_3 formation (Berridge et al., 2000). Recent evidence suggests that IP_3 receptors may also be intimately involved in controlling the Ca^{2+} channels in the plasma membrane that are activated when intracellular Ca^{2+} stores are emptied, the so-called "capacitative Ca^{2+} entry channels" (Zhang et al., 2001). In mammals and birds, three IP_3 receptor genes encode proteins with closely related primary sequences and these are assembled into both homo- and hetero-tetrameric structures to form functional channels (Taylor et al., 1999). Alternative splicing of type 1 and probably type 2 IP_3 receptors contributes further to the diversity of IP_3 receptors. Despite reports of modest differences in the properties of the IP_3 receptor subtypes and clear evidence for their differential expression (Patel et al., 1999; Taylor et al., 1999), the full functional significance of IP_3 receptor diversity remains unclear. Each receptor subunit has an IP_3-binding site made up of two distinct domains lying towards the N-terminal of the protein. Near the carboxy terminal there are six membrane-spanning regions, the last two of which and the intervening loop, together with the same regions from the other subunits, line the pore of the channel. Although these key regions, the IP_3-binding site and the pore, lie at opposite ends of the primary sequence of this very large protein, the two domains appear to be closely associated in the functional channel. In this way, the large conformational change evoked by binding of IP_3 to the N-terminal of one subunit (Mignery and Südhof, 1990) may be directly communicated to the C-terminal pore region of a neighbouring subunit (Boehning and Joseph, 2000a).

IP_3 is not the only regulator of IP_3 receptors. ATP, redox potential, pH, phosphorylation and many accessory proteins (Mackrill, 1999; Patel et al., 1999) modulate IP_3 receptor behaviour, but the most important of these additional regulators is cytosolic Ca^{2+}. Indeed all Ca^{2+} channels, whether in the plasma membrane or in the membranes of intracellular stores, are regulated by cytosolic Ca^{2+} and such regulation is invariably important in allowing feedback control of channel activity. Regulation of IP_3 receptors and of their relatives, the ryanodine receptors, by cytosolic Ca^{2+} is important in co-ordinating the activities of different intracellular Ca^{2+} release processes (see chapter by Galione and Churchill). Furthermore, most mathematical models of the spatial and temporal organisation of intracellular Ca^{2+} signals depend heavily on Ca^{2+} regulation of the gating of IP_3 or ryanodine receptors (see Chap. 8, 10 and 13).

1.2. Biphasic Regulation of IP_3-Evoked Ca^{2+} Mobilization by Cytosolic Ca^{2+}

A few years after IP_3 was first shown to release Ca^{2+} from intracellular stores (Berridge et al., 2000; Streb et al., 1983), Masamitsu Iino showed that when permeabilized smooth muscle cells were pre-treated with cytosol-like media containing different free Ca^{2+} concentrations, the ability of a submaximal concentration of IP_3 to cause Ca^{2+} release was biphasically dependent on the cytosolic $[Ca^{2+}]$ (Iino, 1990). The response to IP_3 was optimal when the free $[Ca^{2+}]$ was about 300nM, but substantially inhibited at lower or higher $[Ca^{2+}]$. Many subsequent studies using essentially similar methods have confirmed that reversible biphasic regulation of IP_3 receptors by cytosolic Ca^{2+} is a common feature shared by each of the three mammalian IP_3 receptor subtypes and the IP_3 receptors of other species (Table 1). Results from hepatocytes, where different bivalent cations differ in their relative abilities to stimulate and inhibit IP_3-evoked Ca^{2+} release (Marshall and Taylor, 1994), suggest that two distinct Ca^{2+}-binding sites mediate the stimulatory and inhibitory effects of cytosolic Ca^{2+}.

There are some reports of Ca^{2+} failing either to stimulate (Combettes et al., 1994) or inhibit (Horne and Meyer, 1995) IP_3-evoked Ca^{2+} release, but the greatest source of controversy has been concerned with whether type 3 IP_3 receptors are inhibited by cytosolic Ca^{2+}. Recordings from the IP_3 receptors of RIN cells (which express largely type 3 IP_3 receptors) (Hagar et al., 1998) after reconstitution into planar lipid bilayers, and Ca^{2+} efflux measurements of permeabilized DT40 cells expressing only type 3 receptors (Miyakawa et al., 1999) have suggested that type 3 receptors are stimulated, but not inhibited, by cytosolic Ca^{2+}. The latter results are perhaps more persuasive because bilayer recordings of type 2 IP_3 receptors also failed to detect Ca^{2+} inhibition (Ramos-Franco et al., 1998b) despite general acceptance that *in situ* this subtype is biphasically regulated by cytosolic Ca^{2+} (Table 1). The lack of Ca^{2+} inhibition observed in these bilayer recordings (Hagar et al., 1998; Ramos-Franco et al., 1998b) may reflect the loss of an essential accessory protein required for Ca^{2+} inhibition (see Sec. 5). Nevertheless, the substantial evidence that Ca^{2+} biphasically regulates IP_3-evoked Ca^{2+} release in cells expressing predominantly type 3 IP_3 receptors (Table 1) must be tempered with the possibility that because these cells also express lower levels of other receptor subtypes, their inclusion in heterotetrameric assemblies may confer the Ca^{2+} sensitivity. That concern is substantially dismissed by the demonstration that homo-tetrameric type 3 receptors heterologously expressed in the nuclear envelope of *Xenopus* oocytes are also biphasically regulated by cytosolic Ca^{2+} (Mak et al., 2000).

Table 1. Biphasic regulation of IP_3-evoked Ca^{2+} release by cytosolic Ca^{2+}. In each of the cell types shown, cytosolic Ca^{2+} biphasically regulates IP_3-evoked Ca^{2+} release. The results with avian DT40 cells, which normally express all three receptor subtypes, were obtained from cells in which the genes encoding the indicated pair of subtypes had been deleted by homologous recombination.

Cell	Reference
Type 1 IP_3 receptors	
Cerebellum	Sugawara et al., 1997
	Bezprozvanny et al., 1997; Finch et al., 1991; Kaftan et al., 1997
A7r5 cells	Missiaen et al., 1998; Sienaert et al., 1997
SH-SY5Y cells	Swatton et al., 1999
DT40 cells ($R3^-$ and $R2^-$)	Miyakawa et al., 1999
Expression in COS cells	Boehning and Joseph, 2000b
Xenopus oocytes	Mak et al., 1998; Parys et al., 1992
Type 2 IP_3 receptors	
Hepatocytes	Dufour et al., 1997; Marshall and Taylor, 1993
Basophilic leukaemia cells	Oancea and Meyer, 1996
DT40 cells ($R1^-$ and $R3^-$)	Miyakawa et al., 1999
Type 3 IP_3 receptors	
16HBE14o-bronchial cells	Missiaen et al., 1998
Expression in *Xenopus* oocytes	Mak et al., 2000; Mak et al., 2001
Express in COS cells	Boehning and Joseph, 2000b
3T6 fibroblasts	Zhang et al., 1993
RINm5F cells	Swatton et al., 1999
Insect IP_3 receptor	
Drosophila and honeybee	Baumann and Walz, 1989; Swatton et al., 2001

We suggest, therefore, that cytosolic Ca^{2+} is capable of both positively and negatively modulating all IP_3 receptor subtypes. It is far from clear, however, that the underlying mechanisms are the same for all receptor subtypes (Sects. 1.3 and 1.5). It is also important to emphasise that IP_3 is *required* for channel opening: Ca^{2+} modulates the ability of IP_3 to activate IP_3 receptors, but it is not alone capable of causing them to open. Very recently, however, members of a family of neuronal Ca^{2+}-binding proteins that bind to the IP_3-binding domain of IP_3 receptors in the presence of Ca^{2+} have been shown to cause channel opening in the absence of IP_3 (Yang et al., 2002) raising the possibility that there may be physiological means of activating IP_3 receptors in the absence of IP_3.

1.3. Effects of Ca^{2+} on IP_3 Binding to its Receptors

Because IP_3 must bind to its receptor for the channel to open and because ^3H-IP_3 binding assays with purified receptors or fragments of them are much easier than functional assays, there have been many studies of the effects of Ca^{2+} on IP_3 binding. Such studies may be useful for revealing differences between receptor subtypes and then, by mutagenesis, identifying sites through which Ca^{2+} exerts its effects, but it is not easy to relate the results to the functional effects of Ca^{2+} on IP_3 receptor gating. An increase in IP_3 receptor affinity, for example, may reflect binding to an inactivated state so that even though more receptors may have bound IP_3, fewer may be active. Another problem is that IP_3 binding assays are often performed in media with high pH and low ionic strength, further diminishing the opportunities for direct comparisons with functional assays.

There is a general consensus that binding of IP_3 to type 1 receptors, including the recombinant receptor expressed in Sf9 cells (Cardy et al., 1997; Yoneshima et al., 1997) is reversibly inhibited by Ca^{2+} (Benevolensky et al., 1994; Lin et al., 2000; Picard et al., 1998; Supattapone et al., 1988; Worley et al., 1987). The inhibition does not require phosphorylation, and several studies have concluded that the inhibition is lost after purification of the receptor (Benevolensky et al., 1994; Danoff et al., 1988; Richardson and Taylor, 1993). The first candidate for the accessory protein proposed to mediate the Ca^{2+} inhibition was "calmedin", a membrane-associated protein whose estimated molecular mass has ranged between 15 and 300kDa (Danoff et al., 1988; Ferris and Snyder, 1992). More than a decade on, the idea that calmedin is the Ca^{2+}-sensor has failed to gain further support (Picard et al., 1998). Nor is Ca^{2+} inhibition of IP_3 binding likely to be mediated by calmodulin because although calmodulin does inhibit IP_3 binding to type 1 receptors, the effect is entirely Ca^{2+}-independent (Patel et al., 1997; Sienaert et al., 2002). We suspect that "calmedin" may have been calmodulin associated with endogenous calmodulin-binding proteins and that the Ca^{2+}-dependence of its effect on IP_3 receptor may simply reflect the Ca^{2+}-regulated redistribution of calmodulin from these proteins (Patel et al., 1997). This explanation could also account for "calmedin" causing Ca^{2+}-independent inhibition of IP_3 binding after storage (Danoff et al., 1988; Ferris and Snyder,

1992; Picard et al., 1998) and the very different estimates of its size in successive studies (Danoff et al., 1988; Ferris and Snyder, 1992). Another possibility is that the neuronal Ca^{2+}-binding proteins recently shown to bind to the N-terminal of IP_3 receptors in a Ca^{2+}-dependent fashion (Yang et al., 2002) may also displace IP_3 from its binding site. Despite rather clear evidence that IP_3 binding to pure receptors is not inhibited by Ca^{2+}, IP_3 binding to N-terminal fragments of the type 1 IP_3 receptor can be partially inhibited by Ca^{2+} binding to a site that lies within the first 225 residues (Sienaert et al., 2002). This enigmatic observation may perhaps be explained if this Ca^{2+}-binding site is inaccessible in the full-length protein.

The results with types 2 and 3 IP_3 receptors are even less clear. IP_3 binding to type 2 receptors in liver (Marshall and Taylor, 1994; Pietri et al., 1990) is stimulated by Ca^{2+}, an effect that may be mediated by a peripheral membrane protein (Picard et al., 1998). By contrast, in basophilic leukaemia cells, which also express predominantly type 2 IP_3 receptors, IP_3 binding is only very modestly stimulated by Ca^{2+} and then profoundly inhibited when the $[Ca^{2+}]$ is further increased (Mohr et al., 1993). Ca^{2+} has been reported to stimulate (Yoneshima et al., 1997), biphasically regulate (Cardy et al., 1997) or to have no effect on IP_3 binding to type 3 IP_3 receptors (Swatton and Taylor, 2002).

Are there any general lessons to be taken from the effects of Ca^{2+} on IP_3 binding? First, the variability between studies. The effects of Ca^{2+} are invariably reversible and must therefore result from Ca^{2+} shifting the equilibrium between conformations of the IP_3 receptor with high and low affinity for IP_3. Whether the lowest affinity sites will be detected experimentally using ^3H-IP_3 binding will, of course, depend upon their affinity, how many low-affinity sites there are, and the concentration of ^3H-IP_3 used. Many experiments may fail to detect heterogeneity within the population of IP_3-binding sites because they fail to detect low-affinity sites. The effect of Ca^{2+} is then seen only through the high-affinity IP_3-binding sites: a Ca^{2+}-regulated switch from a low- to a high-affinity conformation then creates the impression of an increase in the total number of IP_3-binding sites (B_{max}). The important point is that our ability to resolve different affinity states of the IP_3 receptor depends critically upon the relative affinities of the different states, the abundance of each, and the concentration of ^3H-IP_3 used relative to the affinity of the states. These parameters vary widely between different experiments and probably contribute to some of the confusion. Second, the effects of Ca^{2+} on IP_3 binding are plainly different for the different receptor subtypes, even when they are expressed in exactly the same cellular context (Cardy et al., 1997; Yoneshima et al., 1997): Ca^{2+} inhibits IP_3 binding to type 1 receptors, but stimulates binding to types 2 (and perhaps type 3) receptors. Finally, an accessory protein, the identity of which remains a mystery, is almost certainly responsible for Ca^{2+}-inhibition of IP_3 binding to type 1 receptors, but whether accesssory proteins also mediate Ca^{2+}-stimulation of IP_3 binding to the other subtypes is unresolved. The involvement of accessory proteins may go some way towards explaining some of the conflicting results.

1.4. Activation of IP_3 Receptors by Cytosolic Ca^{2+}

There is universal agreement that appropriate concentrations of Ca^{2+} increase the ability of IP_3 to stimulate channel opening, whether the measurements are performed in intact (Parker et al., 1996a) or permeabilized (Taylor, 1998) cells, on microsomes (Finch et al., 1991; Horne and Meyer, 1995), by patch-clamp recording of the nuclear envelope (Mak et al., 1998) or after reconstitution of partially purified (Bezprozvanny et al., 1991; Ramos-Franco et al., 1998a) or purified receptors (Callamaras and Parker, 1994; Michikawa et al., 1999) into lipid bilayers. The agreement extends also to each IP_3 receptor subtype. Because Ca^{2+} stimulation of IP_3 receptors is so robustly preserved in so many experimental settings and purified IP_3 receptors are also stimulated by Ca^{2+}, it seems unlikely that accessory proteins are required.

Although stimulation of IP_3-evoked Ca^{2+} release by Ca^{2+} had been demonstrated by Iino in 1987 (Iino, 1987), the first suggestion that binding of both Ca^{2+} and IP_3 might be *essential* for channel gating was provided by Finch *et al.* in 1991 (Finch et al., 1991). By rapid superfusion of brain microsomes loaded with $^{45}Ca^{2+}$, they demonstrated that cytosolic Ca^{2+} was required for IP_3 to stimulate Ca^{2+} release. Subsequent work with IP_3 receptors in bilayers confirmed that conclusion (Bezprozvanny et al., 1991). IP_3 and Ca^{2+} thus came to be viewed as co-agonists of the IP_3 receptor, and the receptor as a molecular co-incidence detector, responding only when two intracellular stimuli were present.

We approached the question of how Ca^{2+} and IP_3 work together to open the channel using rapid superfusion methods to measure $^{45}Ca^{2+}$ release from permeabilized cells (Taylor and Marchant, 1999) (Fig. 1a). These methods combine high temporal resolution (9ms) with opportunities to rapidly switch the media bathing the cells under conditions that minimise the likelihood of Ca^{2+} released from the stores feeding back to regulate IP_3 receptors (Marchant and Taylor, 1997). For both permeabilized hepatocytes and RIN cells, which express predominantly type 2 and type 3 IP_3 receptors respectively, we observed that after very rapid delivery of a high concentration of IP_3 to the cells in a normal cytosol-like medium, the first detectable $^{45}Ca^{2+}$ release occurred after a delay of about 30ms (Fig. 1b). Even massive further increases in the IP_3 concentration, sufficient to achieve supramaximally effective concentrations within a few ms, were unable to further reduce this absolute latency (Marchant and Taylor, 1997; Swatton and Taylor, 2002). In both cell types, however, changing the cytosolic $[Ca^{2+}]$ affected the latency before the first detectable release of $^{45}Ca^{2+}$ evoked by a supramaximal concentration of IP_3: reducing the $[Ca^{2+}]$ substantially prolonged the latency, while increasing the $[Ca^{2+}]$ shortened and eventually abolished the latency (Fig. 1c). We concluded, in keeping with results from intact cells (Carter and Ogden, 1997; Ogden and Capiod, 1997; Parker et al., 1996b), that binding of IP_3 to its receptor caused a stimulatory Ca^{2+}-binding site to be exposed to which Ca^{2+} had then to bind before the channel could open (Fig. 1d). Our model preserves the idea that Ca^{2+} and IP_3 are obligate co-agonists of the receptor, but extends it by suggesting that the agonists must bind in a strict sequence, first IP_3 and then Ca^{2+}. IP_3, in effect, allows Ca^{2+} to gate the channel.

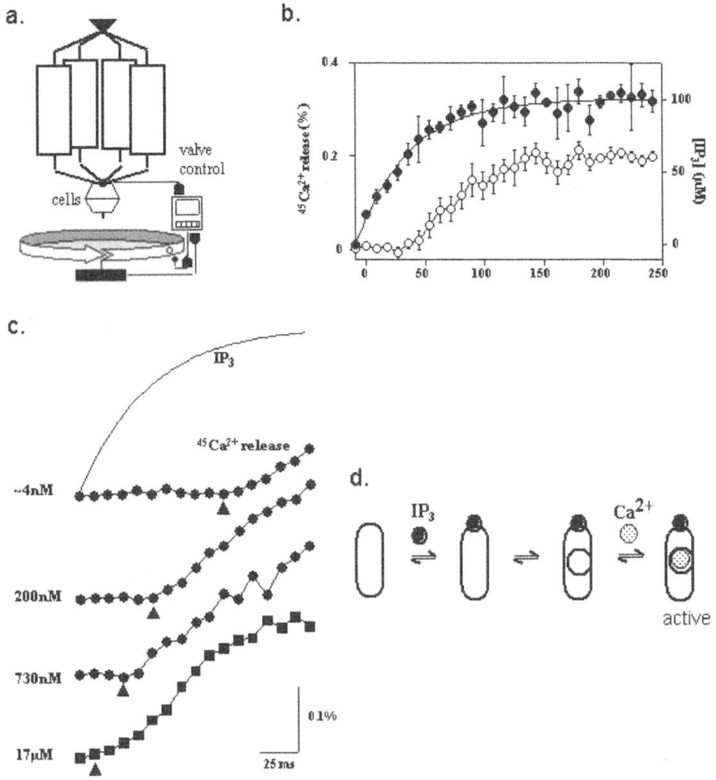

Fig.1. Activation of IP_3 receptors. a. Rapid superfusion of permeabilized hepatocytes loaded with $^{45}Ca^{2+}$ allows rapid switching between the media bathing the cells while collecting the $^{45}Ca^{2+}$ released from them. b. Superfusion of cells with IP_3 (final concentration 100μM, delivered with the time course shown ●) evokes detectable $^{45}Ca^{2+}$ release (○) only after a latency of about 30ms, that cannot be further reduced by increasing the IP_3 concentration. c. Increasing the free $[Ca^{2+}]$ of the medium (from ~4nM to 17μM) reduces the latency between the arrival of a supramaximal concentration of IP_3 (solid line) and the first detectable release of $^{45}Ca^{2+}$ (●). d. Results similar too those shown in *b* and *c* lead to a model in which binding of IP_3 opens a site to which Ca^{2+} must bind before the channel can open. A single subunit is shown, but it is likely that all four subunits of the tetrameric receptor must complete this sequential binding for the channel to open. For clarity, the Ca^{2+}-binding site is shown as an integral part of the receptor, but the involvement of accessory proteins cannot be excluded. Results taken from (Marchant and Taylor, 1997) inhibition by cytosolic Ca^{2+}. It is not immediately obvious how such a scheme can be reconciled with single channel recordings from lipid bilayers where IP_3 clearly gates the channel, but where (for whatever reason) there is no inhibition by cytosolic Ca^{2+} (Hagar et al., 1998; Ramos-Franco et al., 1998b).

From steady-state analyses of the opening of single IP_3 receptors in the nuclear envelope of *Xenopus* oocytes, Foskett and his colleagues (Mak et al., 1998) have suggested an alternative model for gating of IP_3 receptors by IP_3. They suggest that the sole effect of IP_3 binding is to decrease the affinity of the inhibitory Ca^{2+}-binding site, and to thereby promote channel opening by relieving a tonic.

If, as we suspect, the stimulatory Ca^{2+}-binding site is part of the IP_3 receptor itself, we might expect it to be both similar in all IP_3 receptor subtypes and to be essential for channel opening. At least seven cytosolic sites have been identified on the type 1 IP_3 receptor that can bind $^{45}Ca^{2+}$ (Sienaert et al., 1997); none have yet been linked to either the regulation of IP_3-evoked Ca^{2+} release by cytosolic Ca^{2+} or to Ca^{2+} regulation of IP_3 binding. However, a glutamate residue (Glu^{2100} of the type 1 IP_3 receptor) known to be important for Ca^{2+} regulation of ryanodine receptors and conserved in all three IP_3 receptor subtypes has been directly implicated in Ca^{2+} regulation of IP_3 receptors (Miyakawa et al., 2001). Unfortunately none of the fusion proteins used to identify Ca^{2+}-binding sites in the type 1 receptor (Sienaert et al., 1997) included this glutamate residue. Mutagenesis of this glutamate residue shifted the biphasic effects of Ca^{2+} on IP_3-evoked Ca^{2+} release to 10-fold higher Ca^{2+} concentrations, suggesting that this residue may be important in mediating the stimulatory effect of Ca^{2+} on IP_3 -evoked Ca^{2+} release (Miyakawa et al., 2001). It is less clear whether the residue is also involved in Ca^{2+} inhibition of the receptor. Nor is it clear whether this residue actually forms part of a stimulatory Ca^{2+}-binding site, whether it regulates a more distant site, or whether it forms part of a site that recognises an accessory protein through which Ca^{2+} stimulates the receptor. To summarise: it is likely that Ca^{2+} stimulates all IP_3 receptors by binding directly to them, and the prime candidate for this Ca^{2+}-binding site is presently the region surrounding glutamate-2100.

1.5. Inhibition of IP_3 Receptors by Cytosolic Ca^{2+}

We have already seen that the inhibitory effects of Ca^{2+} on IP_3 receptors have proven less robust than the stimulation, with Ca^{2+} often failing to inhibit receptors after their purification (Michikawa et al., 1999) or partial purification (Hagar et al., 1998; Ramos-Franco et al., 1998b, but see Thrower et al., 1998). These results suggest that an accessory protein may mediate Ca^{2+} inhibition, in keeping with our earlier conclusion that different Ca^{2+}-binding sites mediate the stimulatory and inhibitory effects of Ca^{2+} in hepatocytes (Marshall and Taylor, 1994). Any such protein must be associated tightly enough to survive cell permeabilization and the rapid fluid flows associated with superfusion methods (Adkins and Taylor, 1999; Swatton and Taylor, 2002). The involvement of two distinct Ca^{2+}-binding sites is not, of course, required by the scheme proposed by Foskett *et al.* (Mak et al., 1998), which envisages IP_3-regulation of a single inhibitory Ca^{2+}-binding site as a sufficient explanation for the co-ordinate regulation of IP_3 receptors by Ca^{2+} and IP_3.

IP_3-evoked Ca^{2+} release by all three IP_3 receptor subtypes is rapidly and reversibly inhibited when the cells are exposed to a high (>10μM) free [Ca^{2+}] before addition of IP_3 (Adkins and Taylor, 1999; Finch et al., 1991; Swatton and Taylor, 2002). But much to our surprise, we found that once type 2 IP_3 receptors had bound IP_3 they could no longer be inhibited by even massive increases in cytosolic [Ca^{2+}] (Adkins and Taylor, 1999). We proposed, therefore, that IP_3 binding not only caused a stimulatory Ca^{2+}-binding site to be exposed (Sect. 1.4), but simultaneously caused the inhibitory Ca^{2+}-binding site to be concealed (Fig. 2).

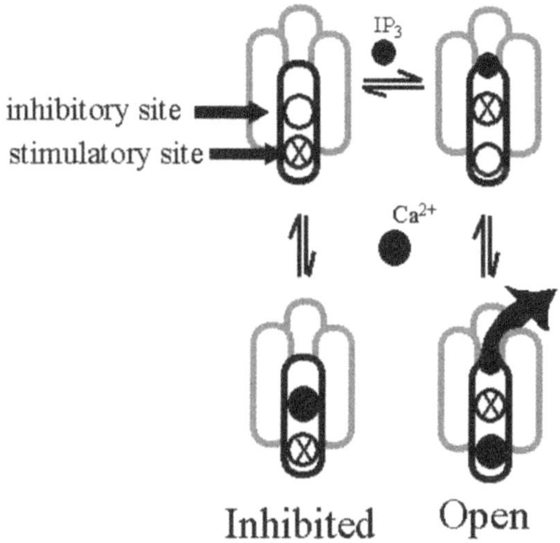

Fig. 2. IP_3 switches type 2 receptors between stimulatory and inhibitory Ca^{2+}-binding sites. Binding of IP_3 to type 2 receptors causes both exposure of a stimulatory Ca^{2+}-binding site (to which Ca^{2+} must then bind to cause channel opening), and occlusion of an inhibitory Ca^{2+}-binding site. For type 3 receptors, IP_3 affects only the stimulatory site; the inhibitory site remains accessible even after IP_3 binding (Swatton and Taylor, 2002). The Ca^{2+}-binding sites are shown as part of the receptor for clarity, though the inhibitory site, at least, is likely to be on another protein. Modified from Adkins and Taylor, 1999.

In this scheme, it is Ca^{2+} that determines whether the channel is open or closed, but IP_3 determines whether it is the inhibitory (no IP_3 bound) or stimulatory (IP_3 bound) Ca^{2+}-binding sites that are available: IP_3 controls what Ca^{2+} does to the IP_3 receptor. This scheme plainly shares at least one feature with that proposed by Foskett *et al.* (Mak et al., 1998), namely that IP_3 relieves the Ca^{2+} inhibition. We refer to this pattern of regulation as "lateral inhibition" to capture the notion that an active IP_3 receptor, by spewing Ca^{2+} over its unliganded neighbours, will inhibit them without attenuating its own activity (Adkins and Taylor, 1999). A receptor that has already bound IP_3, but not yet bound Ca^{2+} to the stimulatory site, would be protected from the inhibitory effect of Ca^{2+} coming from an active neighbour. It would instead be rapidly stimulated to open as Ca^{2+} binds to the exposed stimulatory Ca^{2+}-binding site. Our simple scheme (Fig. 2) therefore provides a plausible means whereby the IP_3 concentration may control the number of IP_3 receptors within a cluster that are rapidly recruited without explosive positive feedback (Adkins and Taylor, 1999).

Whereas sequential binding of IP_3 and then Ca^{2+} appears to underlie activation of each IP_3 receptor subtype (Sect. 1.4), inhibition by cytosolic Ca^{2+} is quite different for types 2 and 3 IP_3 receptors. The latter, unlike type 2 receptors, are similarly inhibited by cytosolic Ca^{2+} whether or not the receptor has IP_3 bound (Swatton and Taylor, 2002). A type 2 IP_3 receptor will therefore close only when IP_3 dissociates from it whatever the local $[Ca^{2+}]$, whereas a type 3 receptor may be prematurely closed by an increase in cytosolic $[Ca^{2+}]$. The shared property of biphasic regulation by cytosolic Ca^{2+} revealed in simple assays of IP_3-evoked Ca^{2+} release (Table 1) therefore masks differences between receptor subtypes in the underlying mechanisms. These differences may be profoundly important in determining the extents to which different IP_3 receptor subtypes contribute to the initiation and propagation of intracellular Ca^{2+} signals.

If, as seems likely, an accessory protein mediates Ca^{2+} inhibition of IP_3 receptors, what might it be? There is no clear answer. Although a Ca^{2+}-regulated phosphorylation/dephosphorylation cycle involving calcineurin (which associates with IP_3 receptors) and Ca^{2+}-calmodulin dependent kinase II has been implicated in Ca^{2+} regulation of IP_3 receptors (Zhang et al., 1993), it cannot be a universal mechanism: there are too many examples of reversible Ca^{2+} regulation in the absence of ATP (Patel and Taylor, 1995; Swatton and Taylor, 2002). At present the only serious candidate for the Ca^{2+}-sensor that mediates inhibition of IP_3 receptors is calmodulin, which was shown to restore Ca^{2+} inhibition to purified type 1 IP_3 receptors in lipid bilayers (Michikawa et al., 1999). Although such a mechanism would provide an attractive parallel with many other Ca^{2+} channels where feedback regulation by Ca^{2+} is mediated by calmodulin (see references in Peterson et al., 1999), there are also problems with this mechanism.

Ca^{2+}-calmodulin antagonists do not generally prevent Ca^{2+} from inhibiting IP_3 receptors (Marshall and Taylor, 1993), although it is recognised that these antagonists may be ineffective once Ca^{2+}-calmodulin has bound to its target. Ca^{2+}-calmodulin does not inhibit IP_3 binding to any IP_3 receptor subtype (Cardy and Taylor, 1998; Patel et al., 1997; Sienaert et al., 2002), but it does inhibit IP_3-evoked Ca^{2+} release from cells expressing predominantly types 1, 2 or 3 IP_3 receptors (Adkins et al., 2000). The same cells are, of course, also inhibited by Ca^{2+} alone (Table 1), implying that if calmodulin is the inhibitory Ca^{2+}-sensor then it must associate with each IP_3 receptor subtype in the absence of Ca^{2+} and then remain associated when the calmodulin binds Ca^{2+}. But whereas full length type 1 IP_3 receptors bind apo-calmodulin, the type 3 receptor does not (Cardy and Taylor, 1998), although IP_3 binding to N-terminal fragments of all three IP_3 receptor subtypes was inhibited by apo-calmodulin (Vanlingen et al., 2000). The only Ca^{2+}-calmodulin binding site identified in the type 1 IP_3 receptor (Yamada et al., 1995) is conserved in type 2, but not in type 3, receptors. Furthermore, mutation of this site, which effectively eliminates Ca^{2+}-calmodulin binding, does not prevent Ca^{2+} from inhibiting the IP_3 receptor (Zhang and Joseph, 2001).

In summary, there is presently insufficient evidence to provide a detailed understanding of the means whereby Ca^{2+} inhibits IP_3 receptors, indeed the details are likely to differ between the subtypes. There are, however, tantalising hints consistent with the sort of mechanism elaborated for other Ca^{2+} channels, including ryanodine receptors (Rodney et al., 2001). A speculative possibility, therefore, is that calmodulin may remain permanently associated with the IP_3 receptor, bound to a Ca^{2+}-independent- (or possibly apo-calmodulin)-binding site when the [Ca^{2+}] is low and flick then to a Ca^{2+}-dependent site, from which it inhibits channel opening, when the [Ca^{2+}] is increased.

1.6. Conclusions

Most, perhaps all, IP_3 receptors are biphasically regulated by cytosolic Ca^{2+} and such regulation is important in allowing the propagation of Ca^{2+} signals within cells. Despite this conserved feature of their behaviour, it seems unlikely that the same mechanisms underlie the biphasic effects of cytosolic Ca^{2+} on all IP_3 receptor subtypes. Nor is the extent to which accessory proteins are required as Ca^{2+}-sensors yet clear. The balance of evidence supports the idea that a Ca^{2+}-binding site on the receptor itself, which becomes accessible only after IP_3 binding, is probably responsible for Ca^{2+} stimulation of IP_3 receptors. Inhibition by cytosolic Ca^{2+} probably requires an additional protein, which may be calmodulin.

References

Adkins, C.E., Morris, S.A., De Smedt, H., Török, K. and Taylor, C.W. (2000) Ca^{2+}-calmodulin inhibits Ca^{2+} release mediated by type-1, -2 and -3 inositol trisphosphate receptors. *Biochem. J.*, **345**, 357-363.

Adkins, C.E. and Taylor, C.W. (1999) Lateral inhibition of inositol 1,4,5-trisphosphate receptors by cytosolic Ca^{2+}. *Curr. Biol.*, **9**, 1115-1118.

Baumann, O. and Walz, B. (1989) Calcium- and inositol polyphosphate-sensitivity of the calcium-sequestering endoplasmic reticulum in the photoreceptor cells of the honeybee drone. *J. Comp. Physiol. A*, **165**, 627-636.

Benevolensky, D., Moraru, I.I. and Watras, J. (1994) Micromolar calcium decreases the affinity of inositol trisphosphate receptor in vascular smooth muscle. *Biochem. J.*, **299**, 631-636.

Berridge, M.J., Lipp, P. and Bootman, M.D. (2000) The versatility and universality of calcium signalling. *Nature Rev. Mol. Cell Biol.*, **1**, 11-21.

Bezprozvanny, I., Watras, J. and Ehrlich, B.E. (1991) Bell-shaped calcium-response curves for Ins(1,4,5)P_3- and calcium-gated channels from endoplasmic reticulum of cerebellum. *Nature*, **351**, 751-754.

Boehning, D. and Joseph, S.K. (2000a) Direct association of ligand-binding and pore domains in homo- and heterotetrameric inositol 1,4,5-trisphosphate receptors. *EMBO J.*, **19**, 5450-5459.

Boehning, D. and Joseph, S.K. (2000b) Functional properties of recombinant type I and type III inositol 1,4,5-trisphosphate receptor isoforms expressed in COS-7 cells. *J. Biol. Chem.*, **275**, 21492-21499.

Callamaras, N. and Parker, I. (1994) Inositol 1,4,5-trisphosphate receptors in *Xenopus laevis* oocytes: localization and modulation by Ca^{2+}. *Cell Calcium*, **15**, 66-78.

Cardy, T.J.A. and Taylor, C.W. (1998) A novel role for calmodulin: Ca^{2+}-independent inhibition of type-1 inositol trisphosphate receptors. *Biochem. J.*, **334**, 447-455.

Cardy, T.J.A., Traynor, D. and Taylor, C.W. (1997) Differential regulation of types 1 and 3 inositol trisphosphate receptors by cytosolic Ca^{2+}. *Biochem. J.*, **328**, 785-793.

Carter, T.D. and Ogden, D. (1997) Kinetics of Ca^{2+} release by InsP_3 in pig single aortic endothelial cells: evidence for an inhibitory role of cytosolic Ca^{2+} in regulating hormonally evoked Ca^{2+} spikes. *J. Physiol.*, **504**, 17-33.

Combettes, L., Hannaert-Merah, Z., Coquil, J.-F., Rousseau, C., Claret, M., Swillens, S. and Champeil, P. (1994) Rapid filtration studies of the effect of cytosolic Ca^{2+} on inositol 1,4,5-trisphosphate-induced $^{45}Ca^{2+}$ release from cerebellar microsomes. *J. Biol. Chem.*, **269**, 17561-17571.

Danoff, S.K., Supattapone, S. and Snyder, S.H. (1988) Characterization of a membrane protein from brain mediating the inhibition of insoitol 1,4,5-trisphosphate receptor binding by calcium. *Biochem. J.*, **254**, 701-705.

Dufour, J.-F., Arias, I.M. and Turner, T.J. (1997) Inositol 1,4,5-trisphosphate and calcium regulate the calcium channel function of the hepatic inositol 1,4,5-trisphosphate receptor. *J. Biol. Chem.*, **272**, 2675-2681.

Ferris, C.D. and Snyder, S.H. (1992) Inositol phosphate receptors and calcium disposition in the brain. *J. Neurosci.*, **12**, 1567-1574.

Finch, E.A., Turner, T.J. and Goldin, S.M. (1991) Calcium as a coagonist of inositol 1,4,5-trisphosphate-induced calcium release. *Science*, **252**, 443-446.

Hagar, R.E., Burgstahler, A.D., Nathanson, M.H. and Ehrlich, B.E. (1998) Type III $InsP_3$ receptor channel stays open in the presence of increased calcium. *Nature*, **296**, 81-84.

Horne, J.H. and Meyer, T. (1995) Luminal calcium regulates the inositol trisphosphate receptor of rat basophilic leukemia cells at the cytosolic side. *Biochemistry*, **34**, 12738-12746.

Iino, M. (1987) Calcium dependent inositol trisphosphate-induced calcium release in the guinea-pig taenia caeci. *Biochem. Biophys. Res. Commun.*, **142**, 47-52.

Iino, M. (1990) Biphasic Ca^{2+} dependence of inositol 1,4,5-trisphosphate-induced Ca^{2+} release in smooth muscle cells of the guinea pig taenia caeci. *J. Gen. Physiol.*, **95**, 1103-1122.

Kaftan, E.J., Ehrlich, B.E. and Watras, J. (1997) Inositol 1,4,5-trisphosphate ($InsP_3$) and calcium interact to increase the dynamic range of $InsP_3$ receptor-dependent calcium signaling. *J. Gen. Physiol.*, **110**, 529-538.

Lin, C., Widjaja, J. and Joseph, S.K. (2000) The interaction of calmodulin with alternatively spliced isoforms of the type-I inositol trisphosphate receptor. *J. Biol. Chem.*, **275**, 2305-2311.

Mackrill, J.J. (1999) Protein-protein interactions in intracellular Ca^{2+}-release channel function. *Biochem. J.*, **337**, 345-361.

Mak, D.-O., McBride, S. and Foskett, J.K. (1998) Inositol 1,4,5-tris-phosphate activation of inositol tris-phosphate receptor Ca^{2+} channel by ligand tuning of Ca^{2+} inhibition. *Proc. Natl. Acad. Sci. USA*, **95**, 15821-15825.

Mak, D.-O., McBride, S., Raghiram, V., Yue, Y., Joseph, S.K. and Foskett, J.K. (2000) Single-channel properties in endoplasmic reticulum membrane of recombinant type 3 inositol trisphosphate receptor. *J. Gen. Physiol.*, **115**, 241-255.

Mak, D.-O., McBride, S. and Foskett, J.K. (2001) Regulation by Ca^{2+} and inositol 1,4,5-trisphosphate ($InsP_3$) of single recombinant type 3 $InsP_3$ receptor channels: Ca^{2+} activation uniquely distinguishes types 1 and 3 $InsP_3$ receptors. *J. Gen. Physiol.*, **117**, 435-446.

Marchant, J.S. and Taylor, C.W. (1997) Cooperative activation of IP_3 receptors by sequential binding of IP_3 and Ca^{2+} safeguards against spontaneous activity. *Curr. Biol.*, **7**, 510-518.

Marshall, I.C.B. and Taylor, C.W. (1993) Biphasic effects of cytosolic calcium on $Ins(1,4,5)P_3$-stimulated Ca^{2+} mobilization in hepatocytes. *J. Biol. Chem.*, **268**, 13214-13220.

Marshall, I.C.B. and Taylor, C.W. (1994) Two calcium-binding sites mediate the interconversion of liver inositol 1,4,5-trisphosphate receptors between three conformational states. *Biochem. J.*, **301**, 591-598.

Michikawa, T., Hirota, J., Kawano, S., Hiraoka, M., Yamada, M., Furuichi, T. and Mikoshiba, K. (1999) Calmodulin mediates calcium-dependent inactivation of the cerebellar type 1 inositol 1,4,5-trisphosphate receptor. *Neuron*, **23**, 799-808.

Mignery, G.A. and Südhof, T.C. (1990) The ligand binding site and transduction mechanism in the inositol-1,4,5-trisphosphate receptor. *EMBO J.*, **9**, 3893-3898.

Missiaen, L., Parys, J.B., Sienaert, I., Maes, K., Kunzelmann, K., Takahashi, M., Tanzawa, K. and De Smet, H. (1998) Functional properties of the type-3 $InsP_3$ receptor in 16HBE14o- bronchial mucosal cells. *J. Biol. Chem.*, **273**, 8983-8986.

Miyakawa, T., Maeda, A., Yamazawa, T., Hirose, K., Kurosaki, T. and Iino, M. (1999) Encoding of Ca^{2+} signals by differential expression of IP_3 receptor subtypes. *EMBO J.*, **18**, 1303-1308.

Miyakawa, T., Mizushima, A., Hirose, K., Yamazawa, T., Bezprozvanny, I., Kurosaki, T. and Iino, M. (2001) Ca^{2+}-sensor region of IP_3 receptor controls intracellular Ca^{2+} signaling. *EMBO J.*, **20**, 1674-1680.

Mohr, F.C., Hershey, P.E.C., Zimányi, I. and Pessah, I.N. (1993) Regulation of inositol 1,4,5-trisphosphate receptors in rat basophilic leukemia cells. I. Multiple conformational states of the receptor in a microsomal preparation. *Biochim. Biophys. Acta.*, **1147**, 105-114.

Oancea, E. and Meyer, T. (1996) Reversible desensitization of inositol trisphosphate-induced calcium release provides a mechanism for repetitive calcium spikes. *J. Biol. Chem.*, **271**, 17253-17260.

Ogden, D.C. and Capiod, T. (1997) Regulation of Ca^{2+} release by InsP_3 in single guinea pig hepatocytes and rat Purkinje neurons. *J. Gen. Physiol.*, **109**, 741-756.

Parker, I., Choi, J. and Yao, Y. (1996a) Elementary events of $InsP_3$-induced Ca^{2+} liberation in *Xenopus* oocytes: hot spots, puffs and blips. *Cell Calcium*, **20**, 105-121.

Parker, I., Yao, Y. and Ilyin, V. (1996b) Fast kinetics of calcium liberation induced in *Xenopus* oocytes by photoreleased inositol trisphosphate. *Biophys. J.*, **70**, 222-237.

Parys, J.B., Sernett, S.W., DeLisle, S., Snyder, P.M., Welsh, M.J. and Campbell, K.P. (1992) Isolation, characterization, and localization of the inositol 1,4,5-trisphosphate receptor protein in *Xenopus laevis* oocytes. *J. Biol. Chem.*, **267**, 18776-18782.

Patel, S., Joseph, S.K. and Thomas, A.P. (1999) Molecular properties of inositol 1,4,5-trisphosphate receptors. *Cell Calcium*, **25**, 247-264.

Patel, S., Morris, S.A., Adkins, C.E., O'Beirne, G. and Taylor, C.W. (1997) Ca^{2+}-independent inhibition of inositol trisphosphate receptors by calmodulin: Redistribution of calmodulin as a possible means of regulating Ca^{2+} mobilization. *Proc. Natl. Acad. Sci. USA*, **94**, 11627-11632.

Patel, S. and Taylor, C.W. (1995) Quantal responses to inositol 1,4,5-trisphosphate are not a consequence of Ca^{2+} regulation of inositol 1,4,5-trisphosphate receptors. *Biochem. J.*, **312**, 789-794.

Peterson, B.Z., DeMaria, C.D. and Yue, D.T. (1999) Calmodulin is the Ca^{2+} sensor for Ca^{2+}-dependent inactivation of L-type calcium channels. *Neuron*, **22**, 549-558.

Picard, L., Coquil, J.-F. and Mauger, J.-P. (1998) Multiple mechanisms of regulation of the inositol 1,4,5-trisphosphate receptor by calcium. *Cell Calcium*, **23**, 339-348.

Pietri, F., Hilly, M. and Mauger, J.-P. (1990) Calcium mediates the interconversion between two states of the liver inositol 1,4,5-trisphosphate receptor. *J. Biol. Chem.*, **265**, 17478-17485.

Ramos-Franco, J., Caenepeel, S., Fill, M. and Mignery, G. (1998a) Single channel function of recombinant type-1 inositol 1,4,5-trisphosphate receptor ligand binding domain splice variants. *Biophys. J.*, **75**, 2783-2793.

Ramos-Franco, J., Fill, M. and Mignery, G.A. (1998b) Isoform-specific function of single inositol 1,4,5-trisphosphate receptor channels. *Biophys. J.*, **75**, 834-839.

Richardson, A. and Taylor, C.W. (1993) Effects of Ca^{2+} chelators on purified inositol 1,4,5-trisphosphate ($InsP_3$) receptors and $InsP_3$-stimulated Ca^{2+} mobilization. *J. Biol. Chem.*, **268**, 11528-11533.

Rodney, G.G., Moore, C.P., Williams, B.Y., Zhang, J.-Z., Krol, J., Pedersen, S.E. and Hamilton, S.L. (2001) Calcium binding to calmodulin leads to an N-terminal shift in its binding site on the ryanodine receptor. *J. Biol. Chem.*, **276**, 2069-2074.

Sienaert, I., Kasri, N.N., Vanlingen, S., Parys, J., Callewaert, G., Missiaen, L. and De Smedt, H. (2002) Localization and function of a calmodulin/apocalmodulin binding domain in the N-terminal part of the type 1 inositol 1,4,5-trisphosphate receptor. *Biochem. J.*, **365**, 269-277

Sienaert, I., Missiaen, L., De Smedt, H., Parys, J.B., Sipma, H. and Casteels, R. (1997) Molecular and functional evidence for multiple Ca^{2+}-binding domains on the type 1 inositol 1,4,5-trisphosphate receptor *J. Biol. Chem.*, **272**, 25899-25906.

Streb, H., Irvine, R.F., Berridge, M.J. and Schulz, I. (1983) Release of Ca^{2+} from a nonmitochondrial store of pancreatic acinar cells by inositol-1,4,5-trisphosphate. *Nature*, **306**, 67-69.

Sugawara, H., Kurosaki, M., Takata, M. and Kurosaki, T. (1997) Genetic evidence for involvement of type 1, type 2 and type 3 inositol 1,4,5-trisphosphate receptors in signal transduction through the B-cell antigen receptor. *EMBO J.*, **16**, 3078-3088.

Supattapone, S., Worley, P.F., Baraban, J.M. and Snyder, S.H. (1988) Solubilization, purification, and characterization of an inositol trisphosphate receptor. *J. Biol. Chem.*, **263**, 1530-1534.

Swatton, J.E., Morris, S.A., Cardy, T.J.A. and Taylor, C.W. (1999) Type 3 inositol trisphosphate receptors in RINm5F cells are biphasically regulated by cytosolic Ca^{2+} and mediate quantal Ca^{2+} mobilization. *Biochem. J.*, **344**, 55-60.

Swatton, J.E., Morris, S.A. and Taylor, C.W. (2001) Functional properties of Drosophila inositol trisphosphate receptors. *Biochem. J.*, **359**, 435-441.

Swatton, J.E. and Taylor, C.W. (2002) Fast biphasic regulation of type 3 inositol trisphosphate receptors by cytosolic calcium. *J. Biol. Chem.*, **277**, 17571-17579.

Taylor, C.W. (1998) Inositol trisphosphate receptors: Ca^{2+}-modulated intracellular Ca^{2+} channels. *Biochim. Biophys. Acta.*, **1436**, 19-33.

Taylor, C.W., Genazzani, A.A. and Morris, S.A. (1999) Expression of inositol trisphosphate receptors. *Cell Calcium*, **26**, 237-251.

Taylor, C.W. and Marchant, J.S. (1999) Measuring inositol 1,4,5-trisphosphate-evoked $^{45}Ca^{2+}$ release from intracellular stores. In Milligan, G. (ed.), *Signal transduction. A practical approach.* IRL Press, Oxford, pp. 361-384.

Thrower, E.C., Lea, E.J.A. and Dawson, A.P. (1998) The effects of free [Ca^{2+}] on the cytosolic face of inositol (1,4,5)-trisphosphate receptor at the single channel level. *Biochem. J.*, **330**, 559-564.

Vanlingen, S., Sipma, H., De Smet, P., Callewaert, G., Missiaen, L., De Smedt, H. and Parys, J.B. (2000) Ca^{2+} and calmodulin differentially modulate *myo*-inositol 1,4,5-trisphosphate (IP_3)-binding to the recombinant ligand-binding domains of the various IP_3 receptor isoforms. *Biochem. J.*, **346**, 275-280.

Worley, P.F., Baraban, J.M., Supattapone, S., Wilson, V.S. and Snyder, S.H. (1987) Characterization of inositol trisphosphate receptor binding in brain. Regulation by pH and calcium. *J. Biol. Chem.*, **262**, 12132-12136.

Yamada, M., Miyawaki, A., Saito, K., Yamamoto-Hino, M., Ryo, Y., Furuichi, T. and Mikoshiba, K. (1995) The calmodulin-binding domain in the mouse type 1 inositol 1,4,5-trisphosphate receptor. *Biochem. J.*, **308**, 83-88.

Yang, J., McBride, S., Mak, D.-O., D., Vardi., N., Palczewski, K., Haeseleer, F. and Foskett, J.K. (2002) Identification of a family of calcium sensors as protein ligands of the inositol trisphosphate receptor Ca^{2+} release channels. *Proc. Natl. Acad. Sci. USA*, In press.

Yoneshima, H., Miyawaki, A., Michikawa, T., Furuichi, T. and Mikoshiba, K. (1997) Ca^{2+} differentially regulates ligand-affinity states of type 1 and 3 inositol 1,4,5-trisphosphate receptors. *Biochem. J.*, **322**, 591-596.

Zhang, B.-X., Zhao, H. and Muallem, S. (1993) Ca^{2+}-dependent kinase and phosphatase control inositol 1,4,5-trisphosphate-mediated Ca^{2+} release. *J. Biol. Chem.*, **268**, 10997-11001.

Zhang, X. and Joseph, S.K. (2001) Effect of mutation of a calmodulin binding sites on Ca^{2+} regulation of inositol trisphosphate receptors. *Biochem. J.*, **360**, 395-400.

Zhang, Z., Tang, J., Tikunova, S., Johnson, J.D., Chen, Z., Qin, N., Dietrich, A., Stefani, E., Birnbaumer, L. and Zhu, M.X. (2001) Activation of Trp3 by inositol 1,4,5-trisphosphate receptors through displacement of inhibitory calmodulin from a common binding domain. *Proc. Natl. Acad. Sci. USA*, **98**, 3168-3173.

2. The Endoplasmic Reticulum: A Central Player in Cell Signalling and Protein Synthesis.

H. Llewelyn Roderick, Michael J. Berridge and Martin D. Bootman

The Babraham Institute, Babraham, Cambridge, UK.

Abstract. In addition to being the principle intracellular Ca^{2+} store, the endoplasmic reticulum (ER) is the initial site of synthesis and folding of membrane and secretory proteins. These two roles of the ER are intimately linked. First, the function of many proteins involved in Ca^{2+} handling are modulated by Ca^{2+} and second, ER lumenal Ca^{2+} modulates protein synthesis and folding. Within the ER, Ca^{2+} is stored by low affinity high capacity Ca^{2+} binding proteins and is maintained at a free concentration between 0.1 and 1 mM relative to 0.1 μM cytosolic Ca^{2+}. This concentration gradient is maintained by the action of the Sarco-Endoplasmic Reticulum ATPases (SERCA) which hydrolyse ATP to pump Ca^{2+} into the ER. Following stimulation Ca^{2+} is released from the ER through several classes of ligand gated channels. The most well characterized of these being the inositol 1,4,5-trisphosphate receptor (IP_3R) and the Ryanodine receptor (RyR) families of proteins. This release of Ca^{2+} results in a drop of ER free Ca^{2+} to levels as low as 10 μM. This decrease in lumenal Ca^{2+} inhibits further release through the channels and increases the rate of re-sequestration of Ca^{2+} into the ER by the SERCA pumps. Under these conditions, in addition to effects on Ca^{2+} handling proteins, protein synthesis is inhibited, chaperones dissociate from their substrates, secondary modifications of proteins are inhibited and the retention of many proteins within the ER is lost. Furthermore, a signalling cascade resulting in the up-regulation of many proteins involved in protein folding and Ca^{2+} homeostasis is initiated. This review will focus on the proteins involved in the regulation ER lumenal Ca^{2+} and the role of ER lumenal Ca^{2+} in cell signalling and protein synthesis.

2.1. ER Structure

The ER is the largest membrane bound organelle within eukaryotic cells, comprising up to 50 % of cellular membrane and 10% of cell volume (Pahl, 1999). It has several major functions within cells. In addition to being the major site of cellular Ca^{2+} storage, it is the location of synthesis, folding and oligomerisation of membrane and secretory proteins and the location of synthesis of lipids and cholesterol (Voelker, 1991; Berridge, 1993; Pozzan et al., 1994b; Hammond and Helenius, 1995;). The ER structure forms a continuous reticular network surrounding and emanating from the nucleus where it is of greatest density (Fig.1). The ER and its muscle specific orthologue the sarcoplasmic reticulum (SR) also comes into very close contact with the plasma membrane (Franzini-Armstrong, 1999; Blaustein and Golovina, 2001) and mitochondria (Sharma et al., 2000; Collins et al., 2002).

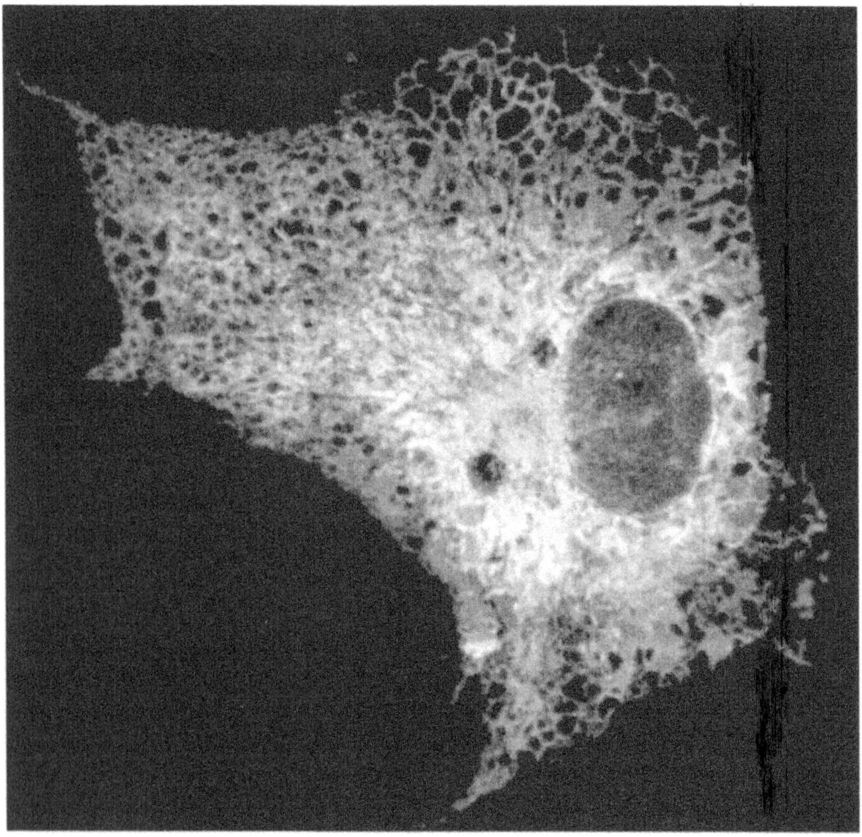

Fig. 1. Confocal image of COS7 cell expressing ER targeted red fluorescent protein. COS7 (a monkey kidney cell line) cells were transfected with an ER targeted red fluorescent protein (dsRed 2) and subsequently imaged with a confocal microscope. A series of Z sections at different focal planes throughout the cell were captured and deconvolved to reveal a 3D representation of the cellular ER network. The ER appears at greatest density around the nucleus which is predominantly devoid of ER. As the ER extends towards the plasma membrane its reticular nature becomes more apparent.

This physical proximity between the ER and the nucleus, mitochondria and plasma membrane has recently attracted much attention and has been shown to be important in intra-organellar signalling. Although the ER is structurally continuous it may contain functionally distinct domains. In non-muscle cells, it can exist as the rough endoplasmic reticulum (RER) and the smooth endoplasmic reticulum (SER). This designation reflects the presence of membrane bound ribosomes on the RER which are responsible for the co-translational translocation of proteins into the lumen. The smooth ER is responsible for lipid and cholesterol synthesis (Voelker, 1991). Specialised regions for Ca^{2+} signalling have also been described in many cell types, for example the PlasmERosomes or terminal cisternae in neurons which are regions of ER close to the plasma membrane (Blaustein and

Golovina, 2001). The architecture of the ER and SR in muscle is a little more complex. The SR is a specialised region of the ER dedicated to Ca^{2+} signalling. In striated muscle this is further divided into the junctional SR which contains the Ca^{2+} release units in proximity to the plasma membrane (Franzini-Armstrong et al., 1998), the corbular SR adjacent to the junctional regions is rich in the Ca^{2+} storage proteins calsequestrin (CSQ) and contains the Ca^{2+} pump. The longititudinal SR is continuous with the corbular SR and ER but CSQ is absent (Meldolesi and Pozzan, 1998).

2.2. Role of the ER in Ca^{2+} Signalling

Whilst extracellular Ca^{2+} concentration varies between 1 and 2 mM, cytosolic Ca^{2+} concentration is maintained at a resting level of between 50-100 nM. Following cellular stimulation cytosolic Ca^{2+} can transiently rise to between 1 and 10 μM (Pozzan et al., 1994a; Berridge et al., 2000b). Although increases in cytosolic Ca^{2+} are necessary for many cellular processes sustained elevations in cytosolic Ca^{2+} are cytotoxic and may lead to cell death. Thus, it is necessary for the cell to strictly regulate cytosolic Ca^{2+} levels within defined limits. In order to achieve this the cell has developed a toolkit of Ca^{2+} channels, binding proteins, pumps and exchangers (Berridge et al., 2000b). Some of these are located on the plasma membrane and allow Ca^{2+} entry into the cell and extrude Ca^{2+} from the cell. However, due to the large volume of the cell and the low mobility of Ca^{2+} in the cytosolic milieu (Allbritton et al., 1992) (0.1 to 0.5 μm per ~ 50 μs) it is necessary for Ca^{2+} to be released from internal stores in order to generate more global Ca^{2+} signals. Following the realization that the SR was the source of mobilisable Ca^{2+} in muscle (MacLennan and Klip, 1979) several lines of evidence indicated that the ER had a similar function in non-muscle cells. The ER contains a number of Ca^{2+} binding proteins which can store and buffer Ca^{2+} (Macer and Koch, 1988), it was shown to accumulate Ca^{2+} in an ATP dependent manner (Heilmann et al., 1983) and Ca^{2+} was released from the ER by the second messenger IP_3 (Streb et al., 1983; Ross et al., 1989). Its functional equivalent, the SR, had previously been established to be the location of mobilisable Ca^{2+} in muscle. This review will primarily discuss the ER although in regions of overlap both will be discussed.

2.3. The ER as a Ca^{2+} Storage Organelle

As described above, the ER/SR is the principle location of Ca^{2+} storage within the cell. The total Ca^{2+} within the ER has been determined to be between 5 and $50mmoll^{-1}$ (Somlyo et al., 1985; Baumann et al., 1991) where it is bound to a number of low affinity Ca^{2+}-binding proteins including calreticulin (CRT), GRP78 and in muscle cells CSQ (MacLennan et al., 1983; Macer and Koch, 1988). In deed, the ER lumen has been described as a gel like matrix of protein coordinated by Ca^{2+} (Tatu and Helenius, 1997; Chevet et al., 1999a). Although the majority of Ca^{2+} is bound within the ER, due to the nature of the Ca^{2+}-binding proteins and the properties of the Ca^{2+} handling machinery, a significant amount of free Ca^{2+} reaching up to 1 mM in concentration is present at rest within this organelle (Meldolesi and Pozzan, 1998). Following stimulation, this level can drop to below 100 µM (Miyawaki et al., 1997; Arnaudeau et al., 2001). The first described Ca^{2+} binding protein was CSQ which is the main Ca^{2+} storage protein in muscle SR (MacLennan, 1974). This protein is characterised by a very acidic domain which binds Ca^{2+} at low affinity and high capacity (k_d = 1 mM, B_{max} ~ 50 mols Ca^{2+} per mol of CSQ). A second protein, calreticulin (CRT) having a very acidic COOH terminus which binds Ca^{2+} with low affinity and high capacity (k_d= 2mM, B_{max} ~ 25 mols Ca^{2+} per mol CRT) was also described in SR (Ostwald and MacLennan, 1974; Baksh and Michalak, 1991). This protein was later shown to be the major Ca^{2+}-binding protein in non-muscle cells (Milner et al., 1991). Although they do not possess long stretches of acidic amino acid within their protein sequence, the glucose regulated proteins GRP78 and 94 (also known as BiP and endoplasmin, respectively) also bind Ca^{2+} with low affinity and have been shown to play an important role in Ca^{2+} storage (Mazzarella and Green, 1987; Lievremont et al., 1997).

2.4. Ca^{2+} Signaling from the ER

In response to one of a plethora of signals at the plasma membrane, Ca^{2+} is released from the ER into the cytosol in the form of global or spatially localised elementary events (Bootman et al., 1997; Bootman et al., 2001). The mechanism of transduction of this signal to the ER is dependent on the nature of the initial stimulus. In all cell types, ligand binding to its receptor at the plasma membrane initiates a cascade of signals which ultimately results in release of Ca^{2+} from the ER. The best characterised of these uses the diffusible messenger IP_3. Activation of G-protein coupled receptors (GPCR) or receptor tyrosine kinases (RTK) results in activation of phospholipase C β or γ, respectively. These enzymes catalyse the cleavage of phosphatidyl inositol 1,4,5-bisphosphate (PIP_2) into diacylglycerol and IP_3. IP_3 subsequently diffuses into the cell binds to its cognate receptor on the endoplasmic reticulum and Ca^{2+} is released (Berridge, 1993). In excitable cells however, i.e. neurons and muscle, an action potential or activation of receptor operated channels results in influx of Ca^{2+} across the plasma membrane which subsequently acts as the messenger to stimulate the RyRs, and perhaps also the IP_3Rs, to release Ca^{2+} through an autocatalytic process of Ca^{2+}-induced Ca^{2+} release (CICR) (Endo et al., 1970; Lechleiter et al., 1991)

Recently, a number of other Ca^{2+} mobilising messengers have been uncovered. These include cyclic ADP ribose and NAADP which are described in greater detail elsewhere in this book. The identity of the ER Ca^{2+} release channel to which these agonists bind remains unclear although both the RyR and the IP_3R have been implicated (Galione and Churchill, 2000; Cancela et al., 2002). Following

Ca^{2+}release, ER Ca^{2+} homeostasis is achieved by the action of the SR/ER Ca^{2+} ATPase (SERCA) which pumps Ca^{2+} from the cytosol back into the ER (MacLennan et al., 1997). With low levels of cell stimulation, leading to modest rates of Ca^{2+} mobilization, the SERCAs can effectively compete with Ca^{2+} release channels to give a graded steady state Ca^{2+} rise (Loomis-Husselbee and Dawson, 1993). However, under conditions of increased stimulation and ER depletion, a Ca^{2+} influx pathway known as store operated Ca^{2+} entry (SOC) is activated at the plasma membrane (Putney, 1990; Berridge et al., 2000a). Ca^{2+} entering the cell via this pathway is subsequently taken up into the ER by the SERCA pump.

2.4.1. Ca^{2+} Release from the ER

In addition to acting as a source of readily releasable intracellular pool, Ca^{2+} within the ER regulates the activity of its own transporters and of many other cellular processes. This regulatory function of Ca^{2+} is particularly evident in the way it controls the RyR and IP_3 receptor channels. IP_3 receptors are the predominant Ca^{2+} release channel in non-electrically excitable cells but are also widely distributed whereas RyRs are predominant in excitable tissues (Iino, 1999). Three genes encoding each of these release channels has been cloned. Alternative splicing of these genes results in several more variants. Since the various isoforms have different Ca^{2+} release properties and they have a tissue-specific distribution, diverse and complex Ca^{2+} signals are possible (Miyakawa et al., 1999). For example, IP_3R1 is the major isoform in brain and IP_3R2, although in low abundance is predominant in the heart. RyR1 is the major isoform in skeletal muscle and RyR2 in cardiac muscle.

Both the RyR and IP_3R are subject to several levels of regulation on their cytosolic face where Ca^{2+} can both promote and inhibit its release from either channel. Calmodulin, PKC, PKA and FKBP12 are also important cytosolic modulators of Ca^{2+} release. At low concentrations Ca^{2+} stimulates Ca^{2+} release through the receptor, whereas at high Ca^{2+} concentration release is inhibited, i.e. a bell shaped dependence to Ca^{2+}. This bell shape is set around the physiological Ca^{2+} concentration, for example, 200-300 nM for the IP_3R1 (Bezprozvanny et al., 1991). Furthermore, at increasing Ca^{2+} concentration the IP_3 receptor becomes more sensitive to ligand and less sensitive to Ca^{2+} dependent inhibition. Thus Ca^{2+} potentiates its own release and can stimulate release from neighbouring receptors. This process of CICR drives Ca^{2+} oscillations and waves (Bootman and Berridge, 1995). The major recognized difference between the regulation of these two channels has been that only in the case of RyR was it thought that Ca^{2+} alone could act to promote its own release (see the article by Taylor in this book) (Sorrentino, 1995; Patel et al., 1999). However, a recent study has elegantly described another form of CICR whereby Ca^{2+}, acting through an EF-hand Ca^{2+} binding protein, could bind to the IP_3R and stimulate Ca^{2+} release in the absence of IP_3 (Yang, J., et al., 2002).

In addition to regulation from the cytosol, Ca^{2+} within the ER lumen may also regulate the IP_3Rs or RyRs (Missiaen et al., 1991; Cheek et al., 1994). Reports in the early 90's described that the amount of Ca^{2+} released from the ER, through either the IP_3R or RyR, was dependent on the state of filling of the store, i.e. greater Ca^{2+} release from a full store. In some cases however, this is only observed when the store is less than 30% full (Parys et al., 1993). In order for Ca^{2+} to regulate its own release from the lumen there has to be a sensor within the ER. This sensor could be on the channel itself or it could lie within accessory proteins attached to the release channel. Evidence for both hypotheses have been presented. IP_3Rs contain a lumenally facing high affinity Ca^{2+}-binding domain (Sienaert et al., 1996). However, since this Ca^{2+} binding domain has such a high affinity (0.3 μM) relative to the Ca^{2+} concentration in the ER, it may not play any physiological role. Calnexin (CLNX) a high affinity Ca^{2+}-binding lectin chaperone of the ER lumen, interacts with the IP_3R, although it was argued that this interaction was purely to assist in the folding of the protein (Joseph et al., 1999). The high affinity of the Ca^{2+} binding region in CLNX makes it unlikely that Ca^{2+} plays a direct role in regulating the interaction between CLNX and IP_3R under physiological conditions. Due to the homology between CLNX and CRT, the IP_3R may also interact with CRT. CRT is a second member of the ER lectin chaperone family of proteins functionally analogous to CLNX (Helenius et al., 1997). However, as described earlier CRT is also the major Ca^{2+} binding protein in non-excitable cells. CRT and CLNX share significant homology in their NH_2-terminal region (N-domain) and their central proline rich domain (P-domain)(Krause and Michalak, 1997). Unlike CLNX, CRT also has a low affinity region (the domain responsible for Ca^{2+} storage) which has been shown to regulate interactions with its substrates (Nigam et al., 1994; Corbett et al., 1999).

Release of Ca^{2+} through the IP_3R is also regulated by the abundance of the receptor in the ER (Tovey et al., 2001). Following ligand binding to the receptor it is targeted for degradation through the ubiquitin/proteosome pathway. Following chronic agonist stimulation over a period of 24 hrs, the abundance of the receptor can decrease to as low as 20 % (Oberdorf et al., 1999). In order for proteins to be degraded through this pathway the presence of CLNX is required (McCracken and Brodsky, 1996). Thus the interaction of the IP_3R with ER lectin chaperones may have significance for both activation of the receptor and regulation of its abundance. Expression levels of the IP_3R are strictly regulated during physiology and pathophysiology. For example, the levels of the IP_3R are increased in end stage heart failure (Go et al., 1995). The RyR is also subject to regulation from the lumen of the ER/SR by Ca^{2+} (Koizumi et al., 1999; Ching et al., 2000). The RyR has been shown to be in quaternary complex with the Ca^{2+} binding protein CSQ, junctin and triadin (Zhang et al., 1997). Triadin and CSQ have opposing roles in regulating the RyR. The opening of the RyR is potentiated by CSQ but inhibited by triadin (Ohkura et al., 1998).

2.4.2. Ca^{2+} Sequestration by the ER

Of equal importance to the regulation of Ca^{2+} release from the ER are the mechanisms of Ca^{2+} clearance from the cytosol. This function is performed by a number of transporters located in the plasma membrane, mitochondria and the ER/SR. In this review, the role of the SERCA pump will be discussed. This is a high affinity Ca^{2+} pump which uses ATP to transport Ca^{2+}into the ER/SR (MacLennan et al., 1997). Three genes encoding different SERCA isoforms have been cloned. Splice variants of each have also been described and are distributed in a tissue specific manner. The SERCA1 isoforms are found predominantly in fast twitch muscle, SERCA2a in cardiac muscle, SERCA2b is ubiquitous and SERCA3 is in platelets (Misquitta et al., 1999). The relative function of each isoforms is ideally suited to the spatio-temporal properties of cytosolic Ca^{2+} transients in the tissue in which they are expressed. SERCA1 and SERCA2a have a very similar K_m for Ca^{2+} (~0.4 μM) (Lytton et al., 1992) and have been shown to be able to substitute for each other in overexpression studies (Loukianov et al., 1998). SERCA2b however has a slower rate of Ca^{2+} transport despite a higher affinity (K_m ~ 0.27 μM) (Lytton et al., 1992) and when expressed in place of SERCA2a in cardiac tissue hypertrophy resulted (Ver Heyen et al., 2001). SERCA3 has a lower affinity than the other isoforms (K_m ~ 1.1 μM) (Lytton et al., 1992). SERCA proteins are regulated by both cytosolic and ER lumenal conditions. Under resting conditions SERCA is relatively inactive but following an increase in cytosolic Ca^{2+} the activity of the pump is increased, resulting in re-sequestration of Ca^{2+} into the ER (Lytton et al., 1992). The K_m and velocities of the different isoforms are appropriate for the Ca^{2+} levels and physiology of their respective tissues. Thus they act to counterbalance Ca^{2+} release and modulate the duration of the Ca^{2+} transient.

SERCA family members are susceptible to regulation by phosphorylation of their cytosolic domains by multiple protein kinases including CamKII and PKC (Misquitta et al., 1999). Furthermore, SERCAs may be modulated by accessory proteins which are themselves phosphorylated in response to various stimuli including an increase in cytosolic Ca^{2+}. In unstimulated cardiac tissue the 52 amino acid ER trans-membrane protein phospholamban (PLB) physically interacts with SERCA2a inhibiting its activity by decreasing its affinity for Ca^{2+} (Simmerman and Jones, 1998). Following release of intracellular Ca^{2+}, or phosphorylation by protein kinase A and casein kinase II, PLB dissociates from the pump and inhibition is lost (Colyer, 1998). This mechanism is used in the heart to upregulate SERCA activity in response to β-adrenergic stimulation (Li et al., 2000). Thus, a rapid increase in SERCA activity under conditions of stress is possible. In cardiac hypertrophy however, an increase in the ratio of PLB to SERCA has been reported to result in a decrease of the Ca^{2+} pumping property of the SR and decreased speed of relaxation (Koss et al., 1997; Brittsan and Kranias, 2000).

A homologous protein to PLB, sarcolipin, has been reported to interact with SERCA1. This is a 31 amino acid proteolipid, which like PLB decreases the affinity of the pump for Ca^{2+}, but unlike PLB it increases the V_{max} of SERCA1. Developmental regulation of expression of sarcolipin is though to be its mode of regulation of SERCA1 (Odermatt et al., 1997).

CLNX and CRT have been shown to inhibit the activity of the SERCA2b isoforms specifically (Camacho and Lechleiter, 1995; John et al., 1998; Roderick et al., 2000). The interaction of these proteins with SERCA2b occurs in the ER lumen and may be modulated by conditions there. However, the interaction of

CLNX with SERCA2b is regulated by the phosphorylation status of its cytosolic domain (Fig. 2, see colorplate) (Roderick et al., 2000). Following an increase in cytosolic Ca^{2+}, CLNX is dephosphorylated and dissociates from SERCA2b resulting in increased pump activity.

The SERCA family of proteins may also functionally interact with a number of other proteins. These include the anti-apoptotic protein Bcl-2 (Kuo et al., 1998) and the insulin receptor substrate, creating a signalling link between the tyrosine phosphorylation signalling cascade and Ca^{2+} signaling (Algenstaedt et al., 1997).

The Ca^{2+} pumping activity of SERCA is also regulated by the Ca^{2+} content of the ER. SERCA activity is maximal when the store is depleted and decreases as the store approaches its maximal capacity (Favre et al., 1996; Mogami et al., 1998). The mechanism of this regulation may be multi-factorial. A direct regulation of the SERCA pump by the binding of Ca^{2+} to the lumenal face of the SERCA pump modulates ATPase activity (Webb et al., 2000). In addition, the Ca^{2+} content of the store regulates the activity of many intralumenal proteins including the interaction of chaperone proteins such as CRT and CLNX with their substrates which include SERCA (Roderick et al., 2000). A kinase activity induced by store depletion has been shown to phosphorylate PLB (Bhogal and Colyer, 1998). Following phosphorylation by this kinase, PLB would dissociate from SERCA and pump activity would increase. This kinase, although not yet cloned, has been called the state of filling kinase (SOF). Furthermore, two ER lumenal cysteine residues in SERCA1 are required for Ca^{2+} dependent activity of this protein. These residues are susceptible to intra-molecular disulphide bond formation and could thus impart regulation by redox conditions in the ER on SERCA activity (Daiho et al., 2001). The ability of the ER to sequester Ca^{2+} is also affected by the expression level of the SERCA pump. Indeed under chronic Ca^{2+} depletion of the ER, SERCA expression levels are upregulated to counteract this deficiency (Caspersen et al., 2000).

2.5. Signalling from the Inside Out

The ER, or portions of it is strategically positioned within the cell to contribute significantly to many signalling events. The most well characterised of these being release of Ca^{2+} into the cytosol, activation of store operated Ca^{2+} entry (SOC) across the plasma membrane and the regulation (Berridge et al., 2000a; Berridge et al., 2000b) and processing of proteins destined for secretion from or resident within the secretory pathway (Ellgaard et al., 1999). The diverse array of cellular processes controlled by Ca^{2+} released from the ER and the regulation of Ca^{2+} influx from the ER have been discussed at length elsewhere and so will not be addressed in this review. Here, the regulation of processes within the ER lumen will be discussed first before describing how conditions within the ER can impact cell physiology in general.

2.6. Role of Ca^{2+} in Protein Folding Within the ER.

Protein folding within the ER is performed by a group of proteins known as chaperones. These chaperones include peptidyl prolyl *cis–trans* isomerases (PPIases), protein disulphide isomerases (PDI), members of the hsp70 family of proteins (GRP78 and 94) and ER lectin chaperones (CRT, CLNX and its testes specific isoforms calmegin) (Chevet et al., 1999b). These assist in the formation of the correct conformation and oligomerisation of proteins and also have a role in quality control (Gething and Sambrook, 1992; Ellgaard et al., 1999). In the latter case, proteins with an incorrect structure are retained within the ER until they are correctly folded or oligomerised. If this is not achieved they are targeted for degradation by the proteosome pathway-a process known as ER associated degradation (ERAD) (McCracken and Brodsky, 1996). Due to the observation that ER Ca^{2+} depletion resulted in the inhibition of protein folding leading to their retention and possible degradation a link between protein folding and Ca^{2+} was made (Lodish et al., 1992). The activity of many chaperones is Ca^{2+} dependent, which imposes a requirement for optimal lumenal Ca^{2+} for protein folding to take place (Corbett et al., 1999). For example, the ability of CRT to assist in protein folding is decreased under conditions of ER Ca^{2+} depletion. The reasons for this are two-fold. First, CRT does not interact with certain of its substrates and second, it also does not recruit thio-reductases, its folding partners, to the site of action (Corbett et al., 1999). Since the activity of the UDP–glucose glycoprotein glucosyltransferase.is dependent on Ca^{2+}, the re-glucosylation of immature N-linked glycoproteins is inhibited (Trombetta and Parodi, 1992). Glycoprotein binding by lectin chaperones is an important part of the protein folding machinery (Trombetta and Helenius, 1998).

Under conditions where protein folding is inhibited malfolded proteins accumulate within the ER (Fig. 2, see colorplate) (Lodish et al., 1992). This may occur as a result of Ca^{2+} depletion, inhibition of glycosylation, expression of mutated proteins or ectopic overexpression of proteins (these are collectively called in this context cell stress). The retention of malfolded proteins as a result of a breakdown of the chaperone machinery was elegantly demonstrated in a study whereby an ATPase mutant of GRP78 was overexpressed. This had a dominant negative effect and resulted in the accumulation of malfolded proteins in the ER leading to its vesicularisation (Hendershot et al., 1995). In order to compensate for an increase in malfolded proteins, cells have developed a mechanism to upregulate the synthesis of proteins involved in protein folding within the ER (Kaufman, 1999). This was initially described for cells which had been deprived of glucose (Shiu et al., 1977). As a result of this stress, expression of a number of ER lumenal low affinity Ca^{2+} binding proteins was shown to increase. Due to their phenotypic regulation they have been named glucose-regulated proteins (GRPs) namely, GRP78 and 94, according to their molecular mass (Little et al., 1994). CRT is also induced under the same stress conditions as the GRPs (Llewellyn et al., 1996). This cellular stress response is known as the unfolded protein response or UPR (Ma and Hendershot, 2001).

The signalling pathway leading to the UPR is a subject of intense investigation. Three ER transmembrane proteins necessary for the initiation of this pathway have been cloned and their mechanism of action characterised. Two of these, Ire1p and PERK (alternatively known as PEK), are integral ER membrane proteins that have a cytosolic kinase domain and an ER lumenal 'environmental' sensor (Fig. 2, see colorplate) (Kaufman, 1999). The substrate for PERK has been demonstrated to be the initiator of translation eIf2α, which when phosphorylated results in inhibition of general translational initiation (Harding et al., 1999) although the translation of a subset of mRNAs including the Activating Transcription Factor 4 (ATF4) is selectively increased. This transcription factor subsequently induces the transcription of the downstream gene CHOP (GADD153) (Harding et al., 2000) with consequences for cell cycle arrest and possible apoptosis (Brewer and Diehl, 2000).

Ire1 has been shown to have endoribonuclease activity. In eukaryotes its target is the messenger RNA encoding the transcription factor XBP1. Following mRNA splicing by Ire1 XBP1 is translated generating the active transcription factor (Yoshida et al., 2001). The synergistic activity of PERK and Ire1 results in the coordinated upregulation of stress responsive genes and the inhibition of general protein synthesis. Although the cytosolic domains of these initiators of the stress response are divergent their lumenal domains are interchangeable suggesting a similar regulatory mechanism (Bertolotti et al., 2000). Indeed in the non-stressed cell the lumenal domains of both PERK and Ire1 are bound to GRP78. Following an increase of malfolded proteins in the ER, GRP78 dissociates from PERK and Ire1. The subsequent oligomerisation and trans-autophosphorylation of PERK and Ire1 leads to their activation (Bertolotti et al., 2000).

Another transducer of the stress response is ATF6 (Yoshida et al., 1998). Like Ire1 and PERK, ATF6 is an ER transmembrane protein containing a lumenal sensor domain and a cytosolic activation domain. Similarly to Ire1 and PERK, ATF6 is bound to GRP78 under non-stressed conditions. Under stressed conditions, ATF6 dissociates from GRP78, moves to the Golgi where it is cleaved by resident proteases thus releasing its cytosolic domain. The ATF6 cytosolic domain subsequently translocates to the nucleus where it activates the transcription of stress-responsive genes (Chen et al., 2002).

Two other stress-signalling pathways have been described in eukaryotes, the ER overload response (EOR) and the sterol response pathway. The ER overload response, like the UPR, results from an accumulation of protein in the ER (Pahl and Baeuerle, 1997). Although the EOR can be activated by certain agents that signal the UPR, for example, tunicamycin and thapsigargin, the EOR can be activated independently from the UPR by Hepatitis B virus truncated middle HB surface antigen and agents that activate NFkB such as TNF-α. Conversely, agents that activate the UPR such as ionophore and β-mercaptoethanol do not induce the EOR. The EOR signal is transduced to the nucleus by the transcription factor NFκB. Translocation of the NFκB occurs following its dissociation from its inhibitory subunit IκB. An increase in cytosolic Ca^{2+} and reactive oxygen intermediates are reported to signal this event (Fig. 2, see colorplate) (Pahl and Baeuerle,

1996). NFκB activates the transcription of cytokines and pro-inflammatory mediators which may be physiologically significant in fighting the viral infection which stimulated the EOR (Pahl, 1999).

As its name indicates, the sterol response pathway is activated by cholesterol deprivation. In response to this stimulus, the full length sterol response element binding protein transcription factor located in the ER is cleaved to yield its transcriptionally active form. This subsequently translocates to the nucleus where it binds to the sterol response element on genes responsible for cholesterol and fatty acid biosynthesis (Brown and Goldstein, 1999).

2.7. ER Plasticity and Distal Consequences

Due to recent advances in our understanding of the physiology of the ER, the interplay between the roles of the ER in Ca^{2+} signalling, protein synthesis and folding and cholesterol and lipid biosynthesis is clearly apparent. Furthermore, the environment within the ER is able to impact on distal signalling events and cellular processes. What emerges is that the ER is a highly plastic signaling organelle capable of rapidly adapting to changing conditions within the cell. The Ca^{2+} signals produced by physiological stimuli or in some cases by stress itself (Pahl, 1999) can regulate the expression of ER proteins involved in Ca^{2+} transport and release. For example expression of the IP_3R is upregulated in neurons in response to electrical stimulation (Carafoli et al., 1999; Genazzani et al., 1999; Graef et al., 1999). Similarly, ER stress can increase the expression of Ca^{2+}-binding chaperones including CRT and GRP78 and the Ca^{2+} transporter SERCA (Fig. 2, see colorplate). This represents a wholesale amplification of many members of the Ca^{2+} signalling machinery within the ER. Through this dual role of Ca^{2+}-binding proteins in Ca^{2+} signalling and as molecular chaperones together with the fact that Ca^{2+} also plays a role as both a signalling molecule and is involved in protein folding, a positive feedback loop is established to counteract the effect of stress upon the cell. In this way the cell is able to rapidly increase the functional capacity of the ER. For example, following differentiation of B-lymphocytes into immunoglobulin-secreting cells, the ER expands up to fourfold to accommodate the increased immunoglobulin secretion (Wiest et al., 1990).

Another way in which the ER can modulate cell signaling is to alter the synthesis and processing of signaling components such as the receptors localized at the plasma membrane. An interesting phenotype of cells that were deficient for calreticulin elegantly demonstrates this hypothesis. These cells were characterised by the lack of a Ca^{2+} signal in response to the agonist bradykinin. In these cells, however, a direct intracellular application of IP_3 resulted in normal Ca^{2+} release indicating that the ER Ca^{2+} signalling machinery was intact. From these data and others the authors concluded that CRT was necessary for surface expression of the bradykinin receptor (Nakamura et al., 2001). Furthermore, as a result of deficiencies in the protein folding machinery or mutations in plasma membrane proteins the surface expression of many proteins is affected. For example, the most common mutation in cystic fibrosis patients is deletion of amino acid phenylalanine 508 (delta (Δ)F508) in the cystic fibrosis chloride transporter protein (CFTR). Although functional, this ΔF508 mutant is not efficiently transported to the plasma membrane and is retained in the ER (Yang et al., 1993). In light of certain examples given and other work not discussed, a possible therapeutic target for the treatment of many diseases may be to functionally inhibit or enhance the protein folding machinery. This may be achieved either directly through the manipulation of the expression levels of ER chaperones or indirectly through ER Ca^{2+} levels. In this way functional proteins retained in the ER as a result of their misfolding may reach their expected destinations where they can perform their normal function.

2.8. Conclusion

The roles of the ER in Ca^{2+} signalling, protein synthesis and folding and cholesterol and lipid biosynthesis have often been regarded in isolation. It is now clear, however, that due to the significant interplay between these functions that manipulation of any one of these functions of the ER may impact on any, or all, of the other known functions. Furthermore, the distribution of the ER throughout the cytosol and its proximity to other cellular organelles including the mitochondria, nucleus, Golgi and plasma membrane strategically positions it to regulate the functions of these organelles and thus to play a major role in intracellular signalling.

Acknowledgements

We greatly appreciate the assistance of Dr. Tony Collins for assistance with the preparation of Fig.1.

References

Algenstaedt, P, Antonetti, DA, Yaffe, MB, and Kahn, CR (1997). Insulin receptor substrate proteins create a link between the tyrosine phosphorylation cascade and the Ca^{2+}-ATPases in muscle and heart, J Biol Chem *272*, 23696-702.

Allbritton, NL, Meyer, T, and Stryer, L (1992). Range of messenger action of calcium ion and inositol 1,4,5-trisphosphate, Science *258*, 1812-15.

Arnaudeau, S, Kelley, WL, Walsh, JV, Jr., and Demaurex, N (2001). Mitochondria recycle Ca^{2+} to the endoplasmic reticulum and prevent the depletion of neighboring endoplasmic reticulum regions, J Biol Chem *276*, 29430-9.

Baksh, S, and Michalak, M (1991). Expression of calreticulin in Escherichia coli and identification of its Ca^{2+} binding domains, J Biol Chem *266*, 21458-65.

Baumann, O, Walz, B, Somlyo, AV, and Somlyo, AP (1991). Electron probe microanalysis of calcium release and magnesium uptake by endoplasmic reticulum in bee photoreceptors, Proc Natl Acad Sci U S A *88*, 741-4.

Berridge, MJ (1993). Inositol trisphosphate and calcium signalling, Nature *361*, 315-325.

Berridge, MJ, Lipp, P, and Bootman, MD (2000a). Signal transduction. The calcium entry pas de deux, Science *287*, 1604-5.

Berridge, MJ, Lipp, P, and Bootman, MD (2000b). The versatility and universality of calcium signalling, Nat Rev Mol Cell Biol *1*, 11-21.

Bertolotti, A, Zhang, Y, Hendershot, LM, Harding, HP, and Ron, D (2000). Dynamic interaction of BiP and ER stress transducers in the unfolded- protein response, Nat Cell Biol *2*, 326-32.

Bezprozvanny, I, Watras, J, and Ehrlich, BE (1991). Bell-shaped calcium-response curves of Ins(1,4,5)P_3- and calcium-gated channels from endoplasmic reticulum of cerebellum, Nature *351*, 751-754.

Bhogal, MS, and Colyer, J (1998). Depletion of Ca^{2+} from the sarcoplasmic reticulum of cardiac muscle prompts phosphorylation of phospholamban to stimulate store refilling, Proc Natl Acad Sci U S A *95*, 1484-9.

Blaustein, MP, and Golovina, VA (2001). Structural complexity and functional diversity of endoplasmic reticulum Ca^{2+} stores, Trends Neurosci *24*, 602-8.

Bootman, MD, and Berridge, MJ (1995). The elemental principles of calcium signaling, Cell *83*, 675-8.

Bootman, MD, Berridge, MJ, and Lipp, P (1997). Cooking with calcium: the recipes for composing global signals from elementary events, Cell *91*, 367-73.

Brewer, JW, and Diehl, JA (2000). PERK mediates cell-cycle exit during the mammalian unfolded protein response, Proc Natl Acad Sci U S A *97*, 12625-30.

Brittsan, AG, and Kranias, EG (2000). Phospholamban and cardiac contractile function, J Mol Cell Cardiol *32*, 2131-9.

Bootman, MD, Lipp, P, and Berridge, MJ (2001). The organisation and functions of local Ca($^{2+}$) signals, J Cell Sci *114*, 2213-22.

Brown, MS, and Goldstein, JL (1999). A proteolytic pathway that controls the cholesterol content of membranes, cells, and blood, Proc Natl Acad Sci U S A *96*, 11041-8.

Camacho, P, and Lechleiter, JD (1995). Calreticulin inhibits repetitive calcium waves, Cell *82*, 765-771.

Cancela, JM, Van Coppenolle, F, Galione, A, Tepikin, AV, and Petersen, OH (2002). Transformation of local Ca^{2+} spikes to global Ca^{2+} transients: the combinatorial roles of multiple Ca^{2+} releasing messengers, Embo J *21*, 909-19.

Carafoli, E, Genazzani, A, and Guerini, D (1999). Calcium controls the transcription of its own transporters and channels in developing neurons, Biochem Biophys Res Commun *266*, 624-32.

Caspersen, C, Pedersen, PS, and Treiman, M (2000). The sarco/endoplasmic reticulum calcium-ATPase 2b is an endoplasmic reticulum stress-inducible protein, J Biol Chem *275*, 22363-72.

Cheek, TR, Berridge, MJ, Moreton, RB, Stauderman, KA, Murawsky, MM, and Bootman, MD (1994). Quantal Ca^{2+} mobilization by ryanodine receptors is due to all-or-none release from functionally discrete intracellular stores, Biochem J *301*, 879-83.

Chen, X, Shen, J, and Prywes, R (2002). The Luminal Domain of ATF6 Senses Endoplasmic Reticulum (ER) Stress and Causes Translocation of ATF6 from the ER to the Golgi, J Biol Chem *277*, 13045-52.

Chevet, E, Jakob, CA, Thomas, DY, and Bergeron, JJ (1999a). Calnexin family members as modulators of genetic diseases, Semin Cell Dev Biol *10*, 473-80.

Chevet, E, Lemaitre, G, Cailleret, K, Dahan, S, Bergeron, JJ, and Katinka, MD (1999b). Identification and characterization of an intracellular protein complex that binds fibroblast growth factor-2 in bovine brain, Biochem J *341*, 713-23.

Ching, LL, Williams, AJ, and Sitsapesan, R (2000). Evidence for Ca^{2+} activation and inactivation sites on the luminal side of the cardiac ryanodine receptor complex, Circ Res *87*, 201-6.

Collins, TJ, Berridge, MJ, Lipp, P, and Bootman, MD (2002). Mitochondria are morphologically and functionally heterogeneous within cells, Embo J *21*, 1616-27.

Colyer, J (1998). Phosphorylation states of phospholamban, Ann N Y Acad Sci *853*, 79-91.

Corbett, EF, Oikawa, K, Francois, P, Tessier, DC, Kay, C, Bergeron, JJ, Thomas, DY, Krause, KH, and Michalak, M (1999). Ca^{2+} regulation of interactions between endoplasmic reticulum chaperones, J Biol Chem *274*, 6203-11.

Daiho, T, Yamasaki, K, Saino, T, Kamidochi, M, Satoh, K, Iizuka, H, and Suzuki, H (2001). Mutations of either or both Cys876 and Cys888 residues of sarcoplasmic reticulum Ca^{2+}-ATPase result in a complete loss of Ca^{2+} transport activity without a loss of Ca^{2+}-dependent ATPase activity. Role of the CYS876-CYS888 disulfide bond, J Biol Chem *276*, 32771-8.

Ellgaard, L, Molinari, M, and Helenius, A (1999). Setting the standards: quality control in the secretory pathway, Science *286*, 1882-8.

Endo, M, Tanaka, M, and Ogawa, Y (1970). Calcium induced release of calcium from the sarcoplasmic reticulum of skinned skeletal muscle fibres, Nature *228*, 34-6.

Favre, CJ, Schrenzel, J, Jacquet, J, Lew, DP, and Krause, KH (1996). Highly supralinear feedback inhibition of Ca^{2+} uptake by the Ca^{2+} load of intracellular stores, J Biol Chem *271*, 14925-30.

Franzini-Armstrong, C, Protasi, F, and Ramesh, V (1998). Comparative ultrastructure of Ca^{2+} release units in skeletal and cardiac muscle, Ann N Y Acad Sci *853*, 20-30.

Franzini-Armstrong, C (1999). The sarcoplasmic reticulum and the control of muscle contraction, Faseb J *13 Suppl 2*, S266-70.

Galione, A, and Churchill, GC (2000). Cyclic ADP ribose as a calcium-mobilizing messenger, Sci STKE *2000*, PE1.

Genazzani, AA, Carafoli, E, and Guerini, D (1999). Calcineurin controls inositol 1,4,5-trisphosphate type 1 receptor expression in neurons, Proc Natl Acad Sci U S A *96*, 5797-801.

Gething, MJ, and Sambrook, J (1992). Protein folding in the cell, Nature *355*, 33-45.

Go, LO, Moschella, MC, Watras, J, Handa, KK, Fyfe, BS, and Marks, AR (1995). Differential regulation of two types of intracellular calcium release channels during end-stage heart failure, J Clin Invest *95*, 888-94.

Graef, IA, Mermelstein, PG, Stankunas, K, Neilson, JR, Deisseroth, K, Tsien, RW, and Crabtree, GR (1999). L-type calcium channels and GSK-3 regulate the activity of NF-ATc4 in hippocampal neurons, Nature *401*, 703-8.

Hammond, C, and Helenius, A (1995). Quality control in the secretory pathway, Curr Opin Cell Biol *7*, 523-9.

Harding, HP, Novoa, I, Zhang, Y, Zeng, H, Wek, R, Schapira, M, and Ron, D (2000). Regulated translation initiation controls stress-induced gene expression in mammalian cells, Mol Cell *6*, 1099-108.

Harding, HP, Zhang, Y, and Ron, D (1999). Protein translation and folding are coupled by an endoplasmic-reticulum- resident kinase, Nature *397*, 271-4.

Heilmann, C, Spamer, C, and Gerok, W (1983). The phosphoprotein intermediate of a Ca^{2+} transport ATPase in rat liver endoplasmic reticulum, Biochem. Biophys. Res. Comm. *114*, 584-592.

Helenius, A, Trombetta, ES, Hebert, Dn, and Simmons, JF (1997). Calnexin, calreticulin and the folding of glycoproteins., Trends Cell Biol *7*, 193-200.

Hendershot, LM, Wei, JY, Gaut, JR, Lawson, B, Freiden, PJ, and Murti, KG (1995). In vivo expression of mammalian BiP ATPase mutants causes disruption of the endoplasmic reticulum, Mol Biol Cell *6*, 283-96.

Iino, M (1999). Dynamic regulation of intracellular calcium signals through calcium release channels, Mol Cell Biochem *190*, 185-90.

John, LM, Lechleiter, JD, and Camacho, P (1998). Differential modulation of SERCA2 isoforms by calreticulin, J. Cell Biol. *142*, 963-73.

Joseph, SK, Boehning, D, Bokkala, S, Watkins, R, and Widjaja, J (1999). Biosynthesis of inositol trisphosphate receptors: selective association with the molecular chaperone calnexin, Biochem J *342*, 153-61.

Kaufman, RJ (1999). Stress signaling from the lumen of the endoplasmic reticulum: coordination of gene transcriptional and translational controls, Genes Dev *13*, 1211-33.

Koizumi, S, Lipp, P, Berridge, MJ, and Bootman, MD (1999). Regulation of ryanodine receptor opening by lumenal Ca($^{2+}$) underlies quantal Ca($^{2+}$) release in PC12 cells., J. Biol. Chem. *274*, 33327-33.

Koss, KL, Grupp, IL, and Kranias, EG (1997). The relative phospholamban and SERCA2 ratio: a critical determinant of myocardial contractility, Basic Res Cardiol *92*, 17-24.

Krause, KH, and Michalak, M (1997). Calreticulin, Cell *88*, 439-43.

Kuo, TH, Kim, HR, Zhu, L, Yu, Y, Lin, HM, and Tsang, W (1998). Modulation of endoplasmic reticulum calcium pump by Bcl-2, Oncogene *17*, 1903-10.

Lechleiter, J, Girard, S, Peralta, E, and Clapham, D (1991). Spiral calcium wave propagation and annihilation in Xenopus laevis oocytes, Science *252*, 123-6.

Li, L, Desantiago, J, Chu, G, Kranias, EG, and Bers, DM (2000). Phosphorylation of phospholamban and troponin I in beta-adrenergic- induced acceleration of cardiac relaxation, Am. J. Physiol. Heart Circ. Physiol. *278*, H769-79.

Lievremont, JP, Rizzuto, R, Hendershot, L, and Meldolesi, J (1997). BiP, a major chaperone protein of the endoplasmic reticulum lumen, plays a direct and important role in the storage of the rapidly exchanging pool of Ca^{2+}, J. Biol. Chem. *272*, 30873-9.

Little, E, Ramakrishnan, M, Roy, B, Gazit, G, and Lee, AS (1994). The glucose-regulated proteins (GRP78 and GRP94): functions, gene regulation, and applications, Crit Rev Eukaryot. Gene Expr. *4*, 1-18.

Llewellyn, DH, Kendall, JM, Sheikh, FN, and Campbell, AK (1996). Induction of calreticulin expression in HeLa cells by depletion of the endoplasmic reticulum Ca^{2+} store and inhibition of N-linked glycosylation, Biochem. J. *318*, 555-60.

Lodish, HF, Kong, N, and Wikstrom, L (1992). Calcium is required for folding of newly made subunits of the asialoglycoprotein receptor within the endoplasmic reticulum, J. Biol. Chem. *267*, 12753-60.

Loomis-Husselbee, JW, and Dawson, AP (1993). A steady-state mechanism can account for the properties of inositol 1,4,5-trisphosphate-stimulated Ca^{2+} release from permeabilized L1210 cells, Biochem. J. *289*, 861-6.

Loukianov, E, Ji, Y, Grupp, IL, Kirkpatrick, DL, Baker, DL, Loukianova, T, Grupp, G, Lytton, J, Walsh, RA, and Periasamy, M (1998). Enhanced myocardial contractility and increased Ca^{2+} transport function in transgenic hearts expressing the fast-twitch skeletal muscle sarcoplasmic reticulum Ca^{2+}-ATPase, Circ. Res. *83*, 889-97.

Lytton, J, Westlin, M, Burk, SE, Shull, GE, and MacLennan, DH (1992). Functional comparisons between isoforms of the sarcoplasmic or endoplasmic reticulum family of calcium pumps, J. Biol. Chem. *267*, 14483-9.

Ma, Y, and Hendershot, LM (2001). The unfolding tale of the unfolded protein response, Cell *107*, 827-30.

Macer, DRJ, and Koch, GLE (1988). Identification of a set of calcium-binding proteins in reticuloplasm, the luminal content of the endoplasmic reticulum, J Cell Science *91*, 61-70.

MacLennan, DH (1974). Isolation of a second form of calsequestrin, J. Biol. Chem. *249*, 980-4.

MacLennan, DH, Campbell, KP, and Reithmeier, RAF (1983). Calsequestrin. In Calcium Cell Function, W Y Cheung, ed. (New York, Academic Press), pp. 151-173.

MacLennan, DH, and Klip, A (1979). Calcium transport and release by sarcoplasmic reticulum: a mini-review, Soc .Gen. Physiol. Ser. *33*, 61-75.

MacLennan, DH, Rice, WJ, and Green, NM (1997). The mechanism of Ca^{2+} transport by sarco(endo)plasmic reticulum Ca^{2+}-ATPases, J. of Biol. Chem. *272*, 28815-8.

Mazzarella, RA, and Green, M (1987). ERp99, an abundant, conserved glycoprotein of the endoplasmic reticulum, is homologous to the 90-kDa heat shock protein (hsp900 and the 94-kDa glucose regulated protein (GRP94), J. Biol. Chem. *262*, 8875-8883.

McCracken, AA, and Brodsky, JL (1996). Assembly of ER-associated protein degradation in vitro: dependence on cytosol, calnexin, and ATP, J. Cell Biol. *132*, 291-8.

Meldolesi, J, and Pozzan, T (1998). The endoplasmic reticulum Ca^{2+} store: a view from the lumen, Trends Biochem Sci *23*, 10-4.

Michalak, M, Baksh, S, and Opas, M (1991). Identification and immunolocalization of calreticulin in pancreatic cells: no evidence for "calciosomes", Exp Cell Res *197*, 91-9.

Milner, RE, Baksh, S, Shemanko, C, Carpenter, MR, Smillie, L, Vance, JE, Opas, M, and Michalak, M (1991). Calreticulin, and not calsequestrin, is the major calcium binding protein of smooth muscle sarcoplasmic reticulum asnd liver endoplasmic reticulum, J. Biol. Chem. *266*, 7155-7165.

Misquitta, CM, Mack, DP, and Grover, AK (1999). Sarco/endoplasmic reticulum Ca^{2+} (SERCA)-pumps: link to heart beats and calcium waves, Cell Calcium *25*, 277-90.

Missiaen, L, Taylor, CW, and Berridge, MJ (1991). Spontaneous calcium release from inositol trisphosphate-sensitive calcium stores, Nature *352*, 241-4.

Miyakawa, T, Maeda, A, Yamazawa, T, Hirose, K, Kurosaki, T, and Iino, M (1999). Encoding of Ca^{2+} signals by differential expression of IP3 receptor subtypes, Embo. J. *18*, 1303-8.

Miyawaki, A, Llopis, J, Heim, R, McCaffery, JM, Adams, JA, Ikura, M, and Tsien, RY (1997). Fluorescent indicators for Ca^{2+} based on green fluorescent proteins and calmodulin, Nature *388*, 882-7.

Mogami, H, Tepikin, AV, and Petersen, OH (1998). Termination of cytosolic Ca^{2+} signals: Ca^{2+} reuptake into intracellular stores is regulated by the free Ca^{2+} concentration in the store lumen, Embo. J. *17*, 435-42.

Nakamura, K, Zuppini, A, Arnaudeau, S, Lynch, J, Ahsan, I, Krause, R, Papp, S, De Smedt, H, Parys, JB, Muller-Esterl, W, *et al.* (2001). Functional specialization of calreticulin domains, J. Cell Biol. *154*, 961-72.

Nigam, SK, Goldberg, AL, Ho, S, Rohde, MF, Bush, KT, and Sherman, M (1994). A set of endoplasmic reticulum proteins possessing properties of molecular chaperones includes Ca^{2+}-binding proteins and members of the thioredoxin superfamily, J. Biol. Chem. *269*, 1744-9.

Oberdorf, J, Webster, JM, Zhu, CC, Luo, SG, and Wojcikiewicz, RJ (1999). Down-regulation of types I, II and III inositol 1,4,5-trisphosphate receptors is mediated by the ubiquitin/proteasome pathway, Biochem. J. *339*, 453-61.

Odermatt, A, Taschner, PE, Scherer, SW, Beatty, B, Khanna, VK, Cornblath, DR, Chaudhry, V, Yee, WC, Schrank, B, Karpati, G, *et al.* (1997). Characterization of the gene encoding human sarcolipin (SLN), a proteolipid associated with SERCA1: absence of structural mutations in five patients with Brody disease, Genomics *45*, 541-53.

Ohkura, M, Furukawa, K, Fujimori, H, Kuruma, A, Kawano, S, Hiraoka, M, Kuniyasu, A, Nakayama, H, and Ohizumi, Y (1998). Dual regulation of the skeletal muscle ryanodine receptor by triadin and calsequestrin, Biochemistry *37*, 12987-93.

Ostwald, TJ, and MacLennan, DH (1974). Isolation of a high affinity calcium-binding protein from sarcoplasmic reticulum, J. Biol. Chem. *249*, 974-9.

Pahl, HL (1999). Signal transduction from the endoplasmic reticulum to the cell nucleus, Physiol. Rev. *79*, 683-701.

Pahl, HL, and Baeuerle, PA (1996). Activation of NF-kappa B by ER stress requires both Ca^{2+} and reactive oxygen intermediates as messengers, FEBS Lett *392*, 129-36.

Pahl, HL, and Baeuerle, PA (1997). The ER-overload response: activation of NF-kappa B, Trends Biochem. Sci. *22*, 63-7.

Parys, JB, Missiaen, L, De Smedt, H, and Casteels, R (1993). Loading dependence of inositol 1,4,5-trisphosphate-induced calcium release in the clonal cell line A7r5, J. Biol. Chem. *268*, 25206-25212.

Patel, S, Joseph, SK, and Thomas, AP (1999). Molecular properties of inositol 1,4,5-trisphosphate receptors, Cell Calcium *25*, 247-64.

Pozzan, T, Rizzuto, R, Volpe, P, and Meldolesi, J (1994a). Molecular and cellular physiology of intracellular calcium stores, Physiol. Rev. *74*, 595-636.

Pozzan, T, Rizzuto, R, Volpe, P, and Meldolesi, J (1994b). Molecular and cellular physiology of intracellular calcium stores., Physiol. Rev. *74*, 595-636.

Putney, JW, Jr. (1990). Capacitative calcium entry revisited, Cell Calcium *11*, 611-24

Roderick, HL, Lechleiter, JD, and Camacho, P (2000). Cytosolic phosphorylation of calnexin controls intracellular Ca^{2+} oscillations via an interaction with SERCA2b, J. Cell Biol. *149*, 1235-48.

Ross, CA, Meldolesi, J, Milner, TA, Satoh, T, Supattapone, S, and Snyder, SH (1989). Inositol 1,4,5-trisphosphate receptor localized to endoplamic reticulum in cerebellar Purkinje neurones, Nature *339*, 468-470.

Sharma, VK, Ramesh, V, Franzini-Armstrong, C, and Sheu, SS (2000). Transport of Ca^{2+} from sarcoplasmic reticulum to mitochondria in rat ventricular myocytes, J Bioenerg Biomembr *32*, 97-104.

Shiu, RP, Pouyssegur, J, and Pastan, I (1977). Glucose depletion accounts for the induction of two transformation- sensitive membrane proteins in Rous sarcoma virus-transformed chick embryo fibroblasts, Proc. Natl. Acad. Sci. U S A *74*, 3840-4.

Sienaert, I, De Smedt, H, Parys, JB, Missiaen, L, Vanlingen, S, Sipma, H, and Casteels, R (1996). Characterization of a cytosolic and a luminal Ca^{2+} binding site in the type I inositol 1,4,5-trisphosphate receptor, J. Biol. Chem. *271*, 27005-12.

Simmerman, HK, and Jones, LR (1998). Phospholamban: protein structure, mechanism of action, and role in cardiac function, Physiol. Rev. *78*, 921-47.

Somlyo, AP, Bond, M, and Somlyo, AV (1985). Calcium content of mitochondria and endoplasmic reticulum in liver frozen rapidly in vivo, Nature *314*, 622-5.

Sorrentino, V (1995). The ryanodine receptor family of intracellular calcium release channels, Adv. Pharmacol. *33*, 67-90.

Streb, H, Irvine, RF, Berridge, MJ, and Schulz, I (1983). Release of Ca^{2+} from a nonmitohondrial intracellular store in pancreatic acinar cells by inositol-1,4,5-trisphosphate, Nature *306*, 67-69.

Tatu, U, and Helenius, A (1997). Interactions between newly synthesized glycoproteins, calnexin and a network of resident chaperones in the endoplasmic reticulum, J Cell Biol *136*, 555-65.

Tovey, SC, de Smet, P, Lipp, P, Thomas, D, Young, KW, Missiaen, L, De Smedt, H, Parys, JB, Berridge, MJ, Thuring, J, *et al.* (2001). Calcium puffs are generic InsP(3)-activated elementary calcium signals and are downregulated by prolonged hormonal stimulation to inhibit cellular calcium responses, J Cell Sci. *114*, 3979-89.

Trombetta, ES, and Helenius, A (1998). Lectins as chaperones in glycoprotein folding, Curr. Opin. Struct. Biol. *8*, 587-92.

Trombetta, SE, and Parodi, AJ (1992). Purification to apparent homogeneity and partial characterization of rat liver UDP-glucose:glycoprotein glucosyltransferase., J. Biol. Chem. *267*, 9236-40.

Ver Heyen, M, Heymans, S, Antoons, G, Reed, T, Periasamy, M, Awede, B, Lebacq, J, Vangheluwe, P, Dewerchin, M, Collen, D, *et al.* (2001). Replacement of the muscle-specific sarcoplasmic reticulum Ca^{2+}-ATPase isoform SERCA2a by the non-muscle SERCA2b homologue causes mild concentric hypertrophy and impairs contraction-relaxation of the heart, Circ. Res. *89*, 838-46.

Voelker, DR (1991). Organelle biogenesis and intracellular lipid transport in eukaryotes, Microbiol. Rev. *55*, 543-60.

Volpe, P, Krause, KH, Hashimoto, S, Zorzato, F, Pozzan, T, Meldolesi, J, and Lew, DP (1988). "Calciosome," a cytoplasmic organelle: the inositol 1,4,5-trisphosphate- sensitive Ca^{2+} store of nonmuscle cells?, Proc. Natl. Acad. Sci. U S A *85*, 1091-5.

Webb, RJ, Khan, YM, East, JM, and Lee, AG (2000). The importance of carboxyl groups on the lumenal side of the membrane for the function of the Ca^{2+}-ATPase of sarcoplasmic reticulum, J. Biol. Chem. *275*, 977-82.

Wiest, DL, Burkhardt, JK, Hester, S, Hortsch, M, Meyer, DI, and Argon, Y (1990). Membrane biogenesis during B cell differentiation: most endoplasmic reticulum proteins are expressed coordinately, J. Cell. Biol. *110*, 1501-11.

Yang, Y, Janich, S, Cohn, JA, and Wilson, JM (1993). The common variant of cystic fibrosis transmembrane conductance regulator is recognized by hsp70 and degraded in a pre-Golgi non-lysosomal compartment, Proc. Natl. Acad. Sci. U S A *90*, 9480-4.

Yang, J., McBride, S., Mak, D.-O., Vardi, N., Palczewski, F.H. and Foskett, J.K. (2002). Identification of a family of calcium sensors as protein ligands of inositol trisphosphate receptor Ca^{2+} release channels. Proc. Natl. Acad. Sci. USA, May 7; [epub ahead of print]

Yoshida, H, Haze, K, Yanagi, H, Yura, T, and Mori, K (1998). Identification of the cis-acting endoplasmic reticulum stress response element responsible for transcriptional induction of mammalian glucose- regulated proteins. Involvement of basic leucine zipper transcription factors, J. Biol. Chem. *273*, 33741-9.

Yoshida, H, Matsui, T, Yamamoto, A, Okada, T, and Mori, K (2001). XBP1 mRNA is induced by ATF6 and spliced by IRE1 in response to ER stress to produce a highly active transcription factor, Cell *107*, 881-91.

Zhang, L, Kelley, J, Schmeisser, G, Kobayashi, YM, and Jones, LR (1997). Complex formation between junctin, triadin, calsequestrin, and the ryanodine receptor. Proteins of the cardiac junctional sarcoplasmic reticulum membrane, J. Biol. Chem. *272*, 23389-97.

3. Mitochondrial and ER Calcium Uptake and Release Fluxes and their Interplay in Intact Nerve Cells

David D. Friel

Dept. of Neurosciences, Case Western Reserve University, Cleveland, Ohio, USA

Abstract. Ionized free Ca (Ca^{2+}) is a ubiquitous signaling ion that serves as the critical link between a variety of physiological stimuli and their intracellular effectors. Previous studies of reduced *in vitro* preparations have provided functional characterizations of various Ca^{2+} channels, pumps and exchangers that regulate cellular Ca^{2+} movements. However, little is known about the functional interplay between transporters that are expressed together in intact cells and orchestrate stimulus-evoked changes in $[Ca^{2+}]$. This review summarizes recent progress in characterizing Ca^{2+} transporters in sympathetic neurons, which provide an ideal model for studying Ca^{2+} dynamics in neurons. Our results show how the functional interplay between Ca^{2+} transport systems that are regulated by Ca^{2+} in quantitatively different ways leads to emergent properties of Ca^{2+} signaling that are expected to play a critical role in defining how Ca^{2+} serves its role as a signaling ion.

3.1. Introduction

Calcium is an important regulator of many cellular processes, serving as the critical link between a variety of physiological stimuli and their intracellular effectors (Berridge, 1998; Meldolesi, 2001). Processes of general biological relevance known to be regulated by Ca include ATP production, protein processing, and gene expression, and specialized Ca^{2+}-sensitive events found in neurons include neurotransmitter release, action potential generation and neurite outgrowth. The constellation of Ca^{2+}-regulated processes found in each cell type depends on the Ca-sensitive effectors it expresses, and where they are expressed in membrane-delimited compartments such as the cytoplasm, mitochondria and endoplasmic reticulum (ER). Since most of the regulatory effects of Ca are mediated by the free ion (Ca^{2+}) through its interactions with Ca^{2+} binding proteins according to the law of mass action, the critical determinant of Ca^{2+} action is the Ca^{2+} concentration ($[Ca^{2+}]$). Thus, $[Ca^{2+}]$ is tightly regulated in each cellular compartment. Regulation is accomplished by specific Ca^{2+} transport pathways (e.g. pumps, exchangers and channels) that can be classified as active or passive depending on whether they lead to Ca^{2+} movement in energetically 'uphill" or "downhill" directions, respectively. Both passive and active Ca^{2+} transport are subject to regulation, and a central regulator is Ca^{2+} itself, which acts both as a substrate for transport and as a modulator of transport activity.

In this review, I will describe the coordinated changes in intracellular calcium concentration that are set into motion by depolarization-evoked Ca^{2+} entry in sympathetic neurons. These cells have been very useful as a model system for studying fundamental properties of Ca^{2+} signal generation that are likely to be relevant to all cells. Inspired by classical studies of membrane electrical phenomena, we used a combination of experimental approaches to separate and characterize the Ca^{2+} fluxes responsible for Ca^{2+} dynamics during and after stimulation. These approaches include measurement of the cytoplasmic free Ca^{2+} concentration, voltage-sensitive Ca^{2+} current, and total calcium concentrations (free + bound) in multiple cellular compartments. Taken together, our measurements provide a picture of the changes in intracellular Ca^{2+} concentration that are initiated by stimulated Ca^{2+} entry, as well as an explanation of these changes in terms of the underlying Ca^{2+} transport and buffering systems. We found that for each compartment, calcium dynamics depend on the functional interplay between multiple active and passive transport systems that are differentially regulated by Ca^{2+}. We also found that stimulus-evoked changes in Ca^{2+} concentration in different compartments are highly interdependent, owing to functional coupling between the systems that regulate Ca^{2+} movements between the compartments. One of the main conclusions is that coupling makes it possible for stimuli that differ only quantitatively to produce qualitatively different Ca^{2+} responses, accounting for the existence of distinct signaling regimes that greatly increase the flexibility of Ca^{2+} signaling. The properties of each regime, and indeed, which regime can be expressed in a given cell, depend critically on the Ca^{2+} transporters and buffers that are expressed, as well as their level of expression and state of modulation. Calcium dynamics thus provides an exemplary case of a complex intracellular signaling system.

This review is not intended to be a comprehensive discussion of the field of calcium signaling, but rather a summary of our work on sympathetic neurons over the last few years. Space limitations make it impossible to cite all the original studies that we have drawn upon, but references to review articles have been included to help guide interested readers to the work of other investigators on a variety of preparations.

3.2. Orientation, Basic Principles and Terminology

We will discuss calcium dynamics in three cellular compartments: the cytoplasm, mitochondria and endoplasmic reticulum (Fig. 1), focusing on changes in free and total calcium concentration that follow depolarization-evoked Ca^{2+} entry across the plasma membrane. This may be thought of as an input/output problem, where the input is a time-varying membrane depolarization, conductance or Ca^{2+} current. A few preliminary remarks are appropriate. First, for each compartment, the total Ca concentration ([Ca]) changes at a rate that depends on the ratio of the net intercompartmental Ca flux (J, e.g. in nmoles/s) to the compartmental volume (v). Second, if Ca^{2+} is uniformly distributed within the compartment, and binding to buffers is always near equilibrium, then $[Ca^{2+}]$ changes at a rate that can be approximated by $\tilde{J}/(v\kappa)$ ($\equiv J$, in nM/s). The factor κ describes how [Ca] changes with small variations in $[Ca^{2+}]$, and $(v\kappa)$ can be interpreted as an effective compartmental volume (Colegrove et al., 2000a). The flux $\tilde{J}$ is a material flux integrated over a compartmental boundary, while J may be thought of as the corresponding 'concentration flux'. Third, each intercompartmental Ca flux is the sum of 'macroscopic' fluxes generated by different populations of transporters. Finally, the macroscopic fluxes are differentially regulated by $[Ca^{2+}]$. Therefore, to understand how $[Ca^{2+}]$ changes in response to a stimulus, it is necessary to examine the macroscopic Ca^{2+} fluxes that collectively define the intercompartmental Ca^{2+} fluxes and how they are regulated by $[Ca^{2+}]$.

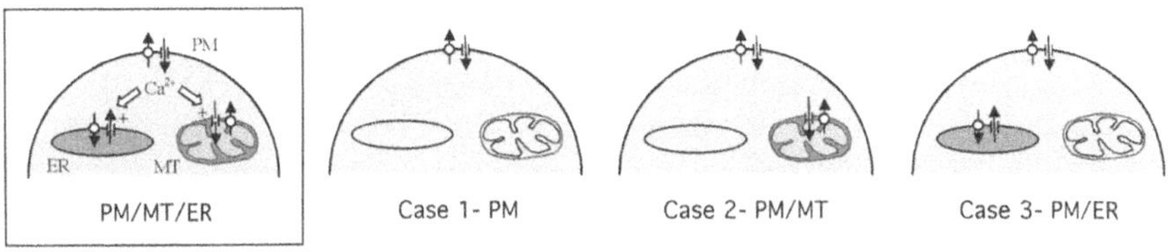

Fig. 1. Major cellular compartments that show regulated Ca^{2+} transport. (Left) Three compartments in which free Ca^{2+} concentration is regulated by specific transport and buffering mechanisms: cytoplasm, mitochondria (MT) and endoplasmic reticulum (ER). Black arrows depict macroscopic Ca^{2+} fluxes generated by active and passive transport systems (shown with and without circles, respectively). White arrows indicate two pathways by which cytoplasmic Ca^{2+} can regulate passive Ca^{2+} movements between the MT, ER and cytoplasm. Right panels depict Ca^{2+} regulation under three simplified conditions. In Case 1, Ca^{2+} dynamics can be described by a single dynamical variable $[Ca^{2+}]_i$ under the influence of plasma membrane transport systems. In Cases 2 and 3 a single organelle is functional and there are two dynamical variables ($[Ca^{2+}]_i$, $[Ca^{2+}]_{MT}$, Case 2), ($[Ca^{2+}]_i$,$[Ca^{2+}]_{ER}$, Case 3). An understanding of $[Ca^{2+}]$ dynamics in Cases 1-3 provides a foundation for analyzing the physiological case where both organelles transport Ca^{2+}.

Stimulus-evoked Ca^{2+} signals in native cells represent the most complicated case, since they involve multiple compartments and interactions between many Ca^{2+} transport systems. Therefore, it is helpful to begin by analyzing Ca^{2+} dynamics in one compartment under the control of a single intercompartmental Ca^{2+} flux. For this purpose, we will start with cytoplasmic $[Ca^{2+}]$ dynamics under the control of surface membrane Ca^{2+} transport in the absence of transport by organelles (Fig. 1, Case 1). This case is important because the cytoplasm provides the context within which both mitochondrial and ER Ca^{2+} transport normally occurs. There is one dynamical variable, $[Ca^{2+}]_i$, whose temporal properties are defined by Ca^{2+} transporters situated at the plasma membrane, cytoplasmic volume and buffering strength. Since this section introduces terms and ideas that are used in later sections, care will be given to their development so they can be used in subsequent sections. Next, we will describe Ca^{2+} dynamics when transport by mitochondria is enabled but ER Ca^{2+} transport is inhibited (Fig. 1, Case 2). This case requires consideration of two dynamical variables ($[Ca^{2+}]_i$, $[Ca^{2+}]_{MT}$) and the interplay between surface membrane and mitochondrial Ca^{2+} transport. We will then discuss the reciprocal case where ER Ca^{2+} transport is enabled and mitochondrial transport is inhibited (Fig.1, Case 3), focusing on the two dynamical variables $[Ca^{2+}]_i$ and $[Ca^{2+}]_{ER}$. Finally, we will discuss the most interesting, and physiologically relevant, case in which both mitochondria and the ER contribute to Ca^{2+} dynamics (Fig. 1, left). This requires consideration of three dynamical variables ($[Ca^{2+}]_i$, $[Ca^{2+}]_{MT}$, $[Ca^{2+}]_{ER}$) that are influenced by all Ca^{2+} transport and buffering systems. As we will see, understanding the components of the Ca^{2+} regulatory network in isolation greatly facilitates an understanding of overall system dynamics when all components are operational.

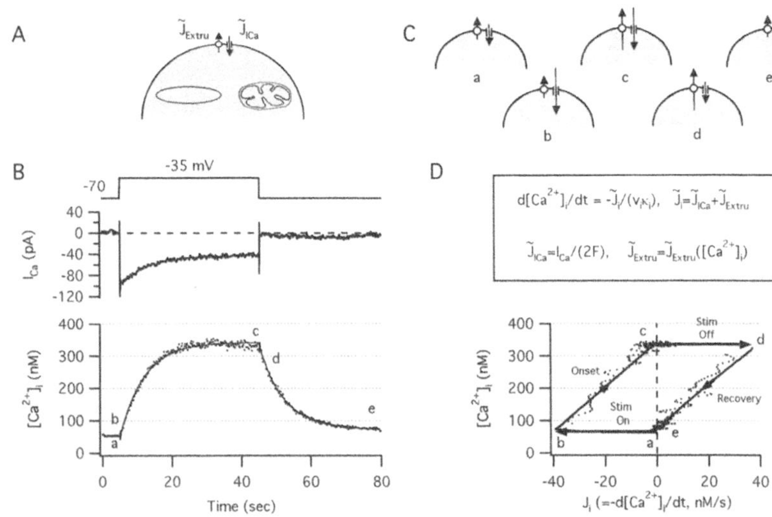

Fig. 2. Cytoplasmic $[Ca^{2+}]$ regulation by surface membrane Ca^{2+} transport systems. (A) Diagram illustrating the one compartment case where $[Ca^{2+}]_i$ is controlled by plasma membrane Ca^{2+} transport and cytoplasmic Ca^{2+} buffers without contributions from mitochondria or the ER. (B) $[Ca^{2+}]_i$ response elicited from a sympathetic neuron by a depolarization-evoked Ca^{2+} current (I_{Ca}) under voltage clamp (perforated patch conditions). (C) Schematic representation of the relationship between the rates of Ca^{2+} entry and extrusion at different times during the response (a-e). (D, Bottom) Measurement of the total cytoplasmic Ca^{2+} flux per unit effective volume (J_i) during the response shown in (B). Dots show the measured flux and lines show the derivative of smooth curves fit to the data in B. By convention, inward fluxes are negative and outward fluxes are positive. (D, Top) Differential equation describing $[Ca^{2+}]_i$ dynamics in terms of the rate of Ca^{2+} transport across the plasma membrane and cytoplasmic volume and buffering strength. The components of the total cytoplasmic Ca^{2+} flux representing Ca^{2+} entry ($\tilde{J}_{ICa}$) and extrusion ($\tilde{J}_{Extru}$) are indicated. In the following, $\tilde{J}$ represents a material flux integrated over a compartmental boundary (e.g. in nmoles/s) while J is the ratio of $\tilde{J}$ to the effective compartmental volume ($v\kappa$). F is the Faraday constant.

3.3. Case 1: Cytoplasmic Ca^{2+} Dynamics under the Control of Plasma Membrane Transport

Net Ca^{2+} transport across the plasma membrane represents the functional interplay between pathways that regulate passive Ca^{2+} entry, and those responsible for active Ca^{2+} extrusion (Fig. 2A). In sympathetic neurons, Ca^{2+} extrusion is controlled by Ca^{2+} pumps (PMCAs, plasma membrane Ca ATPases) and a Na^+/Ca^{2+} exchanger, and Ca^{2+} entry is regulated by voltage-sensitive Ca^{2+} channels (Suzuki et al. 2002). To determine how these transport pathways together contribute to calcium dynamics, we examined stimulus-evoked $[Ca^{2+}]_i$ responses under conditions where surface membrane Ca^{2+} transport dominates. As a stimulus, we depolarized the plasma membrane to open voltage-sensitive Ca^{2+} channels, leading to Ca^{2+} entry. To minimize contributions from organelles, Ca^{2+} transport by mitochondria and the ER was either inhibited pharmacologically (Colegrove et al., 2000) or minimized using weak stimuli, making it possible to consider a single dynamical variable ($[Ca^{2+}]_i$) under the influence of the net Ca^{2+} flux across the plasma membrane. Since this flux depends on the relative rates of passive Ca^{2+} entry and active extrusion, an understanding of $[Ca^{2+}]_i$ dynamics requires consideration of these two fluxes and their interplay.

Interplay between Ca^{2+} entry and extrusion during depolarization-evoked $[Ca^{2+}]_i$ elevations. Figure 2 illustrates a $[Ca^{2+}]_i$ response elicited by membrane depolarization under voltage clamp (perforated patch conditions). Depolarization opens voltage-gated Ca^{2+} channels in the plasma membrane, leading to Ca^{2+} entry and a rise in $[Ca^{2+}]_i$ towards a new steady level (Fig. 2B). Restoring the membrane potential to its initial value leads to channel closure and a recovery of $[Ca^{2+}]_i$ to its prestimulation level. Qualitatively, the changes in $[Ca^{2+}]_i$ produced by this stimulus can be understood in terms of the relationship between Ca^{2+} entry and extrusion rates (Fig. 2C). Under resting conditions, the rate of net Ca^{2+} transport across the plasma membrane is zero, reflecting a balance between Ca^{2+} entry and extrusion rates (a). When Ca^{2+} channels open, there is a sudden increase in the rate of passive Ca^{2+} entry, creating an imbalance between entry and extrusion rates that causes $[Ca^{2+}]_i$ to rise (b). As $[Ca^{2+}]_i$ rises, the rate of extrusion also rises, causing the net flux to decline as $[Ca^{2+}]_i$ approaches a new steady level (c). Membrane repolarization then leads to rapid Ca^{2+} channel closure and an abrupt drop in Ca^{2+} entry rate, which unmasks ongoing Ca^{2+} extrusion, causing $[Ca^{2+}]_i$ to decline (d). Finally, as $[Ca^{2+}]_i$ falls, so does the rate of extrusion, restoring the balance between Ca^{2+} entry and extrusion as $[Ca^{2+}]_i$ returns to its resting level (e).

Quantitative properties of the underlying Ca^{2+} fluxes. The first step in describing $[Ca^{2+}]_i$ dynamics quantitatively is to measure the time derivative of $[Ca^{2+}]_i$ ($d[Ca^{2+}]_i/dt$). If Ca^{2+} is uniformly distributed within the cytoplasm and Ca^{2+} binding to cytoplasmic buffers is always close to equilibrium, then at each instant in time during the response, $d[Ca^{2+}]_i/dt$ gives the net cytoplasmic Ca^{2+} flux per unit effective cytoplasmic volume ($v_i\kappa_i$). We refer to this flux as J_i ($\equiv \tilde{J}_i/(v_i\kappa_i)$). In the present case, where only plasma membrane Ca^{2+} transport occurs, J_i depends on the net Ca^{2+} flux across the plasma membrane, which in turn depends on the relative rates of stimulated Ca^{2+} entry through Ca^{2+} channels ($J_{ICa} = \tilde{J}_{ICa}/(v_i\kappa_i)$) and Ca^{2+} extrusion ($J_{Extru} = \tilde{J}_{Extru}/(v_i\kappa_i)$). Here, '$Ca^{2+}$ extrusion' refers to all Ca^{2+} transport across the plasma membrane in the absence of stimulation, including contributions from the PMCA, Na^+/Ca^{2+} exchanger, and a small basal leak. J_{ICa} can be determined from the measured Ca^{2+} current given estimates of v_i and κ_i (Albrecht et al., 2001, 2002) and J_{Extru} can be obtained directly from $d[Ca^{2+}]_i/dt$ during the recovery, when $J_{ICa}=0$. Since J_i ultimately determines how $[Ca^{2+}]_i$ changes in response to stimulation in our experiments, we will focus on the Ca^{2+}-dependence this flux. Initial work towards converting concentration fluxes to material fluxes is described in Albrecht et al. (2002).

Before proceeding, it is important to ask if cytoplasmic Ca^{2+} is really spatially homogeneous during responses like those shown in Fig. 2. Spatial $[Ca^{2+}]_i$ gradients are observed during strong depolarization (Kuba, 1994), so quantitative analysis of $[Ca^{2+}]_i$ dynamics under these conditions either requires explicit consideration of the spatiotemporal properties of $[Ca^{2+}]_i$, or a demonstration that spatial uniformity is an acceptable approximation for the problem under study. Recoveries that follow repolarization, on the other hand, are typically slow enough compared to the dissipation of radial Ca^{2+} gradients that Ca^{2+} may be treated as being spatially uniform. Therefore, post-depolarization recoveries, or responses to weak depolarization, when Ca^{2+} entry is relatively slow, can be analyzed with these methods, and can serve as a basis for predicting responses elicited by strong depolarization which can then be compared with experimental observations.

Figure 2D (bottom) shows a plot of J_i ($=-d[Ca^{2+}]_i/dt$) vs $[Ca^{2+}]_i$ during the entire $[Ca^{2+}]_i$ response; the minus sign implements the convention that fluxes into the cytoplasm are negative and outward fluxes are positive. This plot is analogous to a current-voltage relation for voltage-sensitive currents, with the axes oriented so that J_i can be read off at each time during the response based on the value of $[Ca^{2+}]_i$ at that time (Fig. 2B,D). During the depolarization, J_i depends on both Ca^{2+} extrusion and stimulated entry, while during the recovery it depends on extrusion alone. Thus, during the onset, J_i is shifted to the left by an amount J_{ICa}. Analyses of responses of this kind lead to two important conclusions regarding J_{Extru}. First, at each point in time during the recovery, J_{Extru} depends on the value of $[Ca^{2+}]_i$ at that time, but not its history. This indicates that the Ca^{2+} transport systems responsible for extrusion adjust rapidly to changes in $[Ca^{2+}]_i$ during the recovery (Colegrove et al., 2000). Second, J_{Extru} has essentially the same $[Ca^{2+}]_i$-dependence during the onset and recovery. This indicates that the mechanisms responsible for Ca^{2+} extrusion have the same sensitivity to Ca^{2+} during both phases of the response and are not very sensitive to membrane potential (Albrecht et al., 2001). Thus, the rate of Ca^{2+} extrusion during weak depolarization can, to a good approximation, be deduced from the $[Ca^{2+}]_i$-dependence of extrusion rate during the recovery following repolarization.

These results indicate that in the absence of contributions from mitochondrial and ER Ca^{2+} transport, $[Ca^{2+}]_i$ dynamics during and after depolarization can be described by the following differential equation:

$$\frac{d[Ca^{2+}]_i}{dt} = -\frac{\tilde{J}_{ICa}(t) + \tilde{J}_{Extru}([Ca^{2+}]_i)}{v_i \kappa_i}$$

$$= J_{ICa}(t) + J_{Extru}([Ca^{2+}]_i)$$

where J_{ICa} and J_{Extru} can be determined experimentally to provide functional descriptions of the underlying Ca^{2+} transport systems. With this information, it is possible to reconstruct depolarization-evoked $[Ca^{2+}]_i$ responses elicited by measured Ca^{2+} currents, and to predict responses to previously untested I_{Ca} inputs in cells where organellar Ca^{2+} transport is weak. Moreover, the properties of J_{Extru} suggest a simple, and powerful, method for measuring the rate of Ca^{2+} transport by organelles in intact cells, when it occurs. Application of this method to the measurement of mitochondrial Ca^{2+} fluxes will be described in the next section.

3.4. Case 2: Coupled Dynamics of Cytoplasmic and Intramitochondrial Calcium Concentrations - Mitochondria as Reversible, $[Ca^{2+}]_i$-Regulated Ca^{2+} Buffers

In this section, we discuss the case where mitochondrial Ca^{2+} transport is enabled and ER Ca^{2+} transport is inhibited, requiring consideration of both cytoplasmic and mitochondrial $[Ca^{2+}]$ dynamics. We first describe the major physiological roles of mitochondrial Ca^{2+} transport, and then discuss the pathways by which these organelles transport Ca^{2+}. We will then turn to experimental observations showing how mitochondria contribute to depolarization-evoked $[Ca^{2+}]_i$ responses in sympathetic neurons. This will be followed by a summary of our functional characterization of mitochondrial Ca^{2+} uptake and release pathways, illustrating how the interplay between these pathways, in a cytoplasmic context, accounts for a number of interesting features of $[Ca^{2+}]$ responses in intact cells.

Cellular functions of mitochondrial Ca^{2+} transport. Mitochondrial Ca^{2+} transport is thought to have three major cellular functions. First, it provides a way to regulate ATP production in anticipation of energy demands through regulation of Ca^{2+}-sensitive dehydrogenases in the Krebs cycle (McCormack and Denton, 1990). Second, it contributes to the regulation of $[Ca^{2+}]_i$, thereby modulating cellular processes that are controlled by cytoplasmic Ca^{2+}. Finally, it appears that excessive mitochondrial Ca^{2+} accumulation during prolonged $[Ca^{2+}]_i$ elevations plays a role in cell death (Stout et al. 1998, Castilho et al., 1998).

Pathways that regulate mitochondrial Ca^{2+} transport. Mitochondria have distinctive pathways for controlling Ca^{2+} uptake and release that were described in classical experiments using isolated organelles (Gunter et al., 2000). Uptake is regulated by a Ca^{2+}-sensitive uniporter (EC_{50} ~ 10-20 μM) that permits Ca^{2+} to flow rapidly into the mitochondrial matrix down a steep electrochemical gradient. In excitable cells, Ca^{2+} release is controlled by a saturable Na^+/Ca^{2+} exchanger that couples energetically downhill Na^+ uptake to uphill Ca^{2+} release. These Ca^{2+} transport pathways impart to mitochondria a number of interesting properties (Hansford, 1985). For example, when Ca^{2+} is added to a solution of fixed volume containing isolated mitochondria, Ca^{2+} accumulation causes the external Ca^{2+} concentration to fall and internal Ca^{2+} levels to rise. If the mitochondrial Ca^{2+} load is small enough that the release rate is not limited ('non-buffering mode'), a steady-state can be reached where the rates of uptake and release are in balance. Under these conditions, further additions of Ca^{2+} lead to progressively larger steady-state elevations of intra- and extramitochondrial $[Ca^{2+}]$. However, when the Ca^{2+} load becomes large enough that release is limited ('buffering mode'), any added Ca^{2+} is taken up by the mitochondria until the extramitochondrial Ca^{2+} concentration falls to a 'set-point' which, at least over a range of experimental conditions, is invariant even with further Ca^{2+} additions.

Buffering and non-buffering modes of mitochondrial Ca^{2+} transport result from the functional interplay between Ca^{2+} uptake and release pathways that are regulated by Ca^{2+} in quantitatively different ways. The rate of Ca^{2+} uptake via the uniporter increases steeply with $[Ca^{2+}]_i$ over most of the physiological range, while the rate of release becomes limited as the mitochondrial Ca load increases. Two explanations have been proposed for this limitation: (1) Saturation of the Na^+/Ca^{2+} exchanger release pathway, and (2) formation of a Ca^{2+} phosphate complex that limits increases in the intramitochondrial free Ca^{2+} concentration ($[Ca^{2+}]_{MT}$), thereby limiting the availability of Ca^{2+} as a substrate for the release pathway. Irrespective of the basis for limited Ca^{2+} release rates, the interplay between mitochondrial Ca^{2+} uptake and release processes would be expected to have a strong impact on depolarization-evoked $[Ca^{2+}]_i$ elevations in intact cells. For example, if $[Ca^{2+}]_i$ is steadily elevated to levels where Ca^{2+} uptake by the uniporter is faster than the maximal rate of release, e.g. during long trains of action potentials, mitochondria would be expected to accumulate Ca^{2+} continuously, with obvious consequences for both cytoplasmic and mitochondrial Ca^{2+} regulation.

Under what conditions of stimulation do mitochondria actually transport Ca^{2+} in intact cells? This has been a source of debate, since the EC_{50} for the uniporter is 10-20 μM, while elevations in $[Ca^{2+}]_i$ to which the majority of mitochondria are exposed during stimulation are generally much lower. Thus, the possibility arises that isolated mitochondria take up Ca^{2+} when exposed to supraphysiological $[Ca^{2+}]$ levels, but not during physiological $[Ca^{2+}]_i$ elevations. This concept received initial support with the inability to demonstrate stimulus-evoked Ca^{2+} accumulation in mitochondria except under conditions of cellular Ca^{2+} overload (Moravec et al., 1997; Horikawa et al., 1998). However, with the finding that mitochondrial Ca^{2+} accumulation occurs even when (bulk) $[Ca^{2+}]_i$ is well below 1 μM (Thayer and Miller, 1990, Friel and Tsien, 1994, Herrington et al., 1996), this concept required re-evaluation. One possible way to account for mitochondrial Ca^{2+} uptake under these conditions is that there are microdomains of high Ca^{2+} to which a subpopulation of mitochondria are exposed during stimulation (Rizzuto et al., 1998). According to this idea, by underestimating local Ca^{2+} levels, bulk $[Ca^{2+}]_i$ measurements overestimate the sensitivity with which mitochondria transport Ca^{2+}.

Another possibility is that in intact cells, mitochondria transport Ca^{2+} with higher sensitivity than *in vitro* (Gunter et al. 2000).

A simple alternative explanation is provided by considering the underlying Ca^{2+} fluxes. If the rate of Ca^{2+} uptake by the uniporter increases with $[Ca^{2+}]_i$, then even small $[Ca^{2+}]_i$ elevations in resting cells would be expected to create an imbalance between uptake and release rates favoring mitochondrial Ca^{2+} accumulation. Moreover, if mitochondria in intact cells transport Ca^{2+} like those studied in isolation, the potential exists for novel buffering and non-buffering modes during stimulation reflecting the interplay between mitochondrial Ca^{2+} uptake and release pathways, but in a cytoplasmic context. We now turn to experimental observations showing that in sympathetic neurons, mitochondria accumulate Ca^{2+} in response to depolarization evoked Ca^{2+} entry.

Mitochondrial Ca^{2+} transport during depolarization-evoked $[Ca^{2+}]_i$ elevations. Figure 3 compares $[Ca^{2+}]_i$ responses elicited by weak and strong depolarization, using high K^+ depolarization, instead of voltage clamp pulses, to stimulate Ca^{2+} entry. Weak depolarization with 30 mM K^+ ($V_m \sim$ -35 mV) elicits small $[Ca^{2+}]_i$ responses that are kinetically simple (Fig. 3A), similar to those seen under voltage clamp (Fig. 2B), while stronger depolarizations (50 mM K^+, $V_m \sim$ -20 mV) elicit larger $[Ca^{2+}]_i$ elevations that are complex (Fig. 3B). In particular, the recovery that

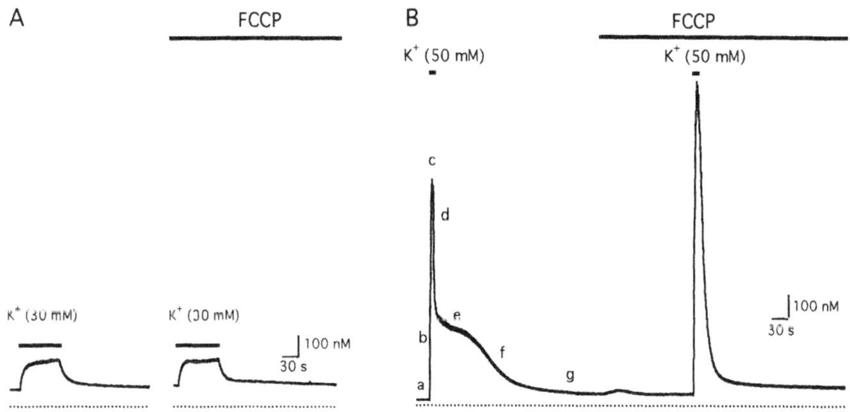

Fig. 3. Mitochondrial Ca^{2+} transport contributes to depolarization-evoked $[Ca^{2+}]_i$ responses in a $[Ca^{2+}]_i$-dependent manner. (A) Responses to weak depolarization (30 mM K^+, $V_m \cong$ -35 mV) before and during exposure to the proton ionophore FCCP (1 µM), which inhibits mitochondrial Ca^{2+} uptake. (B) Responses elicited by stronger depolarization (50 mM K^+, $V_m \cong$ -20 mV) before and during exposure to FCCP (1 µM). FCCP had a much larger effect on responses to strong vs weak depolarization, as expected for inhibition of Ca^{2+} uptake by the mitochondrial Ca^{2+} uniporter, whose activity increases steeply with $[Ca^{2+}]_i$. Phases of the response to strong depolarization (B, a-g) are described in text.

follows repolarization exhibits multiple kinetically distinct phases, including an initial rapid decline (Fig. 3B, d), a slow plateau phase that can last many seconds (e), an accelerated decline (f), and a final slow approach to the prestimulation level (g). Such recoveries are inconsistent with a simple extrapolation of the $J_{Extru}/[Ca^{2+}]_i$ relation described in the preceding section and suggest the involvement of a Ca^{2+} sequestration pathway that becomes strongly activated at high $[Ca^{2+}]_i$.

One hint that such $[Ca^{2+}]_i$ responses are influenced by mitochondrial Ca^{2+} transport comes from the observation that they are greatly modified by inhibitors of mitochondrial Ca^{2+} uptake, like the proton ionophore FCCP. During exposure to FCCP, the same stimulus raises $[Ca^{2+}]_i$ to a higher level, consistent with inhibition of an intracellular Ca^{2+} sequestration process, and the recovery that follows repolarization is faster overall, as expected if Ca^{2+} release from loaded mitochondria no longer occurs. In the presence of FCCP, the $J_{Extru}/[Ca^{2+}]_i$ relation appears to be a simple extrapolation of the relation determined from recoveries following weak depolarization in the absence of FCCP (Fig. 2D, 4C, below). Since FCCP has little effect on the $[Ca^{2+}]_i$ level reached during weak depolarization (Fig. 3A), it is unlikely that the inhibitor produces its effects simply by modifying surface membrane Ca^{2+} transport. Comparison between recoveries following strong depolarization before and after FCCP treatment (Fig. 3B) shows that when mitochondrial Ca^{2+} uptake is enabled, $[Ca^{2+}]_i$ declines more rapidly at high $[Ca^{2+}]_i$, and more slowly at low $[Ca^{2+}]_i$, than it does when uptake is inhibited. This is consistent with known properties of Ca^{2+} transport by isolated mitochondria: strong Ca^{2+} accumulation at high $[Ca^{2+}]_i$, and when loaded by a previous stimulus, net Ca^{2+} release at low $[Ca^{2+}]_i$.

While these observations suggest that mitochondria accumulate Ca^{2+} reversibly during depolarization-evoked Ca^{2+} entry, it was necessary to obtain direct measurements of intramitochondrial calcium levels to test this idea definitively. For this purpose, electron probe microanalysis was used (Pivovarova et al., 1999). Neurons were stimulated and then rapidly frozen, and the total Ca concentration (free + bound) was measured in structurally identified mitochondria from sections of cryofixed cells. These measurements (see Fig. 9C) confirmed that mitochondria do, indeed, accumulate Ca^{2+} in an FCCP-sensitive manner, and that net Ca^{2+} release from loaded mitochondria contributes to the slow plateau phase of the $[Ca^{2+}]_i$ recovery. They also showed, paradoxically, that mitochondria accumulate Ca^{2+} during weak depolarization, despite insensitivity of the steady $[Ca^{2+}]_i$ elevations they produce to FCCP (Fig. 3A). Moreover, during maintained depolarization, mitochondria accumulate Ca^{2+} continuously, even while $[Ca^{2+}]_i$ is steadily elevated at a level far below that seen in the presence of FCCP. To determine if these observations can be explained by the interplay between mitochondrial Ca^{2+} uptake and release pathways operating in a cytoplasmic milieu, we characterized mitochondrial Ca^{2+} uptake and release rates, along with Ca^{2+} extrusion, and their regulation by Ca^{2+} in intact cells.

Characterization of mitochondrial Ca^{2+} uptake and release fluxes in intact cells. Stimulated by the work of Herrington et al. (1996), we performed the following experiments. After inhibiting Ca^{2+} uptake by the ER, cells were depolarized to raise $[Ca^{2+}]_i$ and the recovery following repolarization was analyzed (Fig. 4B). This recovery depends on the combined rates of Ca^{2+} extrusion and mitochondrial Ca^{2+} uptake and release. Individual flux components were resolved by measuring the total cytoplasmic Ca^{2+} flux (J_i) during the recovery under three different conditions. First, the rate of Ca^{2+} extrusion (J_{Extru}) was determined by measuring J_i while mitochondrial Ca^{2+} transport was inhibited (with FCCP, see Fig. 4B,C, Curve 3). Components of the net mitochondrial Ca^{2+} flux were then determined from J_i measured when mitochondrial Ca^{2+} transport was enabled by subtracting J_{Extru} from J_i at corresponding values of $[Ca^{2+}]_i$. This procedure is valid because at each point in time, J_{Extru} is defined by $[Ca^{2+}]_i$ and the fluxes are additive (Colegrove et al. 2000a).

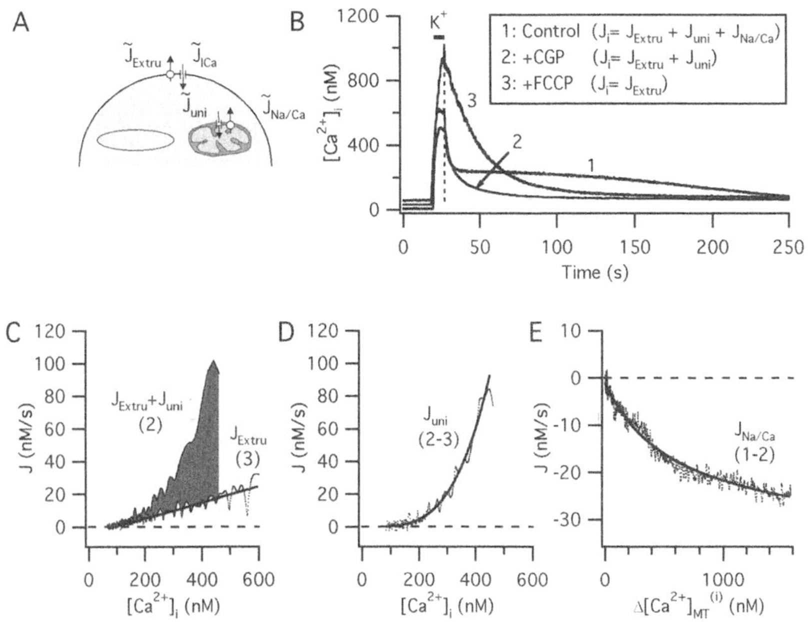

Fig. 4. Dissection of the total cytoplasmic Ca^{2+} flux into components representing Ca^{2+} extrusion and mitochondrial Ca^{2+} uptake and release. (A) Illustration of the transporters responsible for cytoplasmic and mitochondrial Ca^{2+} dynamics during and after depolarization-induced Ca^{2+} entry. (B) $[Ca^{2+}]_i$ response elicited by exposure to 50 mM K^+ under control conditions (Curve 1), or during exposure to CGP 37157 (CGP, Curve 2) or FCCP (Curve 3). Components of the total cytoplasmic Ca^{2+} flux during the recovery were determined as follows. The rate of Ca^{2+} extrusion (J_{Extru}, Panel C) was determined by measuring the FCCP-resistant component of the total Ca^{2+} flux (Curve 3). The net mitochondrial Ca^{2+} flux was then determined in other responses in the same cells as the FCCP-sensitive component of J_i. The rate of Ca^{2+} uptake via the uniporter (J_{uni}, C,D) was taken as the CGP-resistant component of the net mitochondrial Ca^{2+} flux (C, shaded flux component). The CGP-sensitive component of the net mitochondrial Ca^{2+} flux ($J_{Na/Ca}$, E) was taken as a measure of the rate of Ca^{2+} release via the Na^+/Ca^{2+} exchanger. During all responses, ER Ca^{2+} uptake was inhibited with thapsigargin (0.2-1 μM).

To characterize the mitochondrial Ca^{2+} uptake pathway, the recovery was analyzed during exposure to the selective inhibitor of the mitochondrial Na^+/Ca^{2+} exchanger, CGP 37157 (CGP, Fig. 4B, Curve 2). Under these conditions, the recovery is fast and shows little or no plateau like that seen in controls. Here, J_i (Fig. 4C, Curve 2) is the sum of J_{Extru} and the CGP-resistant component of the net mitochondrial Ca^{2+} flux. The latter flux (shaded region in Fig. 4C) represents Ca^{2+} uptake and increases steeply with $[Ca^{2+}]_i$, as expected for the uniporter (Fig. 4D). We attribute this flux to the uniporter and refer to it as J_{uni}. A similar procedure provided a description of the CGP-sensitive component of the net mitochondrial Ca^{2+} flux. This flux represents Ca^{2+} release and depends on cytoplasmic Na^+ and increases with intramitochondrial Ca^{2+} concentration, measured indirectly by the quantity $\Delta[Ca^{2+}]_{MT}^{(i)}$ (Colegrove et al., 2000a,b). We attribute this flux to the mitochondrial Na^+/Ca^{2+} exchanger and refer to it as $J_{Na/Ca}$ (Fig. 4E).

Since spatial $[Ca^{2+}]_i$ gradients dissipate within ~ 1 s of repolarization, these bulk measurements of $[Ca^{2+}]_i$ during the recovery are expected to give a good description of local $[Ca^{2+}]_i$ during the recoveries, and therefore of the $[Ca^{2+}]_i$-dependence of transport by the individual pathways. The observation that J_{uni} is non-zero even when $[Ca^{2+}]_i$ is near resting levels demonstrates that mitochondrial Ca^{2+} transport occurs at low $[Ca^{2+}]_i$. The rate of *net* transport, however, also depends on $J_{Na/Ca}$, which varies with the mitochondrial Ca^{2+} load.

To see if the experimentally determined rate descriptions of J_{Extru}, J_{uni} and $J_{Na/Ca}$ account for changes in $[Ca^{2+}]_i$ and $[Ca]_{MT}$ induced by depolarization, they were used to define rate equations so that they could be tested for their ability to explain the observed effects of Ca^{2+} entry on intracellular Ca^{2+} concentrations (Fig. 5). We did not detect intrinsic time-dependence of J_{Extru}, J_{uni} or $J_{Na/Ca}$, so the rate equations were simply functions of $[Ca^{2+}]_i$ and $[Ca^{2+}]_{MT}$. Figure 5B shows simulated $[Ca^{2+}]_i$ (c_i) and $[Ca^{2+}]_{MT}$ (c_m) responses elicited by Ca^{2+} entry at different rates (s1 and s2). Overall, the simulations agree quite well with $[Ca^{2+}]_i$ responses elicited by weak and strong depolarization (cf Fig. 3). Moreover, they explain why even small $[Ca^{2+}]_i$ elevations (e.g. like those produced by s1) stimulate mitochondrial Ca^{2+} accumulation. In this case, which is the intact cell analog of the 'non-buffering' mode described in isolated organelles, mitochondrial Ca^{2+} transport influences the rate at which $[Ca^{2+}]_i$ changes during stimulation, but not the steady-state level reached during maintained stimulation. This explains why steady $[Ca^{2+}]_i$ elevations elicited by weak depolarization are not influenced by FCCP (Fig. 2A); the *rate* at which $[Ca^{2+}]_i$ changes should, however, be sensitive to FCCP. Strong stimuli, in contrast, which raise c_i to levels where Ca^{2+} uptake by the uniporter is faster than the maximal rate of release via the Na^+/Ca^{2+} exchanger, lead to continuous mitochondrial Ca^{2+} accumulation. In this case, c_i attains a steady-state level when the rate of net Ca^{2+} entry equals the rate of Ca^{2+} accumulation. This is the intact cell analog of the 'buffering' mode described previously in isolated mitochondria. In this mode, mitochondria influence both the rate at which $[Ca^{2+}]_i$ changes during stimulation *and* the steady $[Ca^{2+}]_i$ level attained during continuous stimulation. This explains why steady $[Ca^{2+}]_i$ elevations elicited by strong depolarization are sensitive to the protonophore.

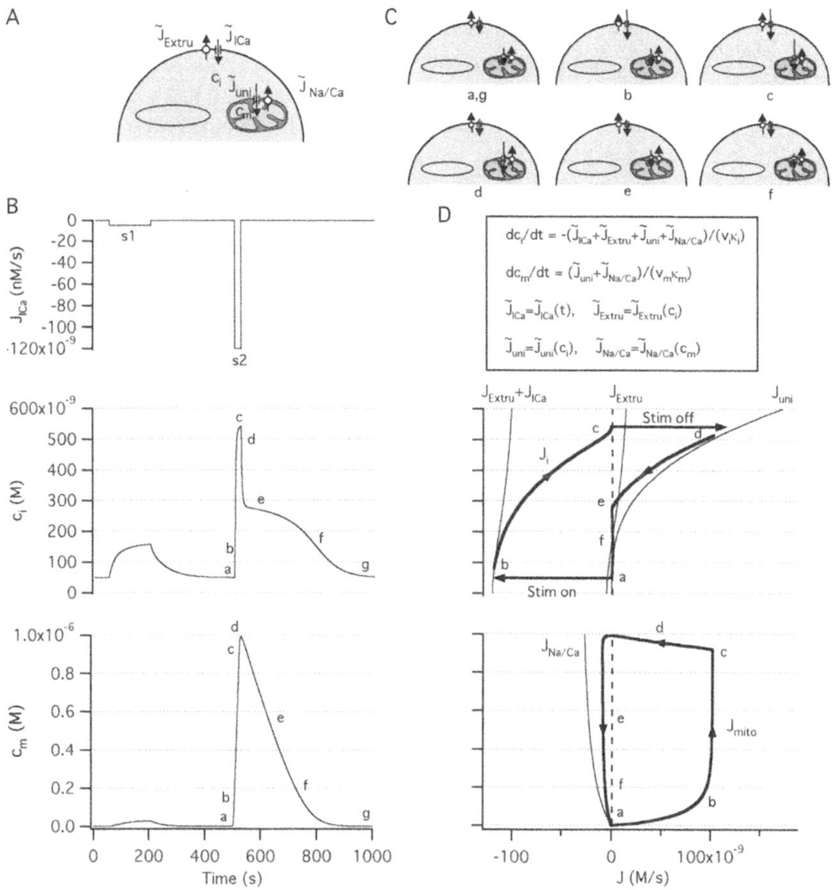

Fig. 5. Quantitative properties of mitochondrial Ca^{2+} uptake and release rates account for qualitative properties of depolarization-evoked changes in cytoplasmic and mitochondrial calcium levels. (A) Schematic showing the dynamical variables ($c_i = [Ca^{2+}]_i$, $c_m = [Ca^{2+}]_{MT}$) and the transport pathways that control them. (B) Comparison between simulated changes in c_i and c_m induced by weak and strong stimuli consisting of step increases in J_{ICa} to different levels (s1 and s2). (C) Qualitative relationship between the Ca^{2+} fluxes during response to the strong stimulus s2 (a-f). (D, second panel) Description of the c_i-dependent fluxes J_{Extru} and J_{uni}. (D, third panel) Description of the c_m-dependent flux, $J_{Na/Ca}$. Both stimuli (s1 and s2) increase c_m, but with the weak stimulus (s1), c_m approaches a steady-state level where the rates of mitochondrial Ca^{2+} uptake and release are equal. During the strong stimulus (s2), c_i rises to a higher level where J_{uni} exceeds the maximal rate of release via the Na^+/Ca^{2+} exchanger, forcing c_m to rise continuously. Dark traces show *net* cytoplasmic and mitochondrial Ca^{2+} fluxes during the response to s2. Thin traces, which describe the c_i- and c_m-dependencies of J_{Extru}, J_{uni} and J_{NaCa}, were determined by fitting equations to flux measurements averaged over multiple cells. Equations describing the two compartment system are shown in D (Top). Dashed vertical lines in D indicate the zero net flux level.

The ability of mitochondria to operate in this mode has important implications for $[Ca^{2+}]_i$ responses elicited by repetitive or excessive stimulation (Friel, 2000).

The flux diagrams in Fig. 5D provide a convenient way for visualizing how the individual Ca^{2+} transport pathways contribute to c_i and c_m dynamics. For example, with the onset of the strong stimulus s2 ('Stim on'), the curve describing plasma membrane Ca^{2+} transport suddenly shifts from J_{Extru} to $J_{Extru}+J_{ICa}$, leading to an inward Ca^{2+} flux that causes c_i to rise (Phase b). As c_i rises, J_{uni} also rises, creating an imbalance between J_{uni} and $J_{Na/Ca}$ that leads to mitochondrial Ca^{2+} accumulation, an increase in c_m, and as a result, an increase in $J_{Na/Ca}$. The rate of Ca^{2+} accumulation (J_{uni} + $J_{Na/Ca}$) approaches the rate of net Ca^{2+} entry ($J_{Extru}+J_{ICa}$) as c_i approaches a new steady-state level, where the net cytoplasmic Ca^{2+} flux is zero and c_m is steadily increasing (Phase c). Terminating the stimulus at this point ('Stim off') unmasks ongoing Ca^{2+} accumulation, which causes c_i to fall rapidly (Phase d) which then largely deactivates the uniporter. This shifts the balance of net mitochondrial Ca^{2+} transport to net release (Phase e), accounting for the transition between fast and slow phases of the recovery. The subsequent recovery then depends on the mitochondrial Ca^{2+} load. If it is large enough to nearly saturate the release pathway, the rate of net Ca^{2+} release is nearly constant, so that c_i hangs up at an elevated level until the release pathway is no longer near saturation. During the plateau phase (Phase e), it is as if Ca^{2+} were injected into the cytosol at a nearly constant rate, forcing c_i to remain at an elevated level where the rate of injection is balanced by extrusion. Of course, since the mitochondrial Ca^{2+} load is exhaustible, c_m declines, as does the rate of release, accounting for the slow c_i decline during the plateau. Similar effects of mitochondrial Ca^{2+} transport on c_i dynamics would result if the rate of Ca^{2+} release were limited by some other mechanism, e.g. intramitochondrial Ca complex formation (Friel and Tsien, 1994), and reversible Ca phosphate formation has been observed experimentally in these cells (Pivovarova et al., 1999). However, the observation that $J_{Na/Ca}$ increases smoothly and saturably with intramitochondrial Ca levels, without hitting an obvious ceiling (Fig. 4E) suggests that Ca-phosphate formation occurs when $[Ca^{2+}]_{MT}$ is above the EC_{50} for the Na^+/Ca^{2+} exchanger, and that the limiting factor during responses to depolarization like those illustrated in Fig. 3B is the Ca^{2+} release pathway, not the rise in $[Ca^{2+}]_{MT}$ (cf David, 1999). However, direct measurements of $[Ca^{2+}]_{MT}$ are required to settle this issue definitively.

Summary. Mitochondria provide a powerful mechanism for modulating depolarization-evoked $[Ca^{2+}]_i$ signals. Two basic signaling regimes can be distinguished based on stimulus strength, for example, defined as the magnitude of a steady depolarization, or as the frequency of repetitive synaptic inputs encountered physiologically. Weak stimuli lead to steady elevations in cytoplasmic and mitochondrial Ca^{2+} levels, or fluctuations around steady elevated levels. Strong stimuli lead to correspondingly larger $[Ca^{2+}]_i$ elevations and continuous mitochondrial Ca^{2+} accumulation. In the aftermath of stimulation, net Ca^{2+} release from loaded mitochondria prolongs the recovery of $[Ca^{2+}]_i$ to its prestimulation level in a way that depends on mitochondrial Ca^{2+} load. The overall effect is to impart to evoked Ca^{2+} signals a far more complex dependence on stimulus strength and history than that seen in Case 1, with far reaching consequences for regulation of Ca^{2+}-sensitive processes within the cell.

3.5. Case 3: Coupled Dynamics of Cytoplasmic and Endoplasmic Reticular Calcium Concentrations - Multiple Modes of Ca^{2+}-Induced Ca^{2+} Release

This section describes the case where ER Ca^{2+} transport is enabled and mitochondrial Ca^{2+} transport is disabled (Fig.1, Case 3). Here, the dynamical variables are $[Ca^{2+}]_i$ and $[Ca^{2+}]_{ER}$, which vary under the influence of plasma membrane and ER Ca^{2+} transport. We begin by describing the major physiological roles of ER Ca^{2+} transport, and then discuss the pathways by which these organelles transport Ca^{2+}. This will be followed by a summary of our functional characterization of ER Ca^{2+} uptake and release pathways, showing how the interplay between these pathways, in a cytoplasmic context, sets the stage for multiple signaling modes during and after depolarization-evoked Ca^{2+} entry.

Cellular functions of ER Ca^{2+} transport. The endoplasmic reticulum (ER) is an important component of Ca^{2+} signaling in most, if not all, non-muscle cells (Meldolesi and Pozzan, 1998). Ca^{2+} transport by the ER is thought to have at least three main functions. First, it permits the ER to act as a Ca^{2+} pool that can be discharged during stimulation. Second, ER Ca^{2+} transport modulates evoked changes in $[Ca^{2+}]_i$, thereby influencing all cellular processes that are regulated by cytoplasmic Ca^{2+}. Finally, high Ca^{2+} levels within the ER are required for proper protein processing, possibly explaining the deleterious effects associated with chronic depletion of Ca^{2+} from the ER (Verkhratsky and Petersen, 2002).

Pathways for ER Ca^{2+} uptake and release. Ca^{2+} uptake by the ER is controlled by specific Ca^{2+} pumps (SERCAs, sarco- and endoplasmic reticulum Ca ATPases) that are distinct from those operating at the plasma membrane. Passive Ca^{2+} release is regulated by Ca^{2+} release channels that open in response to elevations in $[Ca^{2+}]_i$ and permit Ca^{2+} to flow down a steep concentration gradient into the cytoplasm. Ca^{2+} release channels are particularly interesting because, under certain conditions, they can permit regenerative Ca^{2+} release through a process called Ca^{2+}-induced Ca^{2+} release (CICR) (Bezprozvanny et al., 1991; for reviews, see Kuba, 1994, Verkhratsky and Shmigol, 1996, Usachev and Thayer, 1999). Two major classes of Ca^{2+} release channels have been identified, ryanodine receptors (RyRs) and inositol (1,4,5)-trisphosphate receptors ($InsP_3Rs$), that can be distinguished based on Ca^{2+}-sensitivity, pharmacology and expression patterns that define their distinctive contributions to Ca^{2+} signaling in different cells.

Stimulus-evoked changes in ER Ca^{2+} transport. Under resting conditions, Ca^{2+} levels within the ER are much higher than $[Ca^{2+}]_i$, and consequently the ER can serve as a Ca^{2+} reservoir that can be mobilized by various stimuli. For example, agonist-induced production of $InsP_3$ leads to opening of $InsP_3Rs$, which allows Ca^{2+} to flow down its concentration gradient into the cytoplasm, causing $[Ca^{2+}]_i$ to rise and intraluminal Ca^{2+} levels to fall. Depolarization-induced $[Ca^{2+}]_i$ elevations, on the other hand, can either stimulate Ca^{2+} accumulation or net Ca^{2+} release. While these actions are expected to have very different functional effects on intraluminal and cytoplasmic Ca^{2+} binding proteins, the stimulus parameters that determine whether net Ca^{2+} uptake or release occurs have been unclear.

The impact of depolarization-evoked Ca^{2+} entry on net ER Ca^{2+} transport depends on the relationship between ER Ca^{2+} uptake and release rates. Clearly, if Ca^{2+} uptake is stimulated more strongly than release, then the ER will act as a Ca^{2+} sink and evoked $[Ca^{2+}]_i$ elevations will be slowed. This would be expected if the Ca^{2+} permeability of the ER is constant, for in this case, a rise in $[Ca^{2+}]_i$ would increase the availability of Ca^{2+} for transport via SERCAs, and would reduce the rate of passive Ca^{2+} release by diminishing the driving force for release. This would not necessarily occur, however, if the Ca^{2+} permeability is sensitive to $[Ca^{2+}]_i$, e.g. owing to the presence of Ca^{2+} release channels. In this case, an increase in the rate of passive Ca^{2+} release could occur if the permeability increases sufficiently to overcome the effect of a reduced driving force. If release is stimulated less effectively than uptake, a Ca^{2+}-dependent increase in permeability would reduce the rate of ER Ca^{2+} accumulation, thereby influencing both $[Ca^{2+}]_{ER}$ and $[Ca^{2+}]_i$ dynamics. On the other hand, if release is stimulated more effectively than uptake, net Ca^{2+} release would result, speeding, and potentially amplifying, evoked $[Ca^{2+}]_i$ responses. Indeed, it is possible that depending on the size of the evoked $[Ca^{2+}]_i$ elevation, either net Ca^{2+} uptake or net Ca^{2+} release occurs.

The initial motivation for considering these possibilities was our observation that $[Ca^{2+}]_i$ elevations evoked by weak depolarization are accelerated by activation of a ryanodine-sensitive CICR pathway but, paradoxically, are accompanied by ER Ca^{2+} accumulation. Since the direction of net ER Ca^{2+} transport depends on the relative rates of Ca^{2+} uptake and release, it seemed plausible that such $[Ca^{2+}]_i$ elevations stimulate both Ca^{2+} uptake and passive Ca^{2+} release, but stimulate the former more effectively than the latter. In this case, activation of the CICR pathway would speed evoked $[Ca^{2+}]_i$ elevations by reducing the rate of ER Ca^{2+} accumulation. However, if at higher $[Ca^{2+}]_i$ levels, release became more powerful than uptake, the ER could undergo a transition from sink to source. Indeed, such a transition is observed (Hongpaisan et al. 2001). To understand the basis for this transition, information about Ca^{2+}-dependent regulation of the underlying fluxes is essential.

$[Ca^{2+}]$-dependence of ER Ca^{2+} uptake and release rates. We carried out experiments to characterize the functional properties of ER Ca^{2+} uptake and release pathways in sympathetic neurons. The approach is described in Albrecht et al., (2002) and the main results are summarized here. The Ca^{2+} flux representing Ca^{2+} uptake via SERCAs (J_{SERCA}) depends on $[Ca^{2+}]_i$ and increases monotonically with $[Ca^{2+}]_i$. The Ca^{2+} flux representing passive Ca^{2+} release ($J_{Release}$) depends on both the driving force for release, and the macroscopic Ca^{2+} permeability of the ER (P_{ER}), representing the activity of all Ca^{2+} permeable channels expressed by the ER. If Ca^{2+} release channels contribute to P_{ER}, then P_{ER} should depend on $[Ca^{2+}]_i$. and express the pharmacological properties of the underlying channels (RyRs and $InsP_3Rs$). We found that P_{ER} increases with $[Ca^{2+}]_i$ in a strongly caffeine- and ryanodine-sensitive manner, suggesting that RyRs are the most abundant Ca^{2+} re-

lease channels in sympathetic neurons. This is consistent with previous reports indicating that agonist-induced Ca^{2+} release is weak in amphibian neurons (Pfaffinger et al. 1988). However, since our experiments were carried out in the absence of $InsP_3$-generating agonists, a potential role for $InsP_3Rs$ in the presence of agonists cannot be excluded.

Differential Ca^{2+}-dependent regulation of ER Ca^{2+} uptake and release rates accounts for multiple modes of CICR. Our descriptions of ER Ca^{2+} transport pathways were used to address two general questions: (1) How does the functional interplay between ER Ca^{2+} uptake and release pathways define the magnitude and direction of the net ER Ca^{2+} flux (J_{ER}), which define the dynamics of $[Ca^{2+}]_{ER}$? (2) How does the functional interplay between net ER Ca^{2+} transport and surface membrane transport contribute to the magnitude and direction of the net cytoplasmic Ca^{2+} flux (J_i), which defines $[Ca^{2+}]_i$ dynamics when mitochondrial Ca^{2+} transport is weak?

The first step was to describe the measured quantities J_{SERCA}, P_{ER}, and J_{Extru} and their $[Ca^{2+}]_i$-dependence using analytical expressions like those employed previously to describe transport by the corresponding systems *in vitro*. We did not detect intrinsic time-dependence of either J_{SERCA} or P_{ER}, and therefore the rates of transport via these pathways were treated as instantaneous functions of cytoplasmic and ER Ca^{2+} concentrations. Rate descriptions were obtained by fitting equations to the measured fluxes over the Ca^{2+} concentration ranges studied, and continuous extensions of these expressions were used to investigate properties of J_{ER} and J_i and their interplay at higher $[Ca^{2+}]_i$ levels. Thus, the flux descriptions at high $[Ca^{2+}]_i$ are approximations to be evaluated by future experiments. Nevertheless, simulations based on these analytical extensions are in good agreement with observations of evoked $[Ca^{2+}]_i$ responses over the range studied (up to $[Ca^{2+}]_i \sim 1.2\ \mu M$, see Fig.7 below), suggesting that the extended rate descriptions are reasonable.

Figure 6 summarizes our results. Panel A (left) illustrates the system schematically, and panels B and D illustrate how J_{ER} and J_i depend on cytoplasmic and ER Ca^{2+} concentrations (c_i and c_{ER}). One way to get a feeling for J_{ER} and J_i and their regulation by c_i and c_{ER} is to examine how each flux changes with one variable while the other is held constant. For example, when c_{ER} is held at zero, which would represent the case in which the ER is empty, J_{ER} increases with c_i in a way that represents the $[Ca^{2+}]_i$-dependencies of J_{SERCA} and the rate of passive Ca^{2+} uptake into the ER from the cytoplasm. In contrast, increasing c_i while holding c_{ER} constant at its resting level (~130 μM, see slices in panels B and D indicated by dotted lines) provides a description of the net ER Ca^{2+} flux that would be expected if c_i were suddenly increased to different levels without perturbing c_{ER}. Figure 6C shows such a slice from the J_{ER} surface, indicating its components J_{SERCA} and $J_{Release}$ ($=P_{ER}(c_i-c_{ER})$). This illustrates how the c_i-dependence of J_{ER} depends on the interplay between two opposing fluxes with different sensitivities to c_i. When c_i is small, J_{SERCA} exceeds $J_{Release}$ so the ER acts as a Ca^{2+} sink. However, when c_i is large, $J_{Release}$ exceeds J_{SERCA} so the ER is a Ca^{2+} source. The transition from sink to source is marked by a well defined threshold (Fig. 6C, up arrow).

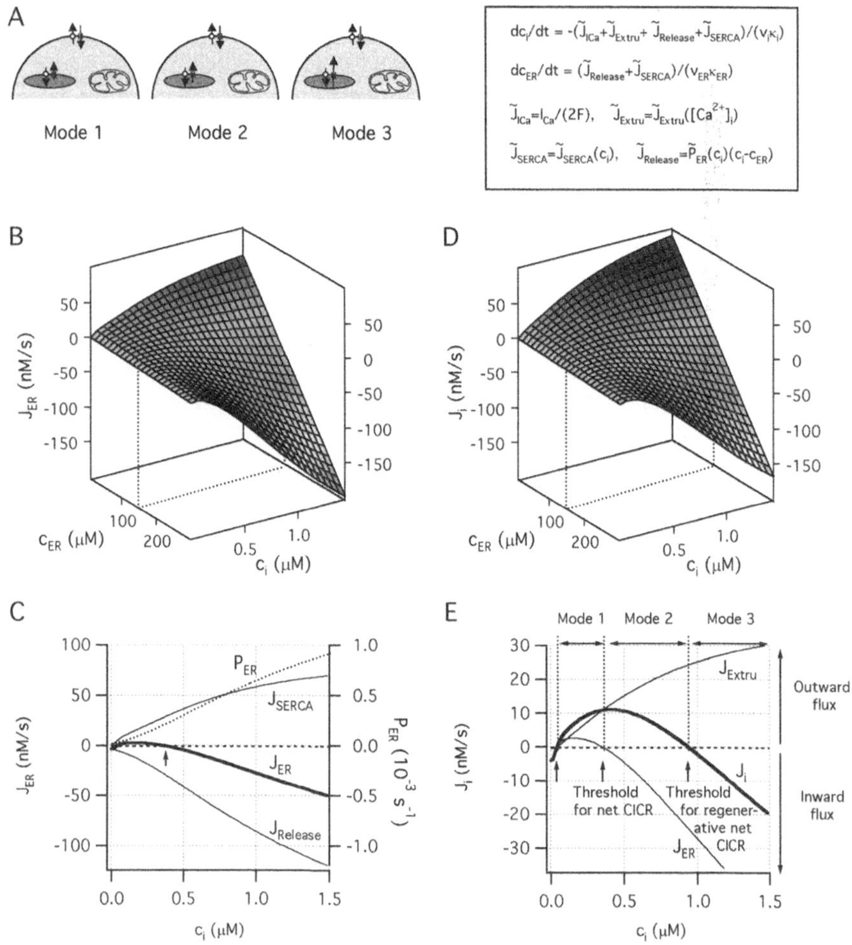

Fig. 6. Quantitative properties of ER Ca^{2+} uptake and release rates account for multiple modes of CICR. (A, left) Schematic showing the dynamical variables (c_i = $[Ca^{2+}]_i$, $c_{ER}=[Ca^{2+}]_{ER}$) and the transport pathways that control them. (B,D) Surfaces describing the c_i and c_{ER} dependencies of the rates of net ER Ca^{2+} transport (J_{ER}, B) and net cytoplasmic Ca^{2+} transport, (J_i, D). (C,E) Instantaneous flux/c_i,c_{ER} relations at constant c_{ER} for J_{ER} (C) and J_i (E) obtained by slicing the surfaces in B and D at fixed (resting) c_{ER} (see dotted lines). Panel C also shows the c_i-dependencies of J_{SERCA} and $J_{Release}$, the components of J_{ER}, along with P_{ER} (dotted curve). Panel E shows the c_i-dependencies of the J_i components J_{ER} and J_{Exrru}. J_{ER} intersects the zero net flux axis at two points (see arrows) corresponding to resting c_i and the threshold for net CICR. There are also two points of interesction for J_i, at resting c_i and at the threshold for regenerative net CICR. Equations describing this system and a schematic illustration of the relationship between fluxes in the three CICR modes are shown in A.

Fig. 6E shows a slice from the J_i surface in Panel D at constant c_{ER} illustrating how the c_i-dependence of J_i reflects the c_i-dependencies of its components, J_{ER} and J_{Extru}. With c_{ER} held at its resting level, J_i first increases with c_i under the influence of J_{Extru} and J_{ER}, crossing the zero net flux axis at a (stable) steady-state value where c_i assumes its resting value (see up arrow, left). If c_i is increased further, J_i continues to rise, but as the permeability of the ER becomes larger, the rate of passive Ca^{2+} release also increases, causing J_i to turn downward, crossing the zero flux axis with negative slope at an (unstable) steady-state to become an inward flux. Thus, there are three ranges of c_i in which J_i and J_{ER} have distinctive signs, revealing three modes of CICR. When c_i is above the resting level but low enough so that RyR activity is weak, J_i and J_{ER} are both positive (Mode 1 CICR). J_{ER} is positive because passive Ca^{2+} release is slower than Ca^{2+} uptake, causing the ER to act as a buffer. In this mode, activation of the c_i-sensitive permeability during stimulation reduces the rate of Ca^{2+} accumulation and accelerates the rise in c_i as compared to the case where P_{ER} is constant at its resting level. Over the intermediate c_i range, J_i is positive but J_{ER} is negative (Mode 2 CICR) because passive Ca^{2+} release is faster than uptake. Net Ca^{2+} release would amplify the effects of stimulated Ca^{2+} entry on c_i and cause c_{ER} to decline to an extent that is graded with c_i. Finally, at higher c_i levels, both J_i and J_{ER} are inward fluxes (Mode 3 CICR). A brief stimulus that brings c_i within this range would stimulate Ca^{2+} release at such a high rate that it overwhelms Ca^{2+} extrusion, leading to a regenerative rise in c_i and fall in c_{ER}. The zero crossing in this case represents the threshold for regenerative net CICR. It should be noted that while this discussion has focused on CICR under the control of RyRs, the general ideas presented here, e.g. regarding multiple modes of CICR, apply equally well to other types of Ca^{2+} release channels, including $InsP_3Rs$.

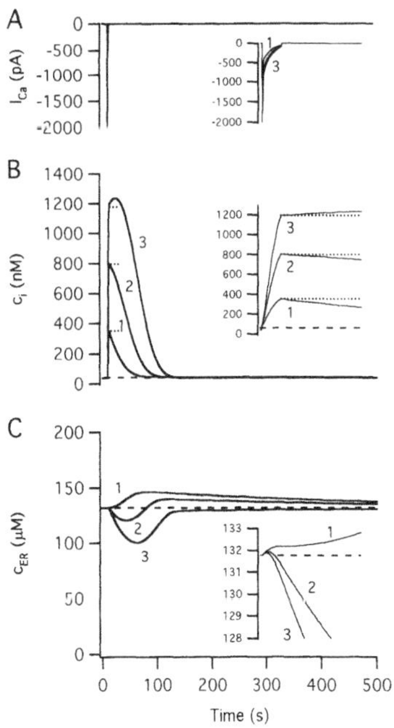

Fig. 7. Simulated responses to Ca^{2+} entry at progressively higher rates reveal three modes of CICR. (A) Ca^{2+} entry waveforms used in the simulations were obtained by fitting measured Ca^{2+} currents with the sum of three exponentials and scaling. Currents were elicited by 2 s depolarizing pulses from -70 mV to -10 mV. (B) Simulated c_i responses elicited by the three currents in A. (C) Simulated c_{ER} responses accompanying the c_i responses shown in B. Insets compare early c_i and c_{ER} responses elicited by stimuli 1-3 on an expanded time scale. Note that stimulus 1 leads to ER Ca^{2+} accumulation, while both stimuli 2 and 3 cause net Ca^{2+} release. However, after terminating stimuli 1 and 2, c_i declined, while a continued rise occurred following stimulus 3, indicating that c_i exceeded the threshold for regenerative CICR.

This analysis describes how c_i and c_{ER} would change if c_i were suddenly increased to different levels at fixed c_{ER}, and then allowed to relax. However, during depolarization, both c_i and c_{ER} would change in response to stimulation, causing the CICR thresholds to be dynamic. Since c_{ER} would depend on the history of stimulation, the conditions required to elicit different modes of CICR would be history-dependent, imparting to Ca^{2+} signaling a form of short term memory. For example, brief priming stimuli that load the ER would lower the threshold for both net CICR and regenerative net CICR (Han et al., 1994).

The properties of J_{SERCA}, P_{ER}, and J_{Extru} provide an explanation for the following experimental findings: (1) The ER accumulates Ca^{2+} at low $[Ca^{2+}]_i$ and releases net Ca^{2+} at high $[Ca^{2+}]_i$ (see Fig. 9B). Simulated responses to Ca^{2+} entry using the same parameters as in Fig. 6 reproduces this behavior as c_i is raised by the stimulus beyond the respective CICR thresholds (see Fig. 7). (2) Inhibition of CICR with ryanodine increases the rate of ER Ca^{2+} accumulation and slows $[Ca^{2+}]_i$ elevations induced by weak depolarization (Albrecht et al., 2001). (3) When sympathetic neurons are depolarized in the presence of caffeine (~5 mM), $[Ca^{2+}]_i$ oscillates. Simulations using rate descriptions of J_{Extru} and J_{SERCA} obtained experimentally in the absence of caffeine, and of P_{ER} in the presence of caffeine, show oscillatory behavior like that observed (Albrecht et al., 2002). (4) The total Ca concentration within the ER ($[Ca]_{ER}$) following a 10 s depolarization varies with distance from the plasma membrane, with $[Ca]_{ER}$ being lower in regions that are closer to the plasma membrane than in more distal regions (Hongpaisan et al. 2001). One simple explanation of this finding is that it represents the simultaneous

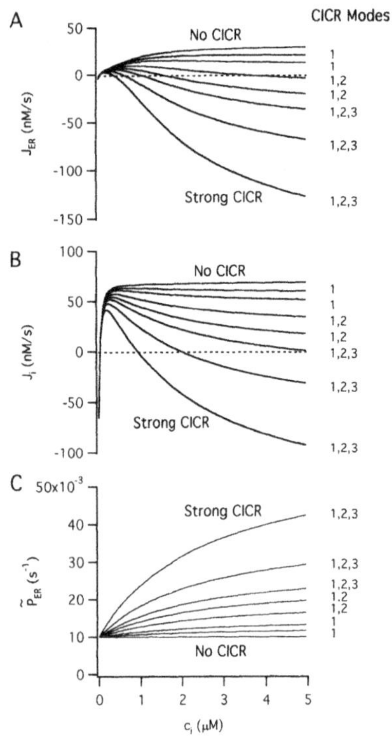

Fig. 8. The allowable modes of CICR depend on quantitative properties of the underlying Ca^{2+} transport systems. Slices from different J_{ER} (A) and J_i (B) surfaces constructed using increasing values of the maximal c_i-sensitive permeability, $P_{max,RyR}$ (C). As this parameter is increased from zero (no CICR) to a large value (strong CICR) new modes of net CICR become available. In each case, c_{ER} was fixed at its resting value, which depended weakly on $P_{max,RyR}$.

expression of different CICR modes when $[Ca^{2+}]_i$ is spatially nonuniform. Portions of the ER near the plasma membrane experience larger increases in $[Ca^{2+}]_i$ due to their proximity to sites of Ca^{2+} entry (Kuba, 1994), which would be more likely to stimulate net Ca^{2+} release. More distant portions of the ER would be exposed to smaller $[Ca^{2+}]_i$ elevations and show little net Ca^{2+} transport.

So far, we have described the modes of CICR that can be observed as the stimulus is varied with a given set of transport parameters obtained from sympathetic neurons. However, which modes are allowable depends critically on the value of these parameters, which are expected to vary with cell type, developmental stage, and state of modulation. For example, Fig. 8 shows how the allowable modes of CICR vary with $P_{ER,max}$, the maximal value of P_{ER} at saturating c_i, which would depend on RyR expression level and unitary permeability. With all other parameters fixed, if $P_{ER,max}$ is small, net Ca^{2+} release cannot occur in response to a brief stimulus-evoked c_i elevation, irrespective of its size. Moderate values permit non-regenerative net release, and large values permit regenerative Ca^{2+} release. Figure 8 gives predicted fluxes following stimuli that do not perturb the intraluminal Ca^{2+} level, but conditioning stimuli that increase or decrease these levels would bias the ER towards acting as a source or sink, respectively.

Summary. Depolarization-evoked $[Ca^{2+}]_i$ elevations can stimulate either net Ca^{2+} uptake or release by the ER. Which form of transport occurs depends on the quantitative properties of the underlying Ca^{2+} transport and buffering systems, the size of the evoked $[Ca^{2+}]_i$ elevation, and the intraluminal Ca^{2+} concentration. In sympathetic neurons, the functional interplay between ER Ca^{2+} uptake and release pathways, representing the properties of SERCAs and ryanodine-sensitive Ca^{2+} release channels, along with the plasma membrane Ca^{2+} extrusion system, leads to distinct modes of CICR during depolarization-evoked $[Ca^{2+}]_i$ elevations. Which modes of CICR prevail during stimulation depends on how high $[Ca^{2+}]_i$ rises, and on the secondary changes in intraluminal Ca^{2+} concentration that accompany these responses. Maintained depolarization can stimulate steady-state Ca^{2+} oscillations. These results show how the ER provides another signaling element that, along with mitochondria, imparts great flexibility to stimulus-evoked Ca^{2+} responses. They also identify cytoplasmic Ca^{2+} concentration as a critical factor in determining whether the ER acts as a Ca^{2+} source or sink during stimulation.

3.6. Case 4: Coupled Dynamics of Cytoplasmic, Mitochondrial and Endoplasmic Reticular Calcium - Mitochondria as Regulators of CICR Modes

We have now examined three components of the Ca^{2+} regulatory network in sympathetic neurons: transport across the plasma membrane, transport by mitochondria, and transport by the ER. We have shown how mitochondria and the ER individually operate in a cytoplasmic Ca^{2+} pool that is ultimately under the control of plasma membrane transport. In this section, we examine the most interesting, and physiologically important case in which all three mechanisms operate simultaneously. As in earlier sections, we assume that Ca^{2+} is uniformly distributed within compartments, and ask how the full system of transporters defines intracellular Ca^{2+} signals during and after depolarization-evoked Ca^{2+} entry. While a description of this system requires three coupled nonlinear differential equations, analysis of its qualitative properties is greatly facilitated by what we have already learned about the components. The following two properties are of particular significance: (1) Mitochondria act as powerful Ca^{2+} buffers whose strength increases steeply with $[Ca^{2+}]_i$. (2) The ER can act as either a Ca^{2+} source or sink depending on cytoplasmic and intraluminal $[Ca^{2+}]$ levels. Together, these two properties point to a simple way of thinking about ER/mitochondrial interactions. In brief, mitochondria can be viewed as devices that indirectly specify the prevailing CICR mode through their ability to regulate $[Ca^{2+}]_i$. In a reciprocal manner, the ER either shares the burden of cytoplasmic Ca^{2+} buffering with mitochondria (at low $[Ca^{2+}]_i$) or contributes to this burden (at high $[Ca^{2+}]_i$).

These concepts follow from the results already presented, but they are also supported by the following experimental observations (Fig. 9B). If cells are depolarized to evoke progressively larger $[Ca^{2+}]_i$ elevations when mitochondrial Ca^{2+} transport is enabled, $[Ca]_{ER}$ is raised by small $[Ca^{2+}]_i$ elevations (~300 nM) but not by intermediate increases (~500 nM). These $[Ca]_{ER}$ measurements were made after prolonged exposure to high K^+ (30 or 50 mM for 45 -120 s) in an effort to permit $[Ca^{2+}]_i$ and $[Ca^{2+}]_{ER}$ to reach steady-state values. That this was accomplished is supported by the observation that $[Ca^{2+}]_i$ was approximately steady, and for the stronger stimulus, $[Ca]_{ER}$ was unchanged if the length of depolarization was increased from 45 s to 120 s (Fig. 9B, points c and d). Mitochondria operate in their buffering mode under these conditions, since mitochondrial total Ca concentration steadily increases under these conditions, as demonstrated by direct measurements of total mitochondrial Ca concentrations using electron probe microanalysis (Fig. 9C, bottom). Mitochondrial Ca^{2+} accumulation depresses $[Ca^{2+}]_i$ and biases the ER away from net Ca^{2+} release. Thus, when cells are exposed to the same depolarizing stimulus after inhibiting mitochondrial Ca^{2+} uptake with FCCP, $[Ca^{2+}]_i$ rises to a higher level, and $[Ca]_{ER}$ falls below its resting level, indicative of net CICR (Fig. 9B, point e). Therefore, it appears that mitochondrial Ca^{2+} transport biases the ER toward Ca^{2+} accumulation (or away from net Ca^{2+} release) by maintaining $[Ca^{2+}]_i$ at a lower level during stimulation. In the absence of this buffering mechanism, $[Ca^{2+}]_i$ rises to a higher level that stimulates net Ca^{2+} release. The reduction in $[Ca]_{ER}$ observed when cells were depolarization in the presence of FCCP appears to be caused by the large $[Ca^{2+}]_i$ increase, not loss of ATP, since $[Ca]_{ER}$ is also reduced by depolarizing stimuli in the absence of FCCP. For example, this occurs in regions that are close to the plasma membrane where $[Ca^{2+}]_i$ is expected to reach its highest levels during depolarization (Fig. 9D).

Additional support for the idea that mitochondria indirectly regulate the direction of net ER Ca^{2+} transport by modulating $[Ca^{2+}]_i$ comes from comparing the time courses of $[Ca]_{ER}$ and $[Ca]_{MT}$ during and after depolarization (Fig. 9C, bottom).

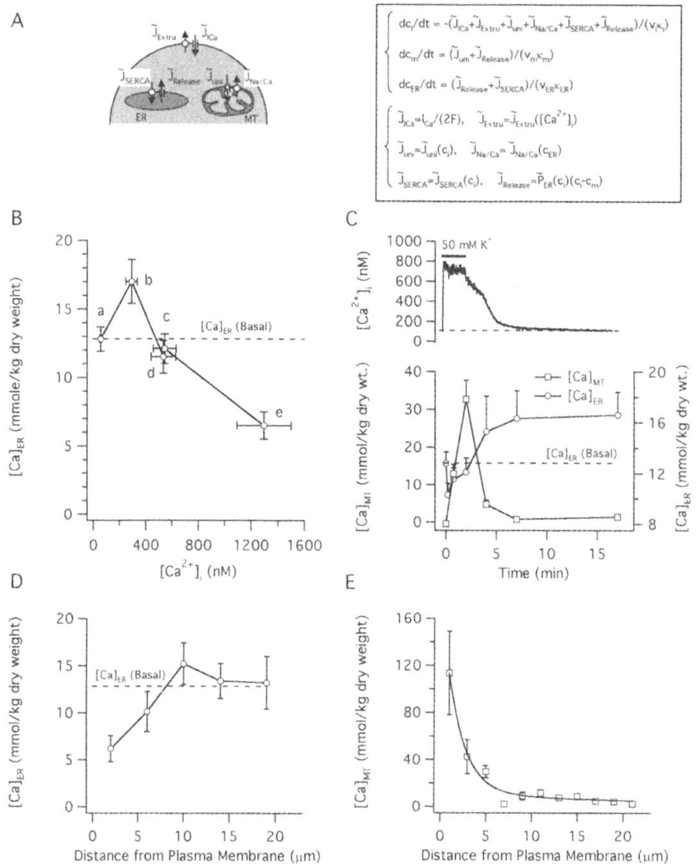

Fig. 9. ER/mitochondrial interactions. (A) Diagram illustrating the three-compartment system in which cytoplasmic, mitochondrial and ER Ca^{2+} concentrations are controlled by plasma membrane, mitochondrial and ER Ca^{2+} transport pathways and intracompartmental Ca^{2+} buffers. The equations used to describe this system are shown at right. (B) Biphasic $[Ca^{2+}]_i$-dependence of total ER Ca^{2+} concentration ($[Ca]_{ER}$) measured under five different experimental conditions: (a) rest, (b) 30 mM K^+, 120 s, (c) 50 mM K^+, 45 s, (d) 50 mM K^+, 120 s, and (e) 50 mM K^+ + 1 μM FCCP, 45 s. Weak stimuli that raise $[Ca^{2+}]_i$ to low levels stimulate net ER Ca^{2+} uptake, while stronger stimuli that raise $[Ca^{2+}]_i$ to higher levels have little effect on $[Ca]_{ER}$. When $[Ca^{2+}]_i$ is raised to much higher levels, net Ca^{2+} release occurs, leading to a reduction in $[Ca]_{ER}$. (C) Comparison between the time courses of $[Ca^{2+}]_i$ (top), $[Ca]_{ER}$ (circles) and $[Ca]_{MT}$ (squares) during a 120 s exposure to 50 mM K^+. Top trace is a representative measurement from a single fura-2 loaded cell, while data in lower panel are collected results from multiple neurons that were stimulated and rapidly frozen at the indicated times in preparation for microanalysis. (D,E) Comparison between the spatial profiles of $[Ca]_{ER}$ measured after a 10 s 50 K^+ depolarization, and of $[Ca]_{MT}$ measured after a 45 s 50 K^+ depolarization. See text for discussion. Measurements of total ER and mitochondrial Ca concentrations were made using electron probe microanalysis and electron energy loss spectroscopy as described in Pivovarova et al. (1999) and Hongpaisan et al. (2001).

When mitochondrial Ca^{2+} transport is enabled, maintained depolarizations that increase $[Ca^{2+}]_i$ to intermediate levels stimulate little net ER Ca^{2+} transport, but there is significant ER Ca^{2+} accumulation during the recovery following repolarization. This occurs specifically during the plateau phase of the $[Ca^{2+}]_i$ recovery, during which mitochondrial Ca^{2+} release temporarily maintains $[Ca^{2+}]_i$ within the permissive range for ER Ca^{2+} accumulation. Thus, it appears that mitochondria regulate the direction of net ER Ca^{2+} transport by modulating $[Ca^{2+}]_i$.

The results presented above suggest that Ca^{2+} released from mitochondria can be taken up by the ER. Does Ca^{2+} release from the ER contribute to mitochondrial Ca^{2+} loading during stimulation? One possible example is illustrated in Fig. 9D,E, where the spatial profiles of $[Ca]_{ER}$ and $[Ca]_{MT}$ are compared at 10 and 45 s during 50 mM K^+ depolarization, respectively. Near the plasma membrane, where $[Ca^{2+}]_i$ is expected to reach its highest levels, the ER releases net Ca^{2+} and mitochondria accumulate Ca^{2+}. Since ER Ca^{2+} release is likely to contribute to the evoked increase in local $[Ca^{2+}]_i$, and since the rate of mitochondrial Ca^{2+} accumulation increases steeply with $[Ca^{2+}]_i$, it seems likely that some of the Ca^{2+} released by the ER would be taken up by mitochondria. One possibility is that net CICR from peripheral ER contributes to large increases in $[Ca^{2+}]_i$ that enhance mitochondrial Ca^{2+} accumulation, accounting for the large increases in peripheral $[Ca]_{MT}$ that are observed (~120 mM/kg dry wt., which can be converted to ~ 80 mM). Given spatial proximity between mitochondria and the ER, it seems likely that Ca^{2+} transfer rates between these organelles could be much higher than those estimated from bulk concentration measurements.

3.7 Summary and Questions for Future Study

Ca^{2+} is an important signaling intermediate in virtually all cells, so there is a strong rationale for understanding $[Ca^{2+}]$ dynamics. Free Ca^{2+} concentration is tightly controlled in all cellular compartments by specific Ca^{2+} transport and buffering systems that have been investigated in isolation, but largely because of technical difficulties in measuring rates of transport *in situ*, have received little attention in intact cells. This review provides an overview of work that we have done to understand the functional properties of Ca^{2+} transport systems expressed in sympathetic neurons and how they operate together as a coupled nonlinear system to orchestrate stimulus-evoked Ca^{2+} signals.

A number of areas are in need of further study. For example, our Ca^{2+} transporter characterizations were obtained under conditions where Ca^{2+} is distributed nearly uniformly within compartments. These characterizations provide a good explanation of $[Ca^{2+}]_i$ dynamics during post-stimulus recoveries, but do they explain $[Ca^{2+}]$ dynamics under conditions where Ca^{2+} is distributed nonuniformly, e.g. during the initial moments of a depolarizing stimulus? Diffusion simulations carried out with rate descriptions obtained from analysis of recoveries account qualitatively for spatially heterogeneous changes in mitochondrial and ER Ca levels during stimulation. However, it is not clear how well they account for these profiles quantitatively, and uncertainties regarding the properties of Ca^{2+} buffers will need to be resolved before this problem can be adequately addressed.

Another question is how elementary properties of Ca^{2+} transporters and buffers contribute to the macroscopic Ca^{2+} fluxes we have described. None of the fluxes showed obvious explicit time dependence and derived their temporal properties from time-dependent substrate availability. This is unlikely to be found for all, or even most transporters, given that transport rates can show time-dependence because of gating or modulation (Schiefer et al. 1995, Bautista et al., 2002). Evidently, such processes do not contribute to extensive changes in transport rates over the limited range of conditions encountered in our experiments. In the case of RyRs, time-dependent gating would impart to P_{ER} an explicit time-dependence, but such changes are likely to be very small when $[Ca^{2+}]_i$ is below several μM (Schiefer et al., 1995), as in our experiments. Thus, more pronounced time-dependence would be expected for this pathway at high $[Ca^{2+}]_i$.

From an evolutionary standpoint, it is important to ask how the intracellular Ca^{2+} signaling mechanisms that we have elucidated in amphibian neurons are conserved in mammals. Recent work reveals an interplay between dendritic ER and mitochondrial Ca^{2+} handling in mammalian neurons following synaptic stimulation (Pivovarova et al., 2002) that is in many ways like that described in the present review. Also, since the underlying transporters undoubtedly show functional variability within animal populations, it will be interesting to determine how such variability contributes to qualitative variations in physiological Ca^{2+} regulation. Finally, what are the implications of derangements in Ca^{2+} signaling in humans? Recent research suggests that pathological changes in intracellular calcium regulation play an important role in neurodegenerative diseases and apoptosis (Verkhratsky and Petersen, 2002), and much work remains to be done to elucidate the mechanisms of normal mammalian CNS Ca^{2+} signaling, its pathophysiology, and potential therapeutic interventions. These and other questions await future experiments and analysis, insuring that the study of Ca^{2+} dynamics will continue to be a source of insights with great relevance to intracellular communication and human physiology for years to come.

Appendix

The equations used to describe Ca^{2+} dynamics were recast from those in Fig. 9 as follows:

$$\frac{dc_i}{dt} = -(J_{Extru} + J_{ICa} + J_{uni} + J_{Na/Ca} + J_{SERCA} + J_{Release})$$

$$\frac{dc_{MT}}{dt} = \left(\frac{J_{uni} + J_{Na/Ca}}{\gamma_{MT}}\right)$$

$$\frac{dc_{ER}}{dt} = \left(\frac{J_{SERCA} + J_{Release}}{\gamma_{ER}}\right)$$

where:

$$J_{Extru} = [k_{leak,pm}(c_i - c_o) + \frac{V_{max,extru}}{1 + (\frac{EC_{50,extru}}{c_i})^{n_{extru}}}]$$

$$J_{ICa} = \frac{I_{Ca}}{2Fv_i\kappa_i}$$

$$J_{uni} = \frac{V_{max,uni}}{1 + (\frac{EC_{50,uni}}{c_i})^{n_{uni}}} c_i$$

$$J_{Na/Ca} = -\frac{V_{max,Na/Ca}}{1 + (\frac{EC_{50,Na/Ca}}{c_m})}$$

$$J_{SERCA} = \left[\frac{V_{max,SERCA}}{1 + (\frac{EC_{50,SERCA}}{c_i})^{n_{SERCA}}}\right]\kappa_i^{-1}$$

$$J_{Release} = [P_{ER}(c_i - c_{ER})]\kappa_i^{-1}$$

$$P_{ER} = P_{basal} + \frac{P_{max,RyR}}{1 + (\frac{EC_{50,RyR}}{c_i})^{n_{RyR}}}$$

$$\kappa_i = \kappa_{endog} + \frac{[\text{Fura}-2]K_{d,Fura2}}{(K_{d,Fura2} + c_i)^2}$$

$$\gamma_{MT} = \frac{v_{MT}\,\kappa_{MT}}{v_i\,\kappa_i}$$

$$\gamma_{ER} = \frac{v_{ER}\,\kappa_{ER}}{v_i\,\kappa_i}$$

where κ_{endog} is the endogenous cytoplasmic buffering strength at rest, which is assumed to be constant, [Fura-2] is the fura-2 concentration and $K_{d,fura}$ is the dissociation constant for Ca^{2+} binding to fura-2. Note that descriptions of ER Ca^{2+} transport explicitly distinguish between the c_i-dependence of transport rate and of κ_i, while for the plasma membrane and mitochondria these two contributions are lumped together. This reflects a difference in form of the equations used to describe the measured fluxes, but has little impact on the simulations.

Table 1 - Representative Parameter Values

Plasma Membrane/Cytoplasm	Mitochondria	Endoplasmic Reticulum
$k_{leak,pm}$=3.7x10^{-7} s^{-1}	$V_{max,uni}$=80 s^{-1}	P_{basal}=0.01 s^{-1}
$V_{max,Extru}$=28 nMs^{-1}	$EC_{50,uni}$=10 µM	$P_{max,RyR}$=0.05 s^{-1}
$EC_{50,Extru}$=379 nM	n_{uni}=2	$EC_{50,RyR}$=2.6 µM
n_{Extru}=1.8		n_{RyR}=1
κ_{endog}=25	$V_{max,Na/Ca}$=35 nMs^{-1}	$V_{max,SERCA}$=2146 nMs^{-1}
[Fura-2]=80 µM	$EC_{50,Na/Ca}$=307 nM	$EC_{50,SERCA}$=30 nM
$K_{d,Fura2}$=224 nM	γ_{MT}=2	n_{SERCA}=2.5
		$v_{ER}\kappa_{ER}/v_i$=1

Acknowledgements

I would like to thank Meredith Albrecht, Brian Andrews, Stephen Colegrove, Jarin Hongpaisan and Natalia Pivovarova for their contributions to this work, and Brian Andrews, Hillel Chiel, Patrick Friel, Barbara Grimpe and James Sneyd for helpful comments and discussion.

References

Albrecht MA, Colegrove SL, Friel DD (2002) Differential regulation of endoplasmic reticulum (ER) Ca^{2+} uptake and release rates accounts for multiple modes of Ca^{2+}-induced Ca^{2+} release. J Gen Physiol 119:211-233.

Albrecht MA, Colegrove SL, Hongpaisan J, Pivovarova NB, Andrews SB, Friel DD (2001). Multiple modes of calcium-induced calcium release in sympathetic neurons I: Attenuation of endoplasmic reticulum Ca^{2+} accumulation at low $[Ca^{2+}]_i$ during weak stimulation. J Gen Physiol 118:83-100.

Bautista DM, Hoth M, Lewis RS (2002). Enhancement of calcium signalling dynamics and stability by delayed modulation of the plasma-membrane calcium-ATPase in human T cells. J Physiol. 541:877-94.

Berridge MJ (1998) Neuronal calcium signaling. Neuron 21:12-26.

Bezprozvanny I, Watras J, Ehrlich BE (1991). Bell-shaped calcium response curves of $Ins(1,4,5)P_3$- and calcium-gated channels from endoplasmic reticulum of cerebellum. Nature 351:751-754.

Castilho RF, Hansson O, Ward MW, Budd SL, Nicholls DG. (1998). Mitochondrial control of acute glutamate excitotoxicity in cultured cerebellar granule cells. J Neurosci. 18:10277-86.

Colegrove SL, Albrecht MA, Friel DD (2000a) Dissection of mitochondrial Ca^{2+} uptake and release fluxes *in situ* following depolarization-evoked $[Ca^{2+}]_i$ elevations in sympathetic neurons. J Gen Physiol 115:351-369.

Colegrove SL, Albrecht MA, Friel DD (2000b). Quantitative analysis of mitochondrial Ca^{2+} uptake and release pathways in sympathetic neurons: reconstruction of the recovery following depolarization-evoked $[Ca^{2+}]_i$ elevations. J Gen Physiol 115:371-388.

David G (1999). Mitochondrial clearance of cytosolic Ca^{2+} in stimulated lizard motor nerve terminals proceeds without progressive elevation of mitochondrial matrix $[Ca^{2+}]$. J Neurosci 19:7495-506.

Friel DD (2000) Mitochondria as regulators of stimulus-evoked calcium signals in neurons. Cell Calcium 28:307-316.

Friel DD, RW Tsien (1994). An FCCP-sensitive Ca^{2+} store in bullfrog sympathetic neurons and its participation in stimulus-evoked changes in $[Ca^{2+}]_i$. J Neurosci 14: 4007-4024.

Gunter TE, Buntinas L, Sparagna G, Eliseev R, Gunter K (2000) Mitochondrial calcium transport: mechanisms and functions. Cell Calcium 28:285-296.

Han S, Schiefer A, Isenberg G (1994). Ca^{2+} load of guinea-pig ventricular myocytes determines efficacy of brief Ca^{2+} currents as trigger for Ca^{2+} release. J Physiol 480:411-21

Hansford RG (1985). Relation between mitochondrial calcium transport and control of energy metabolism. Rev Physiol Pharmacol 102:1-72.

Herrington J, Park YB, Babcock DF, Hille B (1996). Dominant role of mitochondria in clearance of large Ca^{2+} loads from rat adrenal chromaffin cells. Neuron. 16:219-28.

Hongpaisan J, Pivovarova NB, Colegrove SL, Leapman RD, Friel DD, Andrews SB (2001). Multiple modes of calcium-induced calcium release in sympathetic neurons II: A $[Ca^{2+}]_i$-dependent transition from Ca^{2+} accumulation to net release from the endoplasmic reticulum and its spatiotemporal characteristics. J Gen Physiol 118: 101-112.

Horikawa Y, Goel A, Somlyo AP, Somlyo AV (1998) Mitochondrial calcium in relaxed and tetanized myocardium. Biophys J 74:1579-90.

Kuba K (1994). Ca^{2+}-induced Ca^{2+} release in neurons. Jap J Physiol 44:613-650.

McCormack JG, Denton RM (1990). The role of mitochondrial Ca^{2+} transport and matrix Ca^{2+} in signal transduction in mammalian tissues. Biochim Biophys Acta. 1018:287-91.

Meldolesi J (2001) Rapidly exchanging Ca^{2+} stores in neurons: molecular, structural and functional properties. Prog Neurobiol 65:309-38

Meldolesi J, Pozzan T (1998). The endoplasmic reticulum Ca^{2+} store: a view from the lumen. Trends Biochem. Sci 23:10-14

Moravec CS, Desnoyer RW, Milovanovic M, Schluchter MD, Bond M (1997). Mitochondrial calcium content in isolated perfused heart: effects of inotropic stimulation. Am J Physiol 273:H1432-H1439.

Pfaffinger PJ, Leibowitz MD, Subers EM, Nathanson NM, Almers W, Hille B. (1988). Agonists that suppress M-current elicit phosphoinositide turnover and Ca^{2+} transients, but these events do not explain M-current suppression. Neuron 1:477-84.

Pivovarova NB, Hongpaisan J, Andrews SB, Friel DD (1999). Depolarization-induced mitochondrial Ca accumulation in sympathetic neurons: Spatial and temporal characteristics. J Neurosci 19:6372-6384.

Pivovarova NB, Pozzo-Miller LD, Hongpaisan J, Andrews SB (2002). Correlated calcium uptake and release by mitochondria and endoplasmic reticulum of CA3 hippocampal dendrites after afferent synaptic stimulation. J Neurosci. 22:10653-61.

Rizzuto R, Pinton P, Carrington W, Fay FS, Fogarty KE, Lifshitz LM, Tuft RA, Pozzan T (1998). Close contacts with the endoplasmic reticulum as determinants of mitochondrial Ca^{2+} responses. Science 280:1763-6.

Schiefer A, Meissner G, Isenberg G (1995). Ca^{2+} activation and Ca^{2+} inactivation of canine reconstituted sarcoplasmic reticulum Ca^{2+} release channels. J Physiol 489:337-348.

Stout AK, Raphael HM, Kanterewicz BI, Klann E, Reynolds IJ (1998). Glutamate-induced neuron death requires mitochondrial calcium uptake. Nat Neurosci 5:366-373.

Suzuki S, Osanai M, Mitsumoto N, Akita T, Narita K, Kijima H, Kuba K (2002). Ca^{2+}-dependent Ca^{2+} clearance via mitochondrial uptake and plasmalemmal extrusion in frog motor nerve terminals. J Neurophysiol 87:1816-23.

Thayer SA, Miller RJ (1990). Regulation of the intracellular free calcium concentration in single rat dorsal root ganglion neurones in vitro. J Physiol 425:85-115.

Usachev Y, Thayer SA (1999). Controlling the urge for a Ca^{2+} surge: all-or-none Ca^{2+} release in neurons. Bioessays 21:743-750.

Verkhratsky A, Petersen OH (2002). The endoplasmic reticulum as an integrating signaling organelle: from neuronal signalling to neuronal death. Eur J Pharmacol. 447:141-154.

Verkhratsky A, Shmigol A (1996). Calcium-induced calcium release in neurons. Cell Calcium 19:1-14.

4. Global and Local Regulation of Neuronal Differentiation by Calcium Transients

Nicholas C. Spitzer

Neurobiology Section, Division of Biological Sciences and Center for Molecular Genetics, UCSD, La Jolla CA 92093

Abstract. Embryonic spinal neurons express ion channels prior to synaptogenesis that mediate specialized forms of excitability promoting Ca^{2+} influx. Ca^{2+} imaging has revealed that these cells generate three distinct types of elevations of intracellular Ca^{2+}, which appear to encode developmental information in the frequency with which they are produced. Ca^{2+} spikes (~10 s duration) are generated by Ca^{2+}-dependent action potentials and Ca^{2+}-induced Ca^{2+}-release and transcriptionally regulate the developmental appearance of GABA in cultured neurons. Growth cone Ca^{2+} transients (~30 s duration) are generated by influx and release from stores and regulate the rate of axon outgrowth both *in vivo* and in cultured neurons. They inhibit axon extension through activation of the Ca^{2+}-dependent phosphatase, calcineurin (CN), that affects the actin cytoskeleton. Filopodial Ca^{2+} transients (~0.3 s duration) are generated by influx through Ca^{2+} channels activated by integrin receptors in cultured neurons. They promote turning when asymmetrically generated at a substrate border or when artificially imposed in selected filopodia, by activation of the Ca^{2+}-dependent protease calpain. Developmental regulation by Ca^{2+} transients provides function-dependent feedback that may compensate for variation in gene expression and facilitate regulation by environmental stimuli.

4.1. Introduction

The development of fluorescent Ca^{2+}indicators has enabled prodigious advances in understanding Ca^{2+} dynamics. Many studies have identified ligands that stimulate elevations of intracellular Ca^{2+} and revealed the existence of patterns of Ca^{2+} transients. The mechanisms by which Ca^{2+} is elevated have been shown to include influx and release from intracellular stores. Other studies have investigated the roles of Ca^{2+} transients in various aspects of cellular behaviors. It has become clear that many functions are accomplished by this single messenger by multiplexing the parameters of Ca^{2+} transients in space and in time.

We have investigated differentiation of excitability of neurons in the *Xenopus* embryo at early stages of development. Initial studies indicated that ion channels are expressed in spinal neurons at the time of closure of the neural tube, prior to synapse formation, and that the constellations of channels expressed at this time allow Ca^{2+} influx (Spitzer and Ribera, 1998). Significantly, removal of extracellular Ca^{2+} from the medium perturbed the differentiation of neurons in dissociated cell culture (Bixby and Spitzer, 1984; Desarmenien and Spitzer, 1991; Spitzer et al., 1993).

The early expression of Ca^{2+} permeable channels, coupled with effects of growth in Ca^{2+}-free medium, suggested that developing neurons might exhibit spontaneous channel activity leading to changes in intracellular Ca^{2+} that in turn affected neuronal differentiation. To detect such changes we imaged intracellular Ca^{2+}. We observed no steady changes in baseline Ca^{2+} (Holliday and Spitzer, 1990) but fluctuations in Ca^{2+} levels were apparent (Holliday et al., 1991; Gu and Spitzer, 1993). These observations led to characterization of three different classes of Ca^{2+} transients, for each of which a function has been determined. Ca^{2+} spikes were discovered by examining the embryonic form of classical electrical excitability – Ca^{2+}-dependent action potentials. Growth cone Ca^{2+} transients were revealed during experiments aimed at imaging Ca^{2+} dynamics in spikes, and found to be distinct both functionally and mechanistically. Filopodial Ca^{2+} transients were subsequently hypothesized to exist as a link between guidance molecules and cell motility, and were found to be different from spikes and growth cone transients with respect to function and mechanism.

Ca^{2+} imaging has been key to observation of these three forms of excitability that are expressed in spinal neurons prior to synapse formation, and their detection requires image acquisition at appropriate sampling intervals. It has been satisfying to be able to determine the functions of all three of these nontraditional forms of excitability. Further types of Ca^{2+} transients with interesting functions likely remain to be discovered.

4.2. Calcium Spikes

Ca^{2+} spikes were the first class of spontaneous Ca^{2+} transient to be characterized in *Xenopus* spinal neurons and were initially studied in neurons in dissociated cell culture. They are ~10 s in duration, involve an elevation of intracellular Ca^{2+} to ~1 μM from a baseline of 100 nM, and are generated at frequencies of 1-10/hr (Gu and Spitzer, 1993; Gu et al., 1994; Gu and Spitzer, 1995). The triggers for spikes are presently unknown, but are likely to involve a combination of intrinsic membrane properties and extrinsic ligands. They are produced by Ca^{2+}-dependent action potentials, involving both low voltage-activated and high voltage-activated Ca^{2+} currents, which are themselves transiently expressed during development and replaced by Na^{+}-dependent action potentials. Ca^{2+} influx in turn triggers Ca^{2+}-induced Ca^{2+} release (CICR) from intracellular Ca^{2+} stores (Holliday et al., 1991). Spikes have a characteristic waveform with a rapid rise and biphasic decay with time constants of 1 and 10 s. Because they are produced by action potentials, they are propagated rapidly throughout neurons. Remarkably, these low frequencies of Ca^{2+} transients regulate transcription by the frequency with which they are expressed.

4.2.1. In Vitro

Clues to the functions of Ca^{2+} spikes came from several observations. The first originated with the concept that it would be important to examine the natural expression of these Ca^{2+} transients in order to learn whether regular patterns could be discerned. To this end, neurons were imaged for 5 hr periods at various stages of their development during the first day *in vitro* after culturing at the neural plate stage. When the pattern of activity was analyzed it became clear that the frequency of these events, but not their amplitude or duration, varied with developmental age. Initially zero, it rose to a mean value of 3/hr by 5 hr in culture, remained near this value for ~5 hr, and then declined to ~1/hr (Gu and Spitzer, 1995).

The second set of observations stemmed from analysis of the effects of growth of neurons in Ca^{2+}-free medium on expression of neurotransmitters. Forty percent of neurons grown in the presence of extracellular Ca^{2+} express the transmitter GABA, revealed with antibodies to either GABA or to glutamic acid decarboxylase (GAD, the synthetic enzyme). In contrast, only ten percent of neurons grown in the absence of Ca^{2+} express GABA, consistent with the possibility that Ca^{2+} influx could regulate this aspect of neuronal differentiation

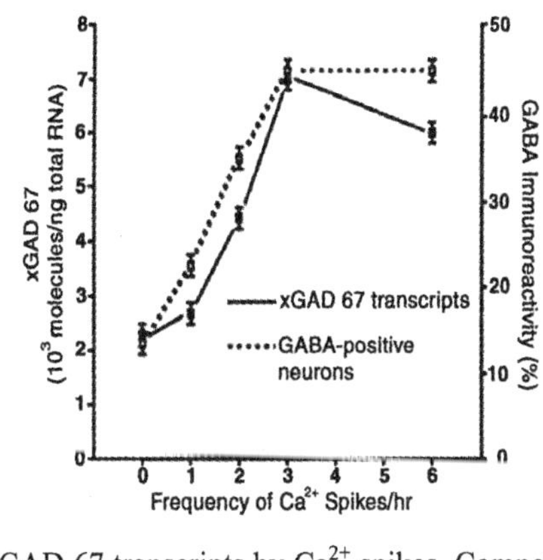

Fig. 1. Regulation of xGAD 67 transcripts by Ca^{2+} spikes. Competitive QRT-PCR analysis of xGAD 67 expression in neurons cultured during stimulation by different frequencies of experimentally imposed Ca^{2+} transients. Numbers of transcripts increase from 2400±0/ng to 7000±200/ng total RNA (2.9±0.1-fold) as the frequency increases in the physiological range from 0-3/hr. The increase in transcript number parallels the percentage of GABA-immunoreactive neurons. N = 5 experiments for each condition. From Watt et al., 2000.

(Spitzer et al., 1993). To test this hypothesis, we grew neurons in the absence of extracellular Ca^{2+} and reimposed Ca^{2+} transients at the natural frequencies with which they are produced. These stimulated Ca^{2+} spikes led to expression of GABA in forty percent of neurons, even though they were grown in Ca^{2+}-free medium. Moreover, chelation of intracellular Ca^{2+} with BAPTA, reducing the amplitude of Ca^{2+} spikes to 25% of their normal amplitude without affecting baseline Ca^{2+}, abolished their effect even though neurons were grown in the presence of extracellular Ca^{2+}.

A third set of observations entailed investigation of the metabolic requirements for GABA expression during development, revealing a requirement for transcription and raising the possibility that Ca^{2+} spikes drive transcription of a gene encoding GAD. To test this idea we cloned xGAD 67, the *Xenopus* isoform of the 67 kD variant of GAD, and were able to show that it is spatially appropriately expressed in the two populations of interneurons previously shown to be GABAergic (Dale et al., 1987; Roberts et al., 1987), and temporally expressed at the time when Ca^{2+} spikes are generated at their highest frequencies (Watt et al., 2000). RNase protection assays and quantitative reverse transcription-polymerase chain reaction (QRT-PCR) were used to determine levels of xGAD 67 transcripts under various growth conditions, and showed that transcript levels were ~3-fold higher when neurons were grown in the presence of Ca^{2+} than in its absence. Moreover the sensitive QRT-PCR method allowed examination of the numbers of transcripts in cultures stimulated with different frequencies of Ca^{2+} spikes. Both the percentage of GABAergic neurons and the levels of xGAD 67 transcripts increased as the frequency of imposed Ca^{2+} spikes increased from 0 to 3/hr, and then plateaued (Fig. 1).

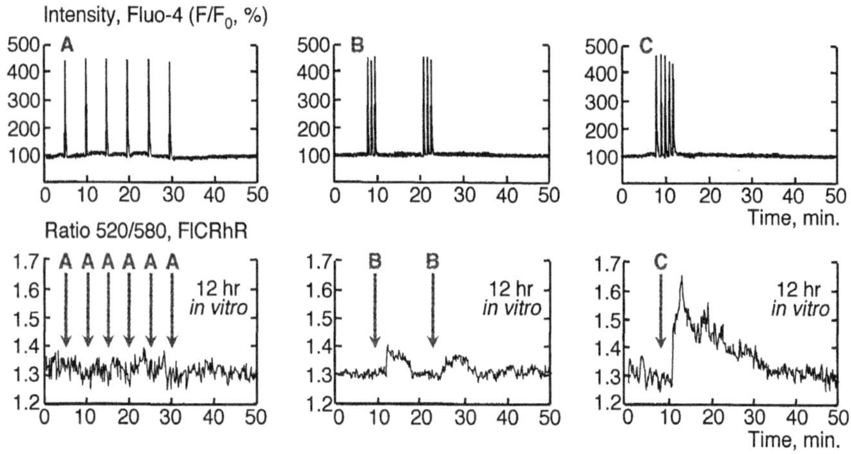

Fig. 2. cAMP transients decode specific patterns of Ca^{2+} spikes. Top, different patterns of stimulated Ca^{2+} spikes. A-single spikes, B-three spikes, C-five spikes. Bottom, cAMP responses to these patterns of Ca^{2+} spike stimulation; arrows mark the times of imposition of the patterns of Ca^{2+} transients indicated by the letter above each one. The pattern of stimulation rather than the number of spikes determines the amplitude and duration of cAMP responses. From Gorbunova and Spitzer, 2002

These observations argue that natural frequencies of Ca^{2+} spikes lead to upregulation of transcripts encoding a transmitter synthetic enzyme, and raise the possibility that other genes may be similarly regulated by this early form of excitability.

Recent work has provided insight into a potential mechanism by which these low frequencies of Ca^{2+} transients can exert their effects. Investigation of the linkage between Ca^{2+} and cAMP signaling in these embryonic spinal neurons revealed that cAMP transients were generated spontaneously but were absent when Ca^{2+} spikes were blocked. Moreover cAMP transients were generated only in response to specific natural patterns of stimulated spikes - brief bursts, indicating that clusters of Ca^{2+} spikes are decoded as cAMP oscillations (Fig. 2) (Gorbunova and Spitzer, 2002). A mathematical model of this module of coupled second messengers produced spontaneous patterns of Ca^{2+} and cAMP oscillations similar to those observed experimentally. Selective elevation of cAMP by distinct patterns of Ca^{2+} transients appears to result from the match of the kinetics of the synthetic enzyme (adenylate cyclase) and the degradative enzyme (phosphodiesterase) with the time scale of Ca^{2+} oscillations. These results suggest that the coding of gene expression by Ca^{2+} spikes may in some cases be driven by cAMP oscillations, because cAMP elevations at similar low frequencies lead to translocation of catalytic subunits of protein kinase A into the nucleus of *Aplysia* neurons and gene expression underlying long-term facilitation (Kang and Kandel, 1993, Bacskai et al., 1993). However elevations of cAMP driven by brief bursts of spikes are not likely to be involved in upregulation of xGAD 67 transcripts by lower spike frequencies.

4.2.2. In Vivo

Ca^{2+} spikes are also observed *in vivo*, in the intact but isolated spinal cord (Gu et al., 1994). The benefit of studying neurons *in situ* lies not only in examining them in their natural environment, but also in the opportunity to identify the spiking behavior of the eight different classes of neurons present in the embryonic spinal cord (Roberts and Clarke, 1982). Preliminary evidence suggests that different types of neurons generate Ca^{2+} spikes at distinct frequencies (Watt et al., 1999; Borodinsky et al., 2002), consistent with differences in terminally differentiated states. It will be useful to examine regulation of cholinergic, glycinergic and glutamatergic phenotypes in spinal neurons and to learn what drives different classes of neurons to fire at different frequencies. Roles for Ca^{2+} spikes in early differentiation of embryonic amphibian spinal neurons are indicated by experiments using embryos in which K^+ channels were unilaterally overexpressed so that spike-mediated Ca^{2+} entry was suppressed. The number of morphologically differentiated neurons in culture derived from the side of the embryo on which K^+ channels are overexpressed is reduced (Jones and Ribera, 1994). Moreover differentiation is rescued when neurons are allowed to differentiate longer *in vivo* prior to culturing; elevations of intracellular Ca^{2+} generated by neurons on the contralateral side, which are not suppressed by BAPTA, may contribute to this recovery (Jones et al., 1995).

4.2.3. Calcium Spikes in Other Cell Types

Signals similar to Ca^{2+} spikes in embryonic *Xenopus* spinal neurons have been ob served in other systems prior to synapse formation. Ca^{2+} transients at frequencies of 1-10/hr are generated in murine neural crest cells and in differentiating sensory neurons in culture during a limited neurogenic period (Carey and Matsumoto, 1999). Spontaneously active neural crest cells produce clones of cells containing neurons, while inactive cells do not. Blocking these Ca^{2+} transients suppresses neuronal differentiation, measured by the expression of Hu RNA-binding protein, voltage-dependent Ca^{2+} channels and caffeine-releasable Ca^{2+} stores. Interestingly, Ca^{2+} transients are also observed in amphibian blastomeres during blastula and gastrula stages, prior to morphological differentiation of the nervous system. These Ca^{2+} transients are generated by L-type Ca^{2+} channels at frequencies of ~3/hr during a 5 hr period of development and can be triggered by the neural inducer, noggin. They are implicated in neural induction, and regulate expression of fos-related protein, the left-right asymmetry specification factor zic3, the DNA replication inhibitor geminin, and the cell adhesion molecule, NCAM (Moreau et al., 1994; Leclerc et al., 1997, 1999, 2000).

4.3. Growth Cone Calcium Transients

Growth cone Ca^{2+} transients were also first identified in cultured neurons. They are ~30 s in duration, entail elevation of intracellular Ca^{2+} to several 100 nM, and are generated at frequencies of 1-10/hr (Gu et al., 1994; Gu and Spitzer, 1995). They appear to be produced by a novel Ca^{2+} entry pathway, since voltage-gated Ca^{2+} channels and several classes of ionotropic receptors are not involved; Ca^{2+} entry triggers CICR. Intriguingly, they are localized to the growth cone, and typically are not propagated back to the cell body. These Ca^{2+} transients regulate the rate of axon outgrowth by affecting the growth cone cytoskeleton.

4.3.1. In Vitro

The first clue to the function of growth cone Ca^{2+} transients came from examination of their pattern of generation, which revealed that they are restricted to growth cones and not propagated back along axons. This observation suggested that they could be involved in some aspect of axon outgrowth. They are generated at a frequency of ~8/hr, from 6 hr in culture when axonogenesis occurs until ~1 day in culture. No changes in mean amplitude or duration were detected during this time (Gu and Spitzer, 1995).

A second set of experiments stemmed from the observation that neurites grew more rapidly and achieved greater lengths when neurons are grown in Ca^{2+}-free medium (Bixby and Spitzer, 1994). Since growth cone transients are abolished by this treatment, we tested the hypothesis that they regulate axon outgrowth. Once again neurons were grown in the absence of extracellular Ca^{2+} and Ca^{2+} transients were reimposed at the natural frequencies with which they are produced in growth cones. These stimulated transients led to restriction of axon outgrowth, even though neurons were grown in Ca^{2+}-free medium. Significantly, stimulation of

Ca^{2+} transients at the lower natural frequencies of Ca^{2+} spikes were ineffective in regulating axonal growth. Moreover, chelation of intracellular Ca^{2+} with BAPTA, reducing the amplitude of growth cone Ca^{2+} transients to 40% of their normal amplitude without affecting baseline Ca^{2+}, caused axon outgrowth to achieve levels seen in the absence of extracellular Ca^{2+}, even though neurons were grown in its presence (Gu and Spitzer, 1995).

Regulation of axon outgrowth by elevations of intracellular Ca^{2+} localized to growth cones suggested the activation of Ca^{2+}-dependent enzymes. Although inhibitors of kinases such as CamKII and PKC had no effect on axon outgrowth (Gu and Spitzer, 1995; Zheng et al., 1994), inhibitors of the phosphatase calcineurin were very effective. Both cyclosporin A and deltamethrin promoted rapid outgrowth and longer neurites in the presence of extracellular Ca^{2+}, that was equivalent to that achieved by growth in the absence of Ca^{2+} (Lautermilch and Spitzer, 2000). To establish more clearly the role of calcineurin, we cloned xCN, the *Xenopus* isoform of the A subunit of calcineurin.

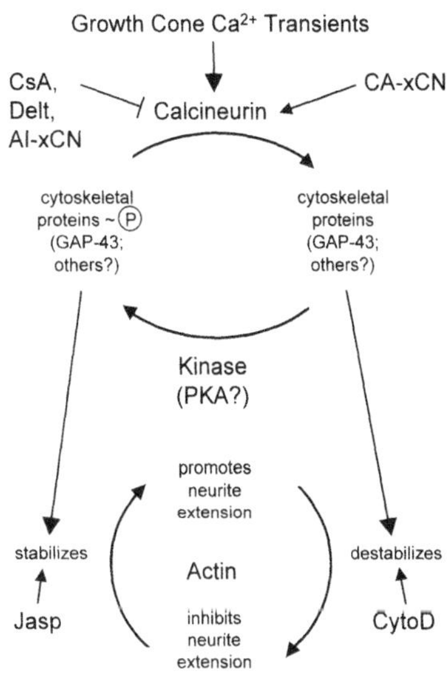

Fig. 3. Model of the signal transduction cascade by which growth cone Ca^{2+} transients act through calcineurin (CN) to regulate axon extension. These transients activate CN, shifting cytoskeletal-regulating proteins to a dephosphorylated state, destabilizing actin filaments and inhibiting axon elongation. CN is not activated in the absence of growth cone Ca^{2+} transients, shifting cytoskeletal-regulating proteins to a phosphorylated state, stabilizing actin filaments and promoting axon outgrowth. Activating or inhibiting this cascade at different points increases or decreases axon extension. CsA, cyclosporin A; Delt, deltamethrin; AI-xCN, autoinhibitory domain of *Xenopus* calcineurin; CA-xCN, constitutively active *Xenopus* calcineurin; jasp, jasplakinolide; cyto D, cytochalasin D. From Lautermilch and Spitzer, 2000

Expression of either the inhibitory domain of xCN or the African swine fever virus protein, another inhibitor of calcineurin, replicated the results obtained with cyclosporin A and deltamethrin. Moreover expression of a constitutively active construct derived from xCN yielded short neurites in the absence of extracellular Ca^{2+} that were not different from those observed when neurons were grown in the presence of Ca^{2+}. The developmental expression pattern of xCN transcripts *in vivo* defined by *in situ* hybridization provided support for the role of calcineurin in regulating Ca^{2+}-dependent neurite extension in the spinal cord. Calcineurin appears to dephosphorylate GAP-43, which in turn destabilizes actin microfilaments in the growth cone. Thus the effect of growth cone transients is to cause a temporary disruption of the local cytoskeleton, slowing axon outgrowth (Fig. 3).

4.3.2. In Vivo

Growth cone Ca^{2+} transients are also observed *in vivo.* Analysis of sensory Rohon-Beard neurons, motoneurons and two classes of interneurons revealed that the rate of axon outgrowth is inversely related to the frequency of spontaneous transients (Gomez and Spitzer, 1999) (Fig. 4). These results showed that Ca^{2+} transients are produced in growth cones migrating through their natural environment, and that the rate of axon extension depends on the frequency with which these transients are generated. Transients appear to be necessary for normal regulation of outgrowth, since suppressing them with the photoactivatable Ca^{2+} chelator diazo-2 AM (caged BAPTA) accelerated the rate of neurite extension. In contrast, diazo-2 had little effect on growth cones that are not generating transients. Transients also appear to be sufficient to slow the rate of axon outgrowth, since generating them with the photoactivatable Ca^{2+} donor NP-EGTA AM (caged Ca^{2+}) at specific frequencies within growth cones that do not generate spontaneous elevations retarded axon extension.

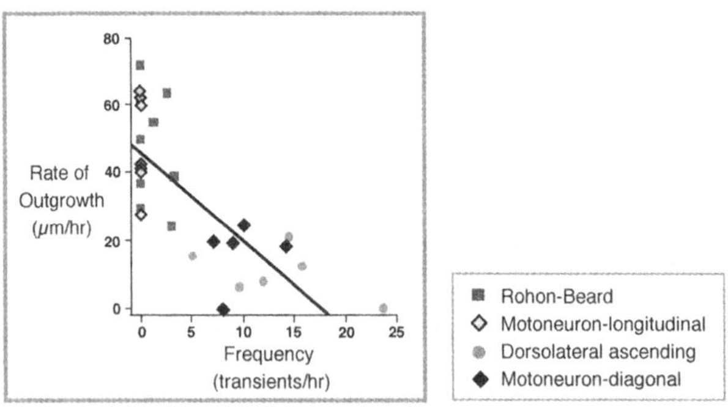

Fig. 4. Control of axon extension by growth cone Ca^{2+} transients in the embryonic spinal cord. The rate of axon outgrowth is inversely proportional to the frequency of spontaneous transients, and the frequency of Ca^{2+} transients depends on cell type. N=43 neurons at stages 21-25. Points at 0/hr are offset to avoid superposition. Slope of fitted line = -3.0, correlation coefficient = 0.78, $p < 0.0001$. From Gomez and Spitzer, 1999

In both sets of experiments the rate of outgrowth was inversely proportional to the frequency of Ca^{2+} transients. These growth cone Ca^{2+} transients may also be involved in axon guidance, since growth cones generated higher frequencies when they stalled at decision regions where they make choices with respect to their trajectories and appeared to sample the local environment for guidance cues. Subsequently the frequency of transients decreased as the growth cones moved off along their routes.

The frequency of growth cone Ca^{2+} transients appears to be determined by environmental cues, since the frequency of spontaneous Ca^{2+} transients is specific to each class of neuron and depends on the position of their growth cones in the spinal cord. All four classes of neurons studied have similar Ca^{2+} buffering and effector machinery, since altering Ca^{2+} dynamics with caged compounds produces similar effects. Thus it seems plausible that there is an endogenous frequency of Ca^{2+} transients that growth cones generate independently, and that this is up or downregulated by different molecules in the spinal cord.

4.3.3. Similar Calcium Transients in Other Cells

Early studies demonstrated that alteration of intracellular Ca^{2+} levels in growth cones of neurons affects their motility (Kater et al., 1988; Kater and Mills, 1991). Additionally, growth cone Ca^{2+} transients have been observed in several types of neurons. Spontaneous increases in intracellular Ca^{2+} are generated in growth cones of neuroblastoma cells (Silver et al., 1990). Higher levels of Ca^{2+} were observed in growth cones that were not advancing, consistent with inhibition of neurite extension. Growth cone transients occurring spontaneously in cultured chick dorsal root ganglion neurons inhibit their migration (Gomez et al., 1995). Ca^{2+} transients elicited from olfactory receptor cell growth cones in response to focal application of 8-Br-cGMP suppress axon extension (Kafitz et al., 2000). Elevations of Ca^{2+} have been shown to affect the local cytoskeleton in several systems (Welnhofer et al., 1999; Wilson et al., 2000). Ca^{2+} transients are generated spontaneously by embryonic myocytes during development *in vitro* and *in vivo* (Ferrari et al., 1996; Ferrari and Spitzer, 1999) during the period of myofibrillogenesis, and blocking transient production disrupts cytoskeletal assembly, blocking the development of myosin thick filament arrays (Ferrari et al., 1998).

4.4. Filopodial Calcium Transients

Many studies have identified molecules that specify axon guidance and numerous investigations have indicated that axon guidance is accomplished by changes in growth cone motility. To examine the mechanism by which molecules affect motility, we tested the hypothesis that this signaling is accomplished by Ca^{2+} transients. Since the responses to guidance cues are focal and rapid, it seemed likely that such signals would be localized and brief. Filopodial Ca^{2+} transients

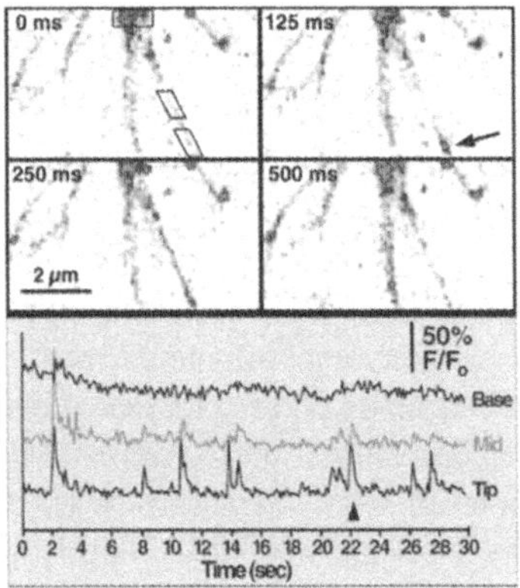

Fig. 5. Spontaneous filopodial Ca^{2+} transients visualized by high speed imaging of Fluo-4 fluorescence. Top: timelapse confocal images of Fluo-4 loaded filopodia. Regions selected for quantification at the tip, middle and base of one filopodium are shown at 0 ms. A local Ca^{2+} transient at the tip occurs at 125 ms (arrow). Bottom: normalized fluorescence changes (scale represents percent over baseline) within regions of this filopodium captured at 125 ms intervals. Ca^{2+} transients occur at high frequency at the tip and move toward the growth cone. Arrowhead indicates transient visualized at top. From Gomez et al., 2001

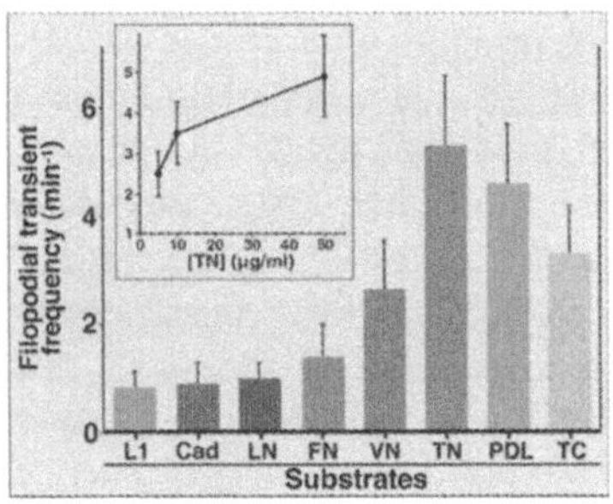

Fig. 6. Frequencies of filopodial Ca^{2+} transients are regulated by the substrate. The average frequency of transients for all filopodia of neurons grown in 2 mM Ca^{2+} on artificial substrates (tissue culture plastic (TC), poly-D-lysine (PDL)), cell adhesion molecules (L1 and N-cadherin (Cad)) and on ECM components laminin (LN), fibronectin (FN), vitronectin (VN) and tenascin (TN) (50 μg/ml). TN stimulated the highest frequency of transients, in a dose-dependent manner (inset). N $\geq$ 10 filopodia for each condition. From Gomez et al., 2001

were identified and observed to be initiated at the tips of single filopodia and propagated back to the growth cone at ~10 μm/s in cultured neurons (Gomez et al., 2001). They are ~300 ms in duration (substantially briefer than spikes and growth cone transients) and involve smaller elevations of intracellular Ca^{2+}. They are produced at frequencies of 1-10/min (higher than those of the other classes of transients) that depend on substrate identity (Fig. 5). They are generated by Ca^{2+} influx through β1 integrin-associated Ca^{2+} channels, and do not involve Ca^{2+} release from intracellular stores. These Ca^{2+} transients signal guidance cues from substrate extracellular matrix molecules to the growth cone and regulate turning.

If filopodial Ca^{2+} transients carry information about the molecular nature of the environment, different substrates could be expected to alter properties such as frequency, amplitude or duration. To test this idea, we imaged filopodial transients in neurons cultured on eight different substrates. Strikingly, the frequency varied with substrate while the amplitude and duration were constant, consistent with a form of frequency code. Extracellular matrix molecules like L1 and cadherin generated lower frequencies of these transients, while vitronectin and tenascin generated higher frequencies, and a dose-response curve could be generated for tenascin (Fig. 6). These substrate molecules are among those implicated in axon guidance; it remains to be determined whether diffusible axon guidance molecules such as netrins and semaphorins also elicit filopodial transients.

How do different molecules trigger distinct frequencies of filopodial transients? Extracellular matrix molecules often signal via RGD (arginine-glycine-aspartate) tripeptide motifs in their sequence that interact with integrin receptors and elevate intracellular Ca^{2+} (Lawson and Maxfield, 1995; Schwartz and Schaller, 1995; Sjaastad et al., 1996). Application of RGDS (S, serine) elevated growth cone Ca^{2+} and increased the frequency of filopodial transients, consistent with activation of integrins, and function-blocking antibodies to β1 integrins blocked these effects. These results demonstrate that integrins on neuronal growth cones can trigger Ca^{2+} elevations and raised the possibility that integrin clusters are the sites of generation of these fast local Ca^{2+} transients. Immunostaining with anti-β1 integrin antibodies indeed showed that filopodial transients are initiated at sites of punctate clusters of integrins. Presentation of different numbers of RGD motifs in various contexts by these extracellular matrix molecules could thus account for different frequencies of filopodial transients.

To test the function of filopodial transients we first examined the relationship between their frequency and filopodial lifetime, measured as the time from extension to retraction. Filopodial lifetime increased with increased frequency, in a transient-dependent manner (blocked by application of BAPTA to abolish transients), likely due to a combination of increased adhesion and decreased motility (Gomez et al., 2001). These results indicated that filopodial transients exert effects autonomous to individual filopodia. Intriguingly, as filopodial lifetime increased, neurite length decreased, consistent with decreased motility and/or increased adhesion (Kuhn et al., 1998).

The local action of these transients on single filopodia, combined with their global action on axon extension, suggested that they might also regulate growth cone turning. On this view, local sampling of heterogeneities in environmental substrates would lead to macroscopic changes in growth cone behavior. Two types of experiments provided evidence supporting this hypothesis. In the first, neurons were confronted with a boundary between two different substrates, on which filopodial transients are generated at different frequencies. Growth cones approaching these boundaries extended filopodia across the border, but turned to remain on the substrate on which filopodial transient frequencies were lower and filopodia were more motile and less adhesive.

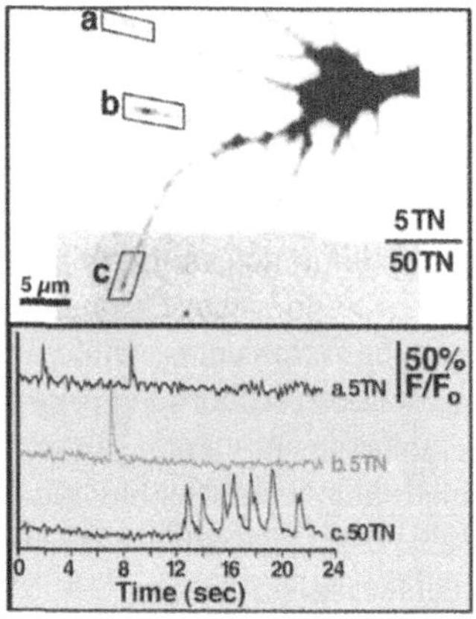

Fig. 7. Local filopodial Ca^{2+} transients regulate growth cone turning. Top: fluo-4 loaded growth cone turning at a border between high and low [TN]. Border is identified by TRITC fluorescence (50 μg/ml TN). Filopodia on either side of the boundary were monitored for Ca^{2+} transient activity (boxes). Bottom: normalized fluorescence changes within regions of the filopodia at top captured separately at 125 ms intervals. The frequency of Ca^{2+} transients is higher in the filopodium on 50 μg/ml TN than those on 5 μg/ml TN. 79% of axons turned to remain on a low [TN] under control conditions (n=28), while only 33% turned when filopodial transients were blocked with 100 nM BAPTA (n=27; $p < .002$, Fisher's exact test). From Gomez et al., 2001

This turning behavior was largely abolished by intracellular loading of BAPTA, eliminating filopodial transients, indicating that these transients are necessary for guidance by matrix molecules *in vitro*. Biologically plausible differences in concentrations of a single molecule, tenascin, were sufficient to elicit this Ca^{2+} transient-dependent turning behavior (Fig. 7). The second set of experiments entailed imposing Ca^{2+} transients on filopodia in which spontaneous transients were suppressed, to learn whether they can drive growth cone turning. NP-EGTA (caged Ca^{2+}) acted as a weak buffer prior to photolysis, suppressing filopodial transients when loaded into neurons; local release of Ca^{2+} by a small flashing UV spot positioned over selected filopodia then generated Ca^{2+} elevations mimicking filopodial transients. Growth cones of NP-EGTA-loaded neurons advancing toward the flashing UV spot turned to avoid entering it, whereas unloaded growth cones grew on through it. These findings indicate that filopodial transients are sufficient to promote growth cone turning in the *in vitro* context. These transients exert their effects via activation of the Ca^{2+}-dependent protease, calpain, that targets proteolyis of cytoskeletal proteins or signaling molecules to regulate growth cone motility and turning (Robles and Gomez, 2002).

4.5. Conclusions

4.5.1. Matching Form to Function

How is each type of Ca^{2+} transient suited to the function it performs in embryonic spinal neurons? In each case the spatial localization appears matched to the function as it is currently understood. Ca^{2+} spikes, because they regulate transcription that may govern long term changes in the state of the cell, are prominent in the cell body and perinuclear region. The involvement of voltage-gated channels and Ca^{2+}-dependent action potentials may provide an avenue for routing stimuli from distant regions of the neuron to the nucleus. Growth cone transients, because they regulate the rate of outgrowth of extending axons, are localized to this region at which axon elongation occurs (Bray, 1996). Filopodial transients, because they regulate turning behavior of growth cones in response to the signals provided by these sensory-motor antennae (Chien et al., 1993; Zheng et al. 1996), are restricted to the filopodia. It is less clear how the durations and amplitudes of these events are matched to their functions, because the molecular targets they activate have yet to be determined. However further analysis of the amplitudes and durations that are necessary to be effective may provide clues to define the identity of the targets.

4.5.2. Multiple Messengers

Elevations of second messengers play key roles in many metabolic processes in fully differentiated cells and tissues. Recently analyzed functions of Ca^{2+} transients in neuronal differentiation are part of this larger picture. It seems likely that both developing and mature cells exhibit patchy oscillations of second messengers. Oscillations of messengers may occur independently, in either nonoverlapping or overlapping domains. Oscillations of messengers may also be linked in networks, since their metabolism is often interconnected (Berridge et al., 2000, Mons et al., 1998; Van Wagenen and Rehder, 1999). This linkage is reminiscent of interactions among plasma membrane components, such as Ca^{2+} channels and Ca^{2+}-activated potassium channels, in which activation of two or more membrane components leads to a different outcome than activation of one alone. Detection of such activity will require further development of optical probes (Zacharias et al., 2000) that will enable simultaneous detection of several messengers simultaneously and implement charting of spatial and temporal dynamics of their fluctuations within different intracellular compartments.

4.5.3. A Broadening View of Excitability

Electrical excitability can be defined as the release of energy stored in ion concentration gradients across biological membranes. There are several prerequisites for this process. A threshold is required, such that only a stimulus substantially greater than noise is adequate to initiate the process. A regenerative process driven by positive feedback is useful in order to generate a large signal rapidly. The signal must then be limited in duration and amplitude by a process that returns the membrane to its resting state, and the components of the system need to recover to allow reexcitation. These are familiar features of classical excitability (Hodgkin and Huxley, 1952) by which action potentials are generated in the plasma membrane.

Studies of excitability now reveal that it is a property of intracellular as well as cell surface membranes. For excitability in the cell interior, threshold is represented by a critical level of stimulation of ryanodine receptors (RyR) or inositol trisphosphate receptors (IP_3R) in the endoplasmic reticulum, below which efflux of Ca^{2+} into the cytoplasm from internal stores is removed by fast Ca^{2+} buffering and extrusion mechanisms (Marchant and Taylor, 1997; Usachev and Thayer, 1997). The regenerative signal is the positive feedback of Ca^{2+} on these receptors that drives large and rapid efflux of Ca^{2+} that can be all-or-none (Missiaen et al., 1991; Hernandez-Cruz et al., 1997; Miyakawa et al., 2001). The elevation of Ca^{2+} is limited by inactivation and desensitization of RyR or IP_3R, store depletion that reduces the driving force on Ca^{2+}, and activation of pumps that return Ca^{2+} to the endoplasmic reticulum (Bezprosvanny et al., 1991; Hirota et al., 1999). These processes contribute to a refractory state; recovery entails reversibility of these processes (Oancea and Meyer, 1996).

4.5.4. Future Directions

The functions of three distinct Ca^{2+} transients in differentiating spinal neurons have been determined and the mechanisms by which they exert their effects have been analyzed. These findings argue that differentiation is driven by gene expression (Nature) that leads to activity (Nurture), which in turn influences gene expression or cytoskeletal architecture (Chicurel et al., 1998; Meyer et al., 2000). This scheme for developmental regulation provides a function-dependent feedback loop to enable flexible adjustment of differentiation and compensation for variation in gene expression, and affords the opportunity for regulation by environmental stimuli. It will be important to decipher the code by which Ca^{2+} spikes and growth cone and filopodial Ca^{2+} transients exert their effects. This will involve a closer analysis of the extent to which information is carried by their frequency or by a more complex patterning of Ca^{2+} transients, and will include determination of the signal transduction pathways. It will also be important to understand the upstream triggering mechanisms that lead neurons to generate different Ca^{2+} transients at specific frequencies. In attempting to develop a general understanding it will be of interest to investigate the extent to which other second messengers follow the paradigms that are becoming understood for Ca^{2+}.

References

Bacskai BJ, Hochner B, Mahaut-Smith M, Adams SR, Kaang BK, Kandel ER, Tsien RY (1993) Spatially resolved dynamics of cAMP and protein kinase A subunits in *Aplysia* sensory neurons. Science 260:222-226

Berridge MJ, Lipp P, Bootman MD (2000) The versatility and universality of calcium signalling. Nature Rev Mol Cell Biol 1:11-21

Bixby JL, Spitzer NC (1984) Early differentiation of vertebrate spinal neurons in the absence of voltage-dependent Ca^{++} and Na^{+} influx. Dev Biol 106:89-96

Borodinsky LN, Root CM, Spitzer NC (2002) The frequency of Ca^{2+} spikes specifies neurotransmitter expression in the embryonic spinal cord. Abstr Soc Neurosci 530.14

Bray D (1996) The dynamics of growing axons. Curr Biol 6:241-243

Carey MB, Matsumoto SG (1999) Spontaneous calcium transients are required for neuronal differentiation of murine neural crest. Dev Biol 215:298-313

Chicurel ME, Singer RH, Meyer CJ, Ingber DE (1998) Integrin binding and mechanical tension induce movement of mRNA and ribosomes to focal adhesions. Nature 392:730-733

Chien CB, Rosenthal DE, Harris WA, Holt CE (1993) Navigational errors made by growth cones without filopodia in the embryonic *Xenopus* brain. Neuron 11:237-251

Dale N, Roberts A, Ottersen OP, Storm-Mathisen J (1987) The morphology and distribution of 'Kolmer-Agduhr cells', a class of cerebrospinal-fluid-contacting neurons revealed in the frog embryo spinal cord by GABA immunocytochemistry. Proc Roy Soc Lond B 232:193-203

Désarménien MG, Spitzer NC (1991) Role of calcium and protein kinase C in development of the delayed rectifier potassium current in *Xenopus* spinal neurons. Neuron 7:797-80

Ferrari MB, Ribbeck K, Hagler DJ Jr, Spitzer NC (1998) A calcium signaling cascade essential for myosin thick filament assembly in *Xenopus* myocytes. J Cell Biol. 141:1349-1356

Ferrari MB, Rohrbough JC, Spitzer NC (1996) Spontaneous calcium transients regulate myofibrillogenesis in embryonic *Xenopus* myocytes. Dev Biol 178:484-497

Ferrari MB, Spitzer NC (1999) Calcium signaling in the developing *Xenopus* myotome. Dev Biol 213:269-282

Gomez TM, Robles E, Poo M-m, Spitzer NC (2001) Filopodial calcium transients promote substrate-dependent growth cone turning. Science 291:1983-1987

Gomez TM, Spitzer NC (1999) *In vivo* regulation of axon extension and pathfinding by growth-cone calcium transients. Nature 397:350-355

Gomez TM, Snow DM, Letourneau PC (1995) Characterization of spontaneous calcium transients in nerve growth cones and their effect on growth cone migration. Neuron 14:1233-1246

Gorbunova YV, Spitzer NC (2002) Dynamic interactions of cyclic AMP transients and spontaneous Ca^{2+} spikes. Nature 418:93-96

Gu X, Olson EC, Spitzer NC (1994) Spontaneous neuronal calcium spikes and waves during early differentiation. J Neurosci 14:6325-6335

Gu X, Spitzer NC (1993) Low-threshold Ca^{2+} current and its role in spontaneous elevations of intracellular Ca^{2+} in developing *Xenopus* neurons. J Neurosci 13:4936-4948

Gu X, Spitzer NC (1995) Distinct aspects of neuronal differentiation encoded by frequency of spontaneous Ca^{++} transients. Nature 375:784-787

Holliday J, Adams RJ, Sejnowski TJ, Spitzer NC (1991) Calcium-induced release of calcium regulates differentiation of spinal neurons. Neuron 7:787-796
Holliday J, Spitzer NC (1990) Spontaneous calcium influx and its roles in differentiation of spinal neurons in culture. Dev. Biol. 141:13-23
Jones SM, Hofmann AD, Lieber JL, Ribera AB (1995) Overexpression of potassium channel RNA: *in vivo* development rescues neurons from suppression of morphological differentiation *in vitro*. J Neurosci 15:2867-2874
Jones SM, Ribera AB (1994) Overexpression of a potassium channel gene perturbs neural differentiation. J Neurosci 14:2789-2799
Kafitz KW, Leinders-Zufall T, Zufall F, Greer CA (2000) Cyclic GMP evoked calcium transients in olfactory receptor cell growth cones. Neuroreport 11:677-681
Kaang BK, Kandel ER, Grant SG (1993) Activation of cAMP-responsive genes by stimuli that produce long-term facilitation in *Aplysia* sensory neurons. Neuron 10:427-435
Kater SB, Mattson MP, Cohan C, Connor J (1988) Calcium regulation of the neuronal growth cone. Trends Neurosci 11:315-321
Kater SB, Mills LR (1991) Regulation of growth cone behavior by calcium. J Neurosci 11:891-899
Kuhn TB, Williams CV, Dou P, Kater SB (1998) Laminin directs growth cone navigation via two temporally and functionally distinct calcium signals. J Neurosci 18:184-194
Lautermilch NJ, Spitzer NC (2000) Regulation of calcineurin by growth cone calcium waves controls neurite extension. J Neurosci 20:315-325
Lawson MA, Maxfield FR (1995) Ca^{2+}- and calcineurin-dependent recycling of an integrin to the front of migrating neutrophils. Nature 377:75-79
Leclerc C, Daguzan C, Nicolas MT, Chabret C, Duprat AM, Moreau M (1997) L-type calcium channel activation controls the *in vivo* transduction of the neuralizing signal in the amphibian embryos. Mech Devel 64:105-110
Leclerc C, Duprat AM, Moreau M (1999) Noggin upregulates Fos expression by a calcium-mediated pathway in amphibian embryos. Dev Growth Diff 41:227-238
Leclerc C, Webb SE, Daguzan C, Moreau M, Miller AL (2000) Imaging patterns of calcium transients during neural induction in *Xenopus laevis* embryos. J Cell Sci 113:3519-3529
Meyer CJ, Alenghat FJ, Rim P, Fong JH, Fabry B, Ingber DE (2000) Mechanical control of cyclic AMP signalling and gene transcription through integrins. Nature Cell Biol 2:666-668
Mons N, Decorte L, Jaffard R, Cooper DM (1998) Ca^{2+}-sensitive adenylyl cyclases, key integrators of cellular signalling. Life Sci 62:1647-1652
Moreau M, Leclerc C, Gualandris-Parisot L, Duprat AM (1994) Increased internal Ca^{2+} mediates neural induction in the amphibian embryo. Proc Nat Acad Sci 91:12639-12643
Robles E, Gomez TM (2002) Filopodial calcium transients regulate growth cone adhesion and motility through local activation of calpain. Abstr Soc Neurosci 628.6
Roberts A, Clarke JDW (1982) The neuroanatomy of an amphibian embryo spinal cord. Phil Trans R Soc Lond B 296:195-212
Roberts A, Dale N, Ottersen OP, Storm-Mathisen J (1987) The early development of neurons with GABA immunoreactivity in the CNS of *Xenopus laevis* embryos. J Comp Neurol 261:435-449.
Schwartz MA, Schaller MD, Ginsberg MH (1995) Integrins: emerging paradigms of signal transduction. Ann Rev Cell Dev Biol 11:549-599

Silver RA, Lamb AG, Bolsover SR (1990) Calcium hotspots caused by L-channel clustering promote morphological changes in neuronal growth cones. Nature 343:751-754

Sjaastad MD, Lewis RS, Nelson WJ (1996) Mechanisms of integrin-mediated calcium signaling in MDCK cells: regulation of adhesion by IP3- and store-independent calcium influx. Mol Biol Cell 7:1025-1041

Spitzer NC, de Baca RC, Allen KA, Holliday J (1993) Calcium dependence of differentiation of GABA immunoreactivity in spinal neurons. J Comp Neurol 337:168-175

Spitzer NC, Ribera AB (1998) Development of electrical excitability in embryonic neurons: mechanisms and roles. J Neurobiol 37:190-197

Van Wagenen S, Rehder V (1999) Regulation of neuronal growth cone filopodia by nitric oxide. J Neurobiol 39:168-185

Watt SD, Gu X, Smith RD, Spitzer NC (2000) Specific frequencies of spontaneous Ca^{2+} transients upregulate GAD 67 transcripts in embryonic neurons. Molec Cell Neurosci 16:376-387

Watt SD, Gu X, Spitzer NC (1999) Regulation of cholinergic phenotype by the frequency of spontaneous calcium spikes in spinal neurons. Abstr Soc Neurosci 25:1285

Welnhofer EA, Zhao L, Cohan CS (1999) Calcium influx alters actin bundle dynamics and retrograde flow in *Helisoma* growth cones. J Neurosci 19:7971-7982

Wilson MT, Kisaalita WS, Keith CH (2000) Glutamate-induced changes in the pattern of hippocampal dendrite outgrowth: a role for calcium-dependent pathways and the microtubule cytoskeleton. J Neurobiol 43:159-172

Zacharias DA, Baird GS, Tsien RY (2000) Recent advances in technology for measuring and manipulating cell signals. Curr Opin Neurobiol 10:416-421

Zheng JQ, Poo MM, Connor JA (1996) Calcium and chemotropic turning of nerve growth cones. Perspect Dev Neurobiol 4:205-213

Zheng JQ, Felder M, Connor JA, Poo M-m (1994) Turning of nerve growth cones induced by neurotransmitters. Nature 368:140-144

5. Role of NAADP in Coordinating Spatiotemporal Aspects of Calcium Signalling

Grant C. Churchill and Antony Galione

Department of Pharmacology, University of Oxford, Mansfield Road, Oxford OX1 3QT,UK

Abstract. We outline the roles of two low molecular weight phosphorylated compounds as intracellular messengers in calcium signaling. These new intracellular messengers (cyclic ADP-ribose–cADPR and nicotinic acid adenine dinucleotide phosphate-NAADP) have been shown to regulate calcium signalling across the plant and animal kingdoms. A central question in cell biology is what are the mechanisms by which calcium ions, arguably most important and universal regulator of cell activation, can encode specificity. The hypothesis that we have been testing is that exist in cells multiple signalling molecules and pathways which give rise to different patterns of calcium signals leading to highly specific cellular responses. We discuss new information about the molecular components of these new Ca^{2+} signalling pathways and their role in generating Ca^{2+} signals.

5.1. The Properties of NAADP are Distinct from IP_3 and cADPR

Nicotinic acid adenine dinucleotide phosphate (NAADP) is an endogenous pyridine nucleotide synthesized from NADP via a base-exchange reaction catalyzed by ADP-ribosyl cyclase (Fig. 1; (Aarhus et al. 1995; Lee 1997)). NAADP potently mobilizes Ca^{2+} from intracellular stores in sea urchin eggs, where it was discovered (Lee and Aarhus 1995), as well as many mammalian, invertebrate and plant systems (Lee 1997; Patel et al. 2001). NAADP mobilizes Ca^{2+} from intracellular stores by a mechanism distinct from those regulated by inositol 1,4,5-trisphosphate (IP_3) and cyclic ADP-ribose (cADPR). This is based on both desensitisation studies whereby a homogenate desensitised to one agent is still responsive to the other agents (Fig. 2A) and pharmacological characterization (Fig. 2B).

Compared to IP_3 and cADPR, in the sea urchin egg, NAADP exhibits at least four unique properties that affect its ability to control spatiotemporal Ca^{2+} signals. First, NAADP-mediated Ca^{2+} release does not exhibit CICR (Chini and Dousa 1996; Genazzani and Galione 1996); therefore, unlike IP_3 (Lechleiter and Clapham 1992; Parker et al. 1996) and cADPR (Churchill and Galione 2000), NAADP cannot turn the cytoplasm into an excitable medium. Second, NAADP potently self-inactivates even when present at concentrations less than those required to

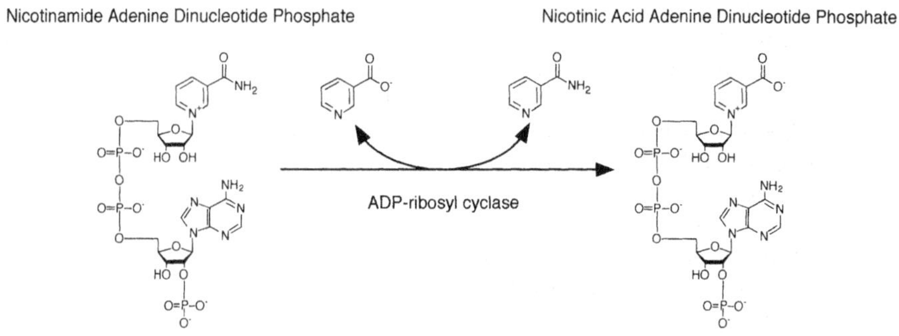

Fig. 1. Structure and synthesis of NAADP. NAADP is synthesized from the coenzyme NADP by a base-exchange reaction catalysed the enzyme ADP-ribosyl cyclase.

activate detectable Ca^{2+} release (Aarhus et al. 1996; Genazzani et al. 1996). Third, once bound to its receptor, radiolabelled NAADP is occluded and cannot be displaced by non-radiolabelled NAADP, which is thought to be related to its self-inactivation mechanism (Aarhus et al. 1996; Billington and Genazzani 2000; Genazzani et al. 1996; Patel et al. 2000). Fourth, NAADP mobilizes Ca^{2+} from a store that is distinct pharmacologically and physically from the IP_3- and cADPR-sensitive Ca^{2+} stores thought to be endoplasmic reticulum (Genazzani and Galione 1996; Lee and Aarhus 1995). We recently identified this organelle as the reserve granule (Churchill et al. 2002), lysosome-related organelles that are the functional equivalent of a lysosome in the sea urchin egg (Armant et al. 1986). Combined, these unique characteristics of NAADP led us to re-evaluate how spatiotemporal Ca^{2+} signals can be generated and brought us to a new understanding of the integration and cooperativity between the different Ca^{2+}-releasing mechanisms.

5.2. Spatial and Temporal Control of Ca^{2+} Signals

Different extracellular signals that all use Ca^{2+} maintain specificity by eliciting different spatial and temporal Ca^{2+} signals. Indeed, in many cell types specific extracellular signals give rise to specific Ca^{2+} signatures defined by the frequency of repetitive Ca^{2+} increases (temporal control) and subcellular location (spatial control). The mechanistic origin of these spatiotemporal Ca^{2+} signals is related to the presence of both multiple Ca^{2+} stores (Churchill and Galione 2001; Churchill et al. 2002; Lee 2000; Lee and Aarhus 2000; Pozzan et al. 1994) and multiple messengers that mobilize them (Cancela et al. 2002; Genazzani and Billington 2002; Lee 2000; Patel et al. 2001). Traditionally, the complex spatiotemporal Ca^{2+} signals have a unifying underlying mechanism based on the principle of positive and negative feedback by Ca^{2+} itself (Berridge et al. 2000). Both the IP_3 receptor and the ryanodine receptor exhibit regulation by Ca^{2+} whereby increases in Ca^{2+} several fold over resting further activate the channel, whereas increases 5-100 fold over resting Ca^{2+} inhibit the channel (Bezprozvanny et al. 1991; Meissner 1994).

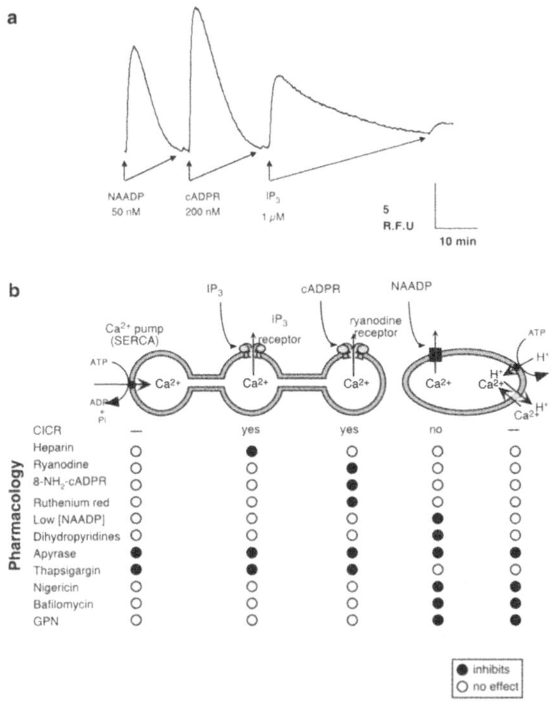

Fig. 2. Overview of Ca^{2+} handling in sea urchin eggs. A, Trace showing the three independent Ca^{2+}-mobilizing pathways present in sea urchin egg homogenate. Note that each pathway shows homologous desensitisation. B, Upper schematic illustrates stores, their overlap and Ca^{2+} uptake and release mechanisms. The lower part of the figure demonstrates the pharmacology of these processes shown in the schematic. CICR, Ca^{2+}-induced Ca^{2+} release. Heparin is a competitive antagonist of the IP_3 receptor. Ryanodine and ruthenium red block the ryanodine receptor. 8-NH_2-cADPR is a competitive antagonist of cADPR. Dihydropyridines at micromalar concentrations block NAADP-mediated Ca^{2+} release. Apyrase hydrolyzes ATP to AMP. Thapsigargin inhibits SERCA. Nigericin is a proton ionophore. Baflilomycin is an inhibitor of V-type H^+-ATPases. GPN is glycylphenylalanine 2-naphthylamide, which disrupts lysosomes via osmotic lysis.

This is commonly termed the bell-shaped response based on the shape of the curve describing first activation and then inactivation of the channel as Ca^{2+} is elevated above its normal resting cytosolic concentration (Bezprozvanny et al. 1991). Spatiotemporal Ca^{2+} signalling can best be understood as of a hierarchy of events (Berridge 1997; Parker et al. 1996). The opening of single channels gives rise to the most elementary events possible, termed blips when mediated by IP_3 receptors (Parker and Yao 1996) and quarks when mediated by ryanodine receptors (Lipp and Niggli 1996). This regenerative autocatalytic positive feedback by Ca^{2+} is termed Ca^{2+}-induced Ca^{2+} release (CICR). The sensitivity to the positive feedback by Ca^{2+} is controlled by other chemicals in the cell, either second messengers or modulators. Specifically, the IP_3 receptor is sensitized by IP_3 (Bezprozvanny et al. 1991; Parker et al. 1996; Patel et al. 1999) and the ryanodine receptor is sensitized by cyclic ADP-ribose (Galione et al. 1991; Lee et al. 1993). How the neighbouring channel responds is dependent on how much Ca^{2+} reaches it (the local increase in free Ca^{2+} concentration) and how sensitive the channel is to the Ca^{2+} (the concentration of IP_3 or cADPR). Thus, a Ca^{2+} increase can remain localized, spread abortively, or propagate throughout an entire cell (Bootman and Berridge 1996; Parker et al. 1996), group of cells (Sanderson et al. 1994) or an entire organ such as the liver (Robb-Gaspers and Thomas 1995).

In many ways Ca^{2+} oscillations are a temporal analogue of Ca^{2+} waves (Berridge and Galione 1988; Berridge 1991; Berridge 1997; Dupont and Goldbeter 1992; Thomas et al. 1991). Several theoretical models have been proposed to explain Ca^{2+} oscillations, and can be classified as to the number of Ca^{2+} stores required (one or two) and whether the level of IP_3 must also oscillate (Berridge and Galione 1988; Berridge 1991; Tsien and Tsien 1990). Additionally, in mammalian cells NAADP exhibits a bell-shaped concentration-response curve for NAADP providing another mechanism for Ca^2 oscillations (Patel et al. 2001). Depending on the cell type and experimental conditions, evidence exists for mechanisms requiring oscillating levels of IP_3 (Harootunian et al. 1991; Hirose et al. 1999; Nash et al. 2001) and control by positive and subsequently negative feedback of Ca^{2+} on Ca^{2+} release channels with a steady level of IP_3 (Hajnoczky and Thomas 1997; Wakui et al. 1989). Additionally, a two-pool mechanism was proposed by Berridge and colleagues (Berridge and Galione 1988; Berridge 1991), which envisages transfer of Ca^{2+} from IP_3-sensitive stores into CICR stores, which overload and spontaneously release Ca^{2+}.

5.3. NAADP Mediates Ca^{2+} Waves by Diffusion

When injected into a sea urchin egg, NAADP gave rise to a Ca^{2+} wave (Perez Terzic et al. 1995), which was puzzling given that NAADP does not exhibit CICR (Chini et al. 1995; Genazzani and Galione 1996). We investigated three possibilities: (1) NAADP released trigger Ca^{2+} and this trigger Ca^{2+} ignited a CICR wave mediated by CICR channels (i.e., the IP_3 receptor and ryanodine receptor); (2) NAADP released Ca^{2+} locally that stimulated the production of either IP_3 or cADPR, which could then diffuse globally, making the entire egg an excitable medium; (3) NAADP diffusion governed the spread of the Ca^{2+} increase (Fig. 3).

When we photoreleased NAADP locally at low concentrations, even though the local increase in Ca^{2+} was very high, the resulting Ca^{2+} wave was not global and appeared diffusive (Fig. 4A). A plot comparing the distance the Ca^{2+} increase reached half maximum versus time revealed that velocity (the slope of this plot) decreased with time (and distance) (Fig. 4B). When the data are transformed so that distance squared (the algebraic equivalent of the square root of time for this purpose) is plotted against time the values fit a straight line (Fig. 4C), consistent with diffusion (Hille 2001), with the higher concentration of NAADP giving a greater slope. The effective diffusion coefficient can be calculated from the slope by dividing it by the dimension term for diffusion, which is, arguably, 6. The dif-

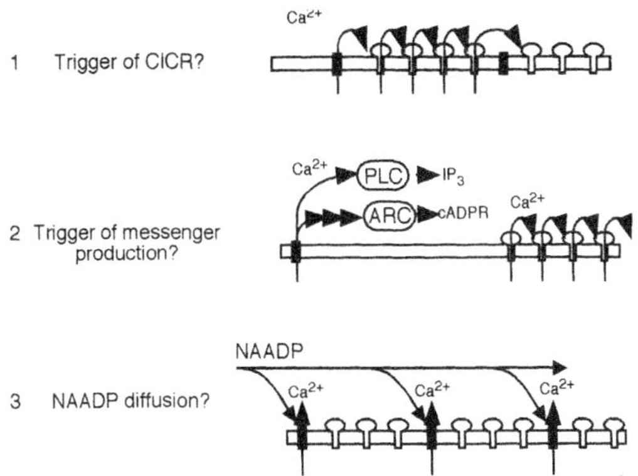

Fig. 3. Possible mechanisms by which NAADP could mediate a Ca^{2+} wave. NAADP-sensitive calcium channels are black rectangles; CICR channels are mushroom shapes. Phospholipase C, PLC; ADP-ribosyl cyclase, ARC.

fusion coefficient is therefore 50-150 $\mu m^2/s$, placing it in the range for Ca^{2+} in a cell (Allbritton et al. 1992). However, Ca^{2+} itself does not spread significantly outside the area of photolysis (Churchill and Galione 2000). Therefore, an NAADP-mediated Ca^{2+} increase is spread by diffusion of NAADP itself. At high concentrations, NAADP exhibits a higher diffusion coefficeint, similar to Ca^{2+} (Allbritton et al. 1992), suggesting that NAADP diffusion is slowed by binding to immobile buffers. Given the occluded state of NAADP when bound to its receptor (Aarhus et al. 1996; Billington and Genazzani 2000; Patel et al. 2000), it is likely that the immobile buffers are the NAADP receptors themselves. Diffusional control of the spreading of a Ca^{2+} increase contrasts markedly to the regenerative Ca^{2+} waves propagated by CICR via either IP_3 receptors or ryanodine receptors.

5.4. NAADP-Mediated Ca^{2+} Responses are Amplified by CICR

We tested the possibility that the Ca^{2+} released by NAADP feeds onto the CICR channels and is thus amplified (Cancela et al. 1999; Santella et al. 2000) by blocking IP_3 receptors with heparin (Taylor and Broad 1998) and the cADPR response with 8-amino-cADPR (Walseth and Lee 1993). Only when both inhibitors were included together was there a dramatic inhibition. The lag time was between the start of photolysis and the beginning of the Ca^{2+} increase lengthened to the extent that the equivalent amount of NAADP to give a response in the control did not elicit any response in the dually inhibited egg (Churchill and Galione 2000). Our interpretation of these data is that an NAADP gradient is always present, but when the CICR amplification system is functioning small amounts of NAADP are sufficient for a maximal response.

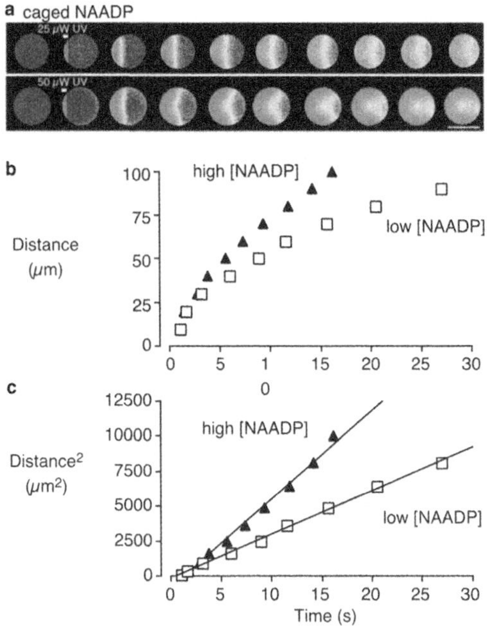

Fig. 4. NAADP-mediated Ca^{2+} waves propagate by diffusion. A, Ca^{2+} waves induced by local photolysis of caged NAADP to a relatively high (50 μW UV) and low (25 μW UV) concentration. B, Plot of distance of the Ca^{2+} wave (half maximum amplitude) by time. C, Plot of distance squared of the Ca^{2+} wave by time.

5.5. Model for the Control of Spatial Ca^{2+} Signalling by NAADP

A model showing how NAADP mediates spatial Ca^{2+} signalling is presented in Fig. 5. A gradient in the concentration of NAADP defines the domain over which the wave propagates. Functional coupling to either IP_3 receptors or ryanodine receptors contributes to the amplitude of the Ca^{2+} increase, but the effect is spatially localized. The lower part of the diagram shows how inhibition of amplification both reduces the Ca^{2+} increase and retards the position of the wavefront, despite the fact that the two waves are dependent on the same underlying NAADP gradient. This is because it is not the NAADP concentration but resulting Ca^{2+} increase that is monitored. The implication of the mechanism by which NAADP mediates spatial Ca^{2+} signalling is that NAADP can be tuned to be either a local or global messenger. In contrast, IP_3 and cADPR may act as global messengers (Allbritton et al. 1992; Petersen 1995), and Ca^{2+} is always a local messenger (Allbritton et al. 1992).

5.6. NAADP-Induced Ca^{2+} Oscillations via a Two-Pool Mechanism

NAADP but not IP_3 or cADPR, induce long-lasting (at least 80 min) Ca^{2+} oscillations when photoreleased in sea urchin eggs (Aarhus et al. 1996; Lee 1997; Lee et al. 1997). To explain the paradox that NAADP potently self-inactivates but induces long-term Ca^{2+} oscillations, Lee and colleagues (Aarhus et al. 1996; Lee 1997; Lee et al. 1997) suggested a two-pool model. We investigated the mechanism of these oscillations. Consistent with recruitment of other Ca^{2+}-releasing mechanisms, the NAADP-induced Ca^{2+} oscillations were affected by interfering with either the IP_3 system with heparin (Taylor and Broad 1998) or the cADPR system with 8-amino-cADPR (Walseth and Lee 1993). Full inhibition of the oscillations required co-application of the antagonists, suggesting both CICR pathways were involved (Churchill and Galione 2001).

We also probed the involvement of CICR stores in these long-term Ca^{2+}oscillations with thapsigargin, an inhibitor of sarco(endo)plasmic recticulum Ca^{2+}-

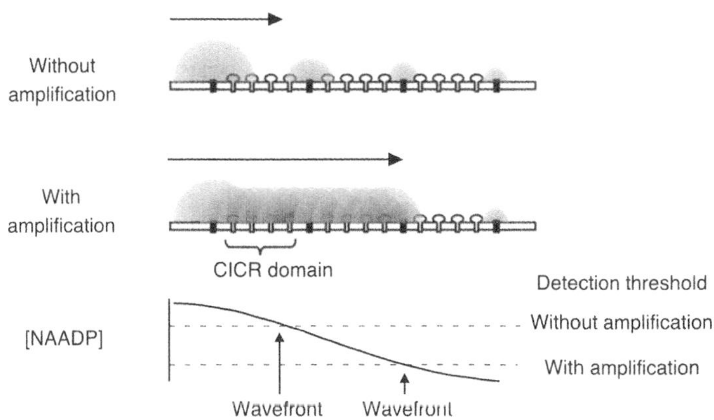

Fig. 5. NAADP mediates a Ca^{2+} wave by diffusion. The distance spread by the Ca^{2+} wave is determined by the diffusion of NAADP. The local amplitude is enhanced by Ca^{2+}-induced Ca^{2+} release.

ATPases (SERCAs)(Taylor and Broad 1998), to selectively deplete the IP_3- and cADPR-sensitive Ca^{2+} stores (Genazzani and Galione 1996). Following exposure to thapsigargin, NAADP still mobilized Ca^{2+} but failed to induce long-term Ca^{2+} oscillations (Churchill and Galione 2001). This provided additional evidence that the thapsigargin-sensitive Ca^{2+} stores are physically distinct from the NAADP-sensitive stores in not only egg homogenates (Lee and Aarhus 1995) but also in the intact egg (Lee and Aarhus 2000). Moreover, the Ca^{2+} transient elicited by thapsigargin is proportional to the Ca^{2+} load of the CICR stores. The size of this Ca^{2+} transient was increased after NAADP-mediated Ca^{2+} mobilization (Churchill and Galione 2001), directly demonstrating transfer of Ca^{2+} from one store to the other, consistent with the prediction of the two-pool model.

The mechanism by which NAADP induces Ca^{2+} oscillations and a summary of the evidence for this mechanism are presented in Fig. 6. The initial Ca^{2+} release mediated by NAADP is largely independent of any additional processes (pathway 1, Fig. 6), except for an amplification by CICR pathways, as described previously (Churchill and Galione 2000). A portion (~one third) of the Ca^{2+} mobilized by NAADP comes from a thapsigargin-resistant store that is distinct from the CICR stores. This released Ca^{2+} is then taken up into CICR stores. Ca^{2+} uptake continues until the stores overload and spontaneously release Ca^{2+} as demonstrated previously (Galione et al. 1991; Missiaen et al. 1991). This mechanism meets the criteria of a two-pool model for Ca^{2+} oscillations (Berridge and Galione 1988; Berridge 1991; Tsien and Tsien 1990) because one Ca^{2+} store is the source of priming Ca^{2+} (NAADP sensitive, which is lysosomal (Churchill et al. 2002)) and a separate Ca^{2+} store mediates the oscillations (CICR), and oscillations can be triggered by any increase in Ca^{2+} that primes the CICR stores. Therefore, this mechanism represents the first direct demonstration of the two-pool model for Ca^{2+} oscillations.

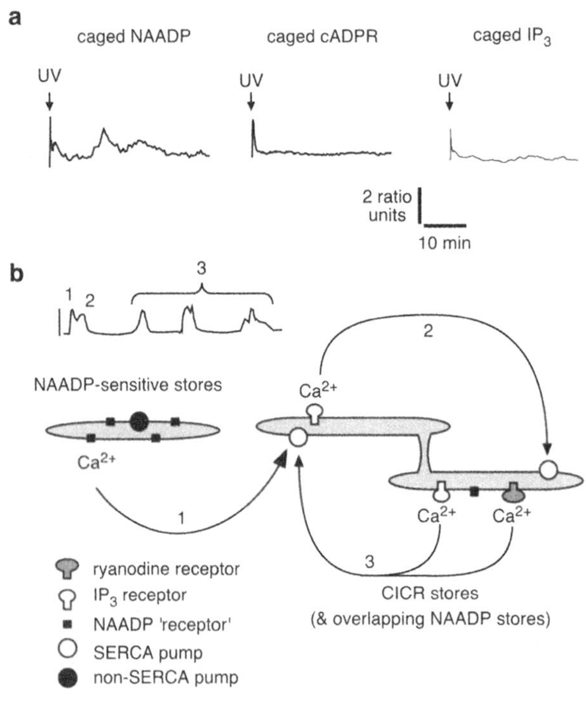

Fig. 6. NAADP induces long-term Ca^{2+} oscillations by a two-pool mechanism. A, Ca^{2+} oscillations are induced by NAADP but not IP_3 or cADPR. B, Model illustrating the mechanism for NAADP-induced Ca^{2+} oscillations. Overlapping NAADP stores are those that are sensitive to IP_3, cADPR as well as NAADP.

5.7. Summary

The characteristics of NAADP-mediated Ca^{2+} signalling differ markedly from the other Ca^{2+} release pathways. Because of this, NAADP plays several unique and potentially critical roles in Ca^{2+} signalling. First, as an initiator of Ca^{2+} signals, NAADP releases a small amount of Ca^{2+} which then stimulates adjacent CICR channels resulting in amplification of the signal. Second, NAADP can also be tuned to be either a local or global messenger depending on its concentration. By varying the concentration of NAADP it can produce localized Ca^{2+} increases, amplified to varying degrees by the global concentrations of cADPR and IP_3, global Ca^{2+} waves, and long-term Ca^{2+} oscillations. In summary, the unique properties of NAADP enable it to control spatiotemporal Ca^{2+} signaling in a fundamentally distinct manner from IP_3 and cADPR.

References

Aarhus, R., Graeff, R. M., Dickey, D. M., Walseth, T. F., and Lee, H. C. (1995). ADP-ribosyl cyclase and CD38 catalyze the synthesis of a calcium-mobilizing metabolite from NADP. J Biol Chem 270: 30327-30333.

Aarhus, R., Dickey, D. M., Graeff, R. M., Gee, K. R., Walseth, T. F., and Lee, H. C. (1996). Activation and inactivation of Ca^{2+} release by $NAADP^{+}$. J Biol Chem 271: 8513-8516.

Allbritton, N. L., Meyer, T., and Stryer, L. (1992). Range of messenger action of calcium ion and inositol 1,4,5-trisphosphate. Science 258: 1812-1815.

Armant, D. R., Carson, D. D., Decker, G. L., Welply, J. K., and Lennarz, W. J. (1986). Characterization of yolk platelets isolated from developing embryos of Arbacia punctulata. Dev Biol 113: 342-355.

Berridge, M., Lipp, P., and Bootman, M. (2000). The versatility and universality of calcium signalling. Nature Mol Cell Biol Rev 1: 11-21.

Berridge, M. J., and Galione, A. (1988). Cytosolic calcium oscillators. Faseb Journal 2: 3074-3082.

Berridge, M. J. (1991). Cytoplasmic calcium oscillations: a two pool model. Cell Calcium 12: 63-72.

Berridge, M. J. (1997). Elementary and global aspects of calcium signalling. J Physiol 499: 291-306.

Bezprozvanny, I., Watras, J., and Ehrlich, B. E. (1991). Bell-shaped calcium-response curves of Ins(1,4,5)P3- and calcium-gated channels from endoplasmic reticulum of cerebellum. Nature 351: 751-754.

Billington, R. A., and Genazzani, A. A. (2000). Characterization of $NAADP^{+}$ binding in sea urchin eggs. Biochem Biophys Res Commun 276: 112-116.

Bootman, M. D., and Berridge, M. J. (1996). Subcellular Ca^{2+} signals underlying waves and graded responses in HeLa cells. Curr Biol 6: 855-865.

Cancela, J. M., Churchill, G. C., and Galione, A. (1999). Coordination of agonist-induced Ca^{2+}-signalling patterns by NAADP in pancreatic acinar cells. Nature 398: 74-76.

Cancela, J. M., Van Coppenolle, F., Galione, A., Tepikin, A. V., and Petersen, O. H. (2002). Transformation of local Ca^{2+} spikes to global Ca^{2+} transients: the combinatorial roles of multiple Ca^{2+} releasing messengers. Embo J 21: 909-919.

Chini, E. N., Beers, K. W., and Dousa, T. P. (1995). Nicotinate adenine dinucleotide phosphate (NAADP) triggers a specific calcium release system in sea urchin eggs. J Biol Chem 270: 3216-3223.

Chini, E. N., and Dousa, T. P. (1996). Nicotinate-adenine dinucleotide phosphate-induced Ca^{2+}-release does not behave as a Ca^{2+}-induced Ca^{2+}-release system. Biochem J 316: 709-711.

Churchill, G. C., and Galione, A. (2000). Spatial control of Ca^{2+} signaling by nicotinic acid adenine dinucleotide phosphate diffusion and gradients. J Biol Chem 275: 38687-38692.

Churchill, G. C., and Galione, A. (2001). NAADP induces Ca^{2+} oscillations via a two-pool mechanism by priming IP_3- and cADPR-sensitive Ca^{2+} stores. Embo J 20: 2666-2671.

Churchill, G. C., Okada, Y., Thomas, J. T., Genazzani, A. A., Patel, S., and Galione, A. (2002). NAADP mobilizes Ca^{2+} from reserve granules, lysosome-related organelles, in sea urchin eggs. Cell In Press.

Dupont, G., and Goldbeter, A. (1992). Oscillations and waves of cytosolic calcium: insights from theoretical models. Bioessays 14: 485-493.

Galione, A., Lee, H. C., and Busa, W. B. (1991). Ca^{2+}-induced Ca^{2+} release in sea-urchin egg homogenates - modulation by cyclic ADP-ribose. Science 253: 1143-1146.

Genazzani, A. A., Empson, R. M., and Galione, A. (1996). Unique inactivation properties of NAADP-sensitive Ca^{2+} release. J Biol Chem 271: 11599-11602.

Genazzani, A. A., and Galione, A. (1996). Nicotinic acid-adenine dinucleotide phosphate mobilizes Ca^{2+} from a thapsigargin-insensitive pool. Biochem J 315: 721-725.

Genazzani, A. A., and Billington, R. A. (2002). NAADP: an atypical Ca^{2+}-release messenger? Trends Pharmacol Sci 23: 165-167.

Hajnoczky, G., and Thomas, A. P. (1997). Minimal requirements for calcium oscillations driven by the IP_3 receptor. EMBO J 16: 3533-3543.

Harootunian, A. T., Kao, J. P., Paranjape, S., and Tsien, R. Y. (1991). Generation of calcium oscillations in fibroblasts by positive feedback between calcium and IP_3. Science 251: 75-78.

Hille, B. (2001) Ion channels of excitable membranes, 3rd edn Sinauer, Sunderland, Mass.

Hirose, K., Kadowaki, S., Tanabe, M., Takeshima, H., and Iino, M. (1999). Spatiotemporal dynamics of inositol 1,4,5-trisphosphate that underlies complex Ca^{2+} mobilization patterns. Science 284: 1527-1530.

Lechleiter, J. D., and Clapham, D. E. (1992). Molecular mechanisms of intracellular calcium excitability in X. laevis oocytes. Cell 69: 283-294.

Lee, H. (2000). Multiple calcium stores: separate but interacting. Science STKE www.stke.org/cgi/content/full/OC_sigtrans;2000/40/pe1.

Lee, H. C., Aarhus, R., and Walseth, T. F. (1993). Calcium mobilization by dual receptors during fertilization of sea urchin eggs. Science 261: 352-355.

Lee, H. C., and Aarhus, R. (1995). A derivative of NADP mobilizes calcium stores insensitive to inositol trisphosphate and cyclic ADP-ribose. J Biol Chem 270: 2152-2157.

Lee, H. C. (1997). Mechanisms of calcium signaling by cyclic ADP-ribose and NAADP. Physiol. Rev 77: 1133-1164.

Lee, H. C., Aarhus, R., Gee, K. R., and Kestner, T. (1997). Caged nicotinic acid adenine dinucleotide phosphate. Synthesis and use. J Biol Chem 272: 4172-4178.

Lee, H. C., and Aarhus, R. (2000). Functional visualization of the separate but interacting calcium stores sensitive to NAADP and cyclic ADP-ribose. J Cell Sci 113: 4413-4420.
Lipp, P., and Niggli, E. (1996). Submicroscopic calcium signals as fundamental events of excitation-- contraction coupling in guinea-pig cardiac myocytes. J Physiol 492: 31-38.
Meissner, G. (1994). Ryanodine receptor/Ca^{2+} release channels and their regulation by endogenous effectors. Annu Rev Physiol 56: 485-508.
Missiaen, L., Taylor, C. W., and Berridge, M. J. (1991). Spontaneous calcium release from inositol trisphosphate-sensitive calcium stores. Nature 352: 241-244.
Nash, M. S., Young, K. W., Challiss, R. A., and Nahorski, S. R. (2001). Intracellular signalling. Receptor-specific messenger oscillations. Nature 413: 381-382.
Parker, I., Choi, J., and Yao, Y. (1996). Elementary events of $InsP_3$-induced Ca^{2+} liberation in Xenopus oocytes: hot spots, puffs and blips. Cell Calcium 20: 105-121.
Parker, I., and Yao, Y. (1996). Ca^{2+} transients associated with openings of inositol trisphosphate- gated channels in Xenopus oocytes. J Physiol 491: 663-668.
Patel, S., Joseph, S. K., and Thomas, A. P. (1999). Molecular properties of inositol 1,4,5-trisphosphate receptors. Cell Calcium 25: 247-264.
Patel, S., Churchill, G. C., and Galione, A. (2000). Unique kinetics of nicotinic acid-adenine dinucleotide phosphate (NAADP) binding enhance the sensitivity of NAADP receptors for their ligand. Biochem J 352 Pt 3: 725-729.
Patel, S., Churchill, G. C., and Galione, A. (2001). Coordination of Ca^{2+} signalling by NAADP. Trends Biochem Sci 26: 482-489.
Perez Terzic, C. M., Chini, E. N., Shen, S. S., Dousa, T. P., and Clapham, D. E. (1995). Ca^{2+} release triggered by nicotinate adenine dinucleotide phosphate in intact sea urchin eggs. Biochemical Journal 312: 955-959.
Petersen, O. H. (1995). Inositol trisphosphate and cyclic ADP-ribose as long range messengers generating local subcellular calcium signals. J Physiol 89: 125-127.
Pozzan, T., Rizzuto, R., Volpe, P., and Meldolesi, J. (1994). Molecular and cellular physiology of intracellular calcium stores. Physiol Rev 74: 595-636.
Robb-Gaspers, L. D., and Thomas, A. P. (1995). Coordination of Ca^{2+} signaling by intercellular propagation of Ca^{2+} waves in the intact liver. J Biol Chem 270: 8102-8107.
Sanderson, M. J., Charles, A. C., Boitano, S., and Dirksen, E. R. (1994). Mechanisms and function of intercellular calcium signaling. Mol Cell Endocrinol 98: 173-187
Santella, L., Kyozuka, K., Genazzani, A. A., De Riso, L., and Carafoli, E. (2000). Nicotinic acid adenine dinucleotide phosphate-induced Ca^{2+} release. Interactions among distinct Ca^{2+} mobilizing mechanisms in starfish oocytes. J Biol Chem 275: 8301-8306.
Taylor, C. W., and Broad, L. M. (1998). Pharmacological analysis of intracellular Ca^{2+} signalling: problems and pitfalls. Trends Pharmacol Sci 19: 370-375.
Thomas, A. P., Renard, D. C., and Rooney, T. A. (1991). Spatial and temporal organization of calcium signalling in hepatocytes. Cell Calcium 12: 111-126.
Tsien, R. W., and Tsien, R. Y. (1990). Calcium channels, stores, and oscillations. Annu Rev Cell Biol 6: 715-760.
Wakui, M., Potter, B. V., and Petersen, O. H. (1989). Pulsatile intracellular calcium release does not depend on fluctuations in inositol trisphosphate concentration. Nature 339: 317-320.
Walseth, T. F., and Lee, H. C. (1993). Synthesis and characterization of antagonists of cyclic ADP-ribose-induced calcium release. Biochim Biophys Acta 1178: 235-242.

6. Models of Cardiac Ca^{2+}-Induced Ca^{2+} Release and Ca^{2+} Sparks

Eric A. Sobie[1,2], M. Saleet Jafri[3,1], W. J. Lederer[1]

[1] Medical Biotechnology Center University of Maryland Biotechnology Institute, 725 W. Lombard Street, Baltimore, MD 21201, USA; Tel (410) 706-4895; FAX: (253) 660-4449; email: lederer@umbi.umd.edu, or sobie@umbi.umd.edu [2] Nora Eccles Harrison CVRTI, University of Utah, [3] Department of Bioinformatics, George Mason University, email: sjafri@gmu.edu

Abstract. This chapter will review computer models of calcium (Ca^{2+}) signaling in cardiac myocytes; specifically, we will discuss mathematical treatments of cardiac Ca^{2+}-induced Ca^{2+} release. Although this process is critical for proper functioning of the heart, early attempts to model it with traditional methods (i.e. a system of non-linear ordinary differential equations) proved to be unable to reproduce experimental data. Nonetheless, these failures led to novel hypotheses about how this process is regulated, and experimental advances in the past decade have confirmed these hypotheses and provided additional information about the role of Ca^{2+} in normal and pathological cardiac physiology. We will review the consequences of these findings for models of cardiac calcium signaling and discuss several recent endeavors aimed at understanding the fundamental units of Ca^{2+} release, Ca^{2+} sparks. We conclude by mentioning unanswered questions that are the focus of current work by us and other groups.

6.1. Introduction

Several inter- and intracellular signaling steps must occur in a synchronous and controlled manner each time the heart beats. An electrical signal, known as an action potential, originates in the sinoatrial node and travels to each cell in the myocardium. Propagation of the action potential into a ventricular cell activates ionic currents that then depolarize the cell's membrane potential (V_m) from a resting value of around -80 mV to a peak value of approximately +40 mV. In general, Na^+ and Ca^{2+} currents flow inward and act to make V_m more positive whereas K^+ currents flow outward and make V_m more negative. Membrane depolarization underlies a large increase in intracellular Ca^{2+} concentration ($[Ca^{2+}]_i$), and contraction of the cell results from Ca^{2+} ions binding to myofilaments. Ca^{2+} then falls as the membrane potential returns to its resting value and the cell relaxes, allowing the cycle to begin again. Figure 1 displays a schematic of these signaling steps, with solid arrows denoting a causal relationship between one signal and a subsequent signal and dashed arrows indicating feedback or subtle modulation.

The increase in $[Ca^{2+}]_i$, known as a Ca^{2+} transient, is therefore the chemical signal that transduces electrical depolarization to mechanical movement, a sequence of events collectively known as excitation-contraction (EC) coupling. Experiments performed nearly 20 years ago (13) established that this key step in EC coupling occurs because a small influx of trigger Ca^{2+}, primarily that flowing through L-type Ca^{2+} channels (30), causes the release of a much larger amount of Ca^{2+} from the sarcoplasmic reticulum (SR). This process, known as Ca^{2+}-induced Ca^{2+}-release (CICR) therefore displays both high gain and positive feedback. These characteristics can, in principle, lead to regenerative, all-or-none behavior and instability, since the Ca^{2+} released from the SR can continue to trigger additional release. However, in healthy hearts, CICR is well-controlled, and Ca^{2+} release is a graded rather than an all-or-none phenomenon. For instance, voltage clamp experiments have demonstrated that small changes in the amount of trigger Ca^{2+} that induces CICR, due to changes in pulse duration (5; 63) or membrane potential (2; 3; 11; 12), cause correspondingly small changes in the amount of Ca^{2+} that is released. However, other positive feedback signaling events, such as action potentials, display relatively constant amplitude for all inputs exceeding a threshold level.

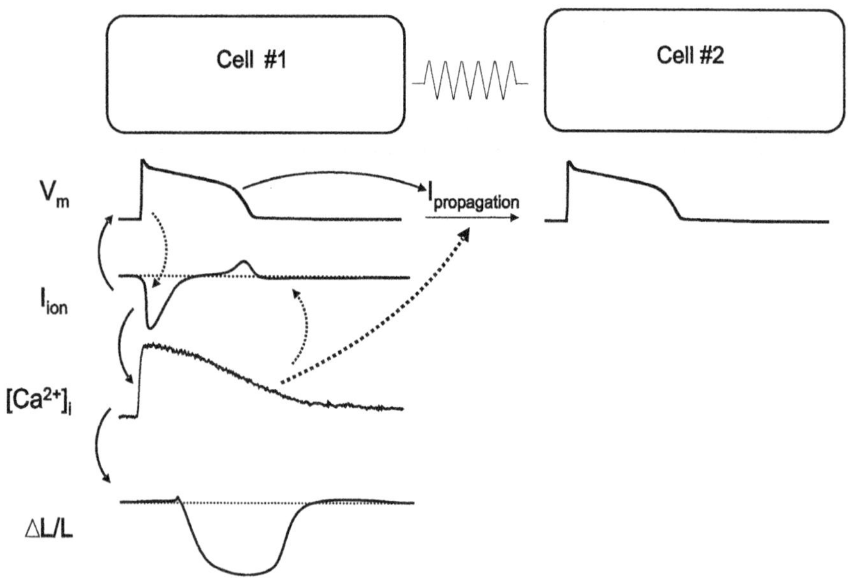

Fig.1. Overview of cardiac signaling. Schematic diagram of the intracellular and intercellular signaling steps that occur each time the heart beats. An electrical signal (action potential), which propagates to each cell in the myocardium, is coupled to mechanical movement via a transient increase in Ca^{2+} concentration with each heartbeat.

6.2. Why Model Cardiac CICR?

The preceding discussion alludes to the two primary reasons for studying, and developing mathematical models of, cardiac Ca^{2+}-induced Ca^{2+} release. First, given the central role of Ca^{2+} signaling in cardiac physiology, one might expect that defects in Ca^{2+} regulation will cause pathological cardiac function. Indeed, the two clinical hallmarks of heart failure, decreased cardiac output and an increased risk of ventricular arrhythmia, are both thought to be related to impaired Ca^{2+} signaling (1; 22; 40; 41; 64). Thus, a thorough, quantitative understanding of this process can provide insight into disease mechanisms and aid in the identification of new targets for therapies. Second, cardiac CICR is interesting on a more abstract level because it represents an example of a high gain, positive feedback system that is nonetheless tightly-controlled and can be modulated in subtle ways. As we will discuss in more detail in the remainder of this chapter, this paradox presents significant challenges to those interested in building mathematical models of CICR and confounded early attempts by modelers to reproduce experimental data. However, theoretical and computational examinations of this problem have also led to novel hypotheses and insights. These advances have inspired important experiments and increased our overall understanding of cardiac physiology.

6.3. A Straightforward, but Problematic, Model of CICR

Figure 2 illustrates the main pathways of Ca^{2+} transport in cardiac myocytes: i.e., the processes that must be considered at a minimum to develop a simple, Hodgkin-Huxley type model of cardiac Ca^{2+} signaling. Ca^{2+} entry across the cell membrane and Ca^{2+} release from the SR act to increase $[Ca^{2+}]_i$ whereas extrusion from the cell via Na^+-Ca^{2+} exchange and resequestration of Ca^{2+} in the SR by the Ca^{2+} ATPase (SERCA) combine to decrease $[Ca^{2+}]_i$. To keep the model simple, we assume that Ca^{2+} removal from the cytosol by either transport mechanism obeys first order kinetics. Ca^{2+} entry in this model occurs only via the L-type Ca^{2+} channel. Other ionic currents, and "reverse-mode" Na^+-Ca^{2+} exchange can also contribute to the Ca^{2+} transient, but these pathways are less important for CICR than the L-type Ca^{2+} channel (30; 47-49). We use the simple model of L-type Ca^{2+} current from Luo and Rudy (32; 33), but more detailed models can be used without fundamentally changing the results. CICR is implemented by a novel mathematical formulation whereby $[Ca^{2+}]_i$ can both activate and inactivate Ca^{2+} release, and these phenomena are modeled with simple, Ca^{2+}-dependent Hodgkin-Huxley type state variables m^{RyR} and h^{RyR}. To improve the stability of Ca^{2+} release we further assume that inactivation is more sensitive to $[Ca^{2+}]_i$ than activation, but the time course of activation is more rapid than that of inactivation, as others have done (16; 51).

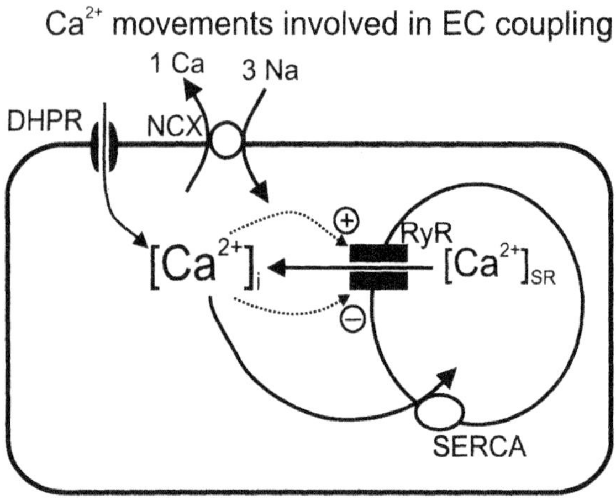

Fig. 2. Ca^{2+} fluxes involved in EC coupling. Diagram of typical movements of Ca^{2+} ions during each cardiac cycle. Ca^{2+} entry through L-type Ca channels (DHPR) causes release of additional Ca^{2+} from the sarcoplasmic reticulum. Ca^{2+} then exits the cell on the Na^+-Ca^{2+} exchanger and is resequestered in the SR by the SR Ca^{2+} ATPase (SERCA). These Ca^{2+} movements must be considered to develop a minimal model of Ca^{2+}-induced Ca^{2+} release.

Figure 3, which displays simulation results, illustrates the problems inherent in models of this type. In this simulation of a voltage-clamp experiment, the cell membrane is depolarized from rest (-80 mV) to 0 mV for a specified duration before a second depolarization clamps V_m at +120 mV. The depolarization to 0 mV activates inward L-type Ca^{2+} current that can trigger release from the SR; the second depolarization to +120 mV, near the Ca^{2+} reversal potential, prevents additional Ca^{2+} entry. A 2 ms depolarization to 0 mV (Fig. 3A) triggers a small Ca^{2+} influx via I_{Ca} (middle), but very little release occurs, and only a minimal change in free $[Ca^{2+}]_i$ (bottom) results. However, a slightly longer (4 ms) depolarization (Fig. 3B) induces enough Ca^{2+} entry to trigger SR release, and a large Ca^{2+} transient is produced. A plot of peak $[Ca^{2+}]_i$ versus the duration of the depolarization to 0 mV (Fig. 3C) reveals that depolarizations shorter than 3 ms produce very little change in $[Ca^{2+}]_i$ whereas those longer than 3 ms produce Ca^{2+} transients with similar amplitudes. SR Ca^{2+} release in this model therefore displays a clear threshold. However, this result is in conflict with experiments that instead show that Ca^{2+} release is a smoothly increasing function of the trigger duration (5). As a possible solution to this dilemma, Niggli and Lederer suggested in 1990 that a spatially non-uniform, and locally regulated, distribution of SR release sites could account for this paradox (38). In support of this idea, Stern predicted in 1992 that CICR in all "common pool" cellular models would be all-or-none and not reproduce experimental data adequately (56). In this context the term "common pool" refers to models in which Ca^{2+} released from the SR enters the same physical space as the Ca^{2+} that crosses the cell membrane to initiate release. Because released Ca^{2+} can trigger additional release, CICR in these models tends to be regenerative once a threshold concentration is reached. This is analogous to the response of excitable cells to electrical stimuli, in which the positive feedback of V_m on the Na^+ current ensures that action potentials are all-or-none phenomena.

Of course, the model presented above represents just one of an infinite number of potential common pool models, and the possibility therefore remains that a properly tuned model will successfully reproduce the phenomenon of graded release. However, the computer modeling studies published to date support Stern's idea that common pool models will generally be inadequate for simulating cardiac CICR. These models of CICR generally fall into one of two classes.

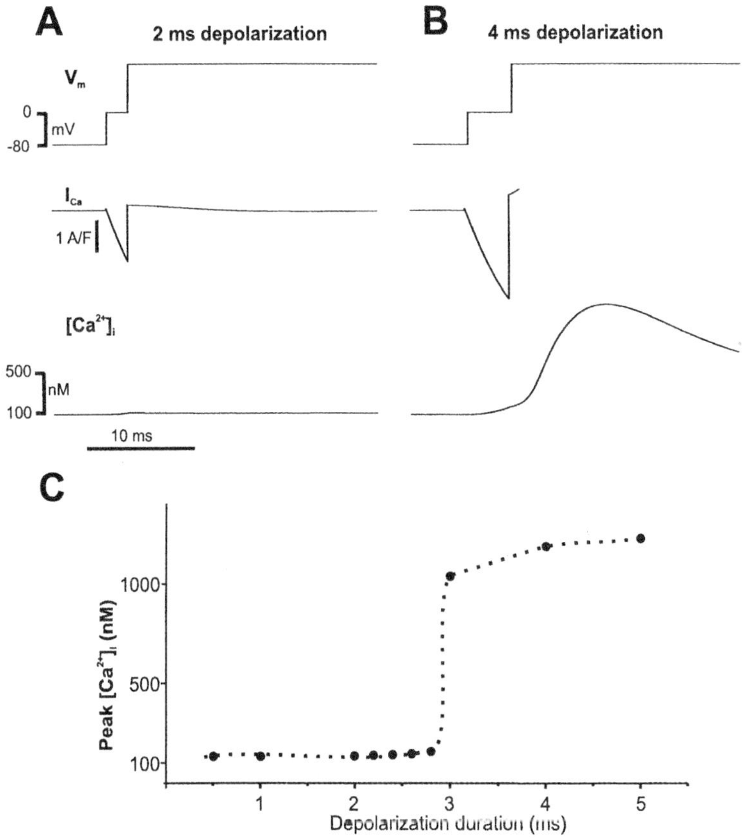

Fig. 3. Simulations with a straightforward model of CICR. Results of simulations with the whole-cell Ca^{2+}-induced Ca^{2+} release model presented in the text. The simulation is of an experiment in which a cardiac myocyte is voltage clamped at 0 mV for a short period of time. **A.** A 2 ms pulse to 0 mV (top) causes a small inward L-type Ca^{2+} current (middle) and a miniscule Ca^{2+} transient (bottom). **B.** Increasing the depolarization to 4 ms causes a large increase in the amplitude of the Ca^{2+} transient. **C.** The plot of Ca^{2+} transient amplitude versus the duration of the depolarization to 0 mV demonstrates that Ca^{2+}-induced Ca^{2+} release in this model is an all-or-none phenomenon displaying a clear threshold.

Several models have employed a straightforward approach such as that presented here in which Ca^{2+} entering the cell triggers SR release by binding to RyRs and altering RyR gating. These models have consistently displayed all-or-none release rather than the gradedness required by experiments (15; 25; 39; 51; 59). Many of these models incorporate a more accurate cellular geometry by assuming that release is controlled not by global intracellular Ca^{2+}, but by Ca^{2+} concentration in a "fuzzy space" (27) between the cell and SR membranes. The use of this fuzzy space reflects that fact that RyRs tend to be located very close to the cell membrane. This approach leads to more accurate gating in terms of the Ca^{2+}-sensitivity of RyRs, but it does not solve the fundamental problem of smoothly graded versus all-or-none release. On the other hand, models that have explicitly attempted to reproduce graded release have been forced to use mathematical tricks which are unjustified by experimental data. For instance, the state variables that control SR release in the model of Hilgemann and Noble (21) depend on V_m as well as on $[Ca^{2+}]_i$, even though most evidence indicates that membrane potential does not influence CICR directly (36; 38). The model developed by Wong et al. (65) assumes that Ca^{2+} released from the SR enters a different physical space than that from which CICR is triggered, even though no anatomical evidence to support such a mechanism has been found. In the model of Luo and Rudy, Ca^{2+} release depends only on the Ca^{2+} that enters the cell in the first 2 ms (33). Biophysically, however, there is no justification for treating a Ca^{2+} ion entering the cell at 4 ms different from one entering at 1 ms.

6.4. Why is Cardiac CICR Graded?

If the gradedness of CICR resulted from a finely-tuned release mechanism, it is likely that the appropriate choice of parameters in a relatively simple, deterministic mathematical model would allow the model to reproduce experimental data showing graded release. Thus, the failure of straightforward models suggests that CICR occurs by a different mechanism from that envisioned. Indeed, experimental results have demonstrated that the amount of Ca^{2+} released from the SR does not depend simply on the total amount of Ca^{2+} crossing the cell membrane. Instead, CICR is controlled locally, as first suggested by Niggli and Lederer in 1990 (38) and mathematically formalized by Stern in 1992 (55). This means that the overall SR release flux represents a collection of discrete release events, known as Ca^{2+} sparks (8), each of which is triggered by local entry of Ca^{2+} through one or more closely associated L-type Ca^{2+} channels.

Figure 4 displays a schematic of how local control works. Figure 4A illustrates two anatomical features of heart cells that are necessary for local control: 1) RyRs are arranged in clusters, and 2) these clusters are spatially separated. Because of this separation between Ca^{2+} release sites (Fig. 4A), under physiological conditions Ca^{2+} release at a particular site will only occur if Ca^{2+} enters through a neighboring L-type Ca^{2+} channel. The distance between release sites (estimated to be about 0.5 to 1.5 μm) ensures that Ca^{2+} release from one site will not be able to trigger Ca^{2+} release from a neighboring site (indicated by the large X). Thus, in this example the right-most site remains silent because the L-type channel associated with that cluster of RyRs does not open. Figure 4B shows that the restricted spatial spread of Ca^{2+} after release from a point source allows this scheme to work. This panel displays a semi-log plot of the Ca^{2+} concentration at the end of a 10 ms,

2 pA current source (similar to the total amount of Ca^{2+} released during a Ca^{2+} spark) as calculated with a spherically-symmetric model that simulates diffusion of Ca^{2+} in the cytoplasm and binding of Ca^{2+} to endogenous buffers. This result demonstrates that $[Ca^{2+}]_i$ falls off rapidly with distance. Although this release flux causes a very large increase in $[Ca^{2+}]_i$ very close to the source (> 1 mM), Ca^{2+} concentration exceeds 10 μM only within ±0.15 μm and 3 μM within ±0.26 mm of the source (see more detailed calculations by Soeller and Cannell (53)). Thus, as long as RyRs only respond to changes in Ca^{2+} above a few μM (which most bilayer experiments indicate is the case), Ca^{2+} release from one RyR cluster will not trigger release from the neighboring clusters, and regenerative, cell-wide, all-or-none release will not occur.

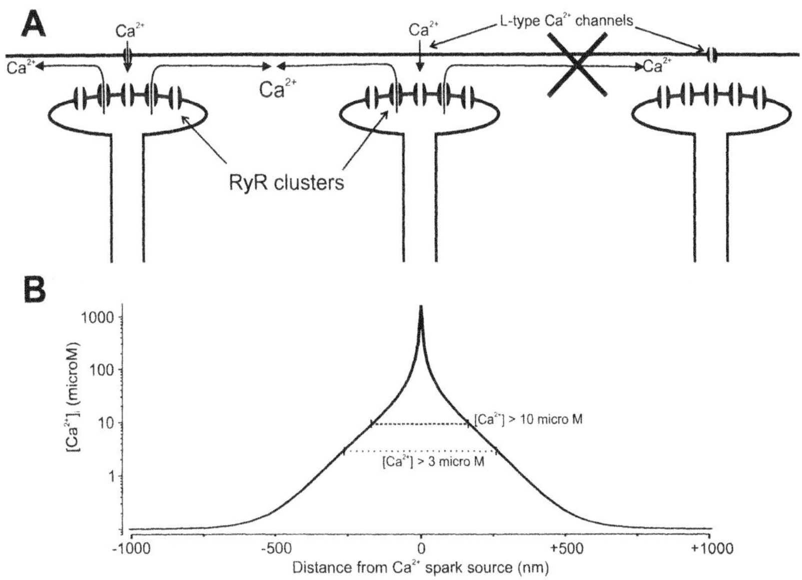

Fig. 4. Local control of cardiac Ca^{2+}-induced Ca^{2+} release. **A.** Because Ca^{2+} release sites (RyR clusters) are spatially separated, Ca^{2+} release from one cluster cannot trigger release from neighboring clusters. Release at a given site therefore only occurs if Ca^{2+} enters the cell through an L-type Ca^{2+} channel situated nearby. **B.** Ca^{2+} concentration in the vicinity of a point source at the end of a 10 ms, 2 pA flux. This calculation was performed with a published model of cellular Ca^{2+} diffusion and buffering (52). The fact that Ca^{2+} falls off dramatically with distance from the source allows the local control mechanism of CICR to work

The mathematical consequences of local control are dramatic. Because the triggering of each Ca^{2+} spark depends on events occurring in the immediate vicinity of the RyR cluster, cell-wide Ca^{2+} release cannot be described by state variables that depend on quantities such as total Ca^{2+} entry or global average $[Ca^{2+}]_i$. To illustrate this, consider again the situation in Fig. 4A. First, imagine that all three L-type Ca^{2+} channels open for the same period of time, pass the same amount of current, and each opening produces an increase in local Ca^{2+} concentration sufficient to trigger a spark. Next, imagine that only one of the three channels opens, but three times as many Ca^{2+} ions pass through the channel (due to an increase in channel conductance and/or open time). Most deterministic, whole-cell models of CICR would not distinguish between these two scenarios because identical amounts of Ca^{2+} enter the cell in either case. However, the two hypothetical events clearly produce different consequences for CICR as three Ca^{2+} sparks are triggered in the former case but only one in the latter.

From the perspective of the modeler, local control means that the behavior of the fundamental units of Ca^{2+} release, Ca^{2+} sparks, must be understood before substantial insight can be gained into the dynamics of CICR at the level of the whole cell. Thus, the remainder of this chapter will discuss computer modeling efforts aimed at understanding these local events.

6.5. Models of Cardiac Ca^{2+} Sparks

Several computer modeling investigations at the level of the Ca^{2+} spark have provided considerable insight into cardiac CICR. These studies can be divided into two categories: 1) models that examine the factors that influence Ca^{2+} spark characteristics, and 2) models that simulate gating of the RyRs that generate the spark. Models in the former category generally assume that a certain local Ca^{2+} release flux underlies the Ca^{2+} spark and then calculate the spatiotemporal fluorescent signal that would be measured experimentally (23; 24; 43; 50; 54). These simulations therefore provide insight into the processes of diffusion and buffering that convert a local release flux into a fluorescent signal, as well as experimental considerations that also affect the measurement such as blurring by the microscope, sampling, noise, and automatic detection algorithms. Studies of this sort have generated novel predictions about the factors influencing Ca^{2+} spark characteristics such as: 1) non-uniform experimental sampling of scan lines can artificially cause Ca^{2+} sparks to appear to be from two separate populations (43), 2) mobility of the fluorescent Ca^{2+} indicator plays a large role in determining the decay of Ca^{2+} sparks (50), and 3) partial saturation of the dye affects the width of the sparks that are measured (23). Thus, these investigations have helped experimentalists to interpret their data with more rigor and avoid erroneous conclusions based on artifacts. However, because the local Ca^{2+} release flux is assumed in these simulations, they cannot provide insight into the factors that control Ca^{2+} spark triggering or the time course of local Ca^{2+} release.

The latter category of models describes studies that have sought to understand the control of Ca^{2+} release in cells from the standpoint of RyR gating. A relatively early endeavor by Soeller and Cannell simulated Ca^{2+} movements in the restricted space between the SR and cell membranes and examined how L-type channel gating and physical spacing between the channels involved affected the triggering of Ca^{2+} sparks (6; 53). This influential work led to the prediction that the release system is "tuned" to respond efficiently to the relatively short (< 1 ms) L-type channel openings that generally occur (45). However, these simulations only modeled

RyR gating up to the moment of first opening and therefore provided no predictions about how the Ca^{2+} release flux evolves with time. Two studies published by different groups in 1999 used Monte-Carlo methods to investigate both the triggering and termination of Ca^{2+} sparks (44; 57). In both cases, stochastic simulations of the L-type Ca^{2+} channel(s) triggering the spark and the RyR cluster producing the Ca^{2+} spark were performed.

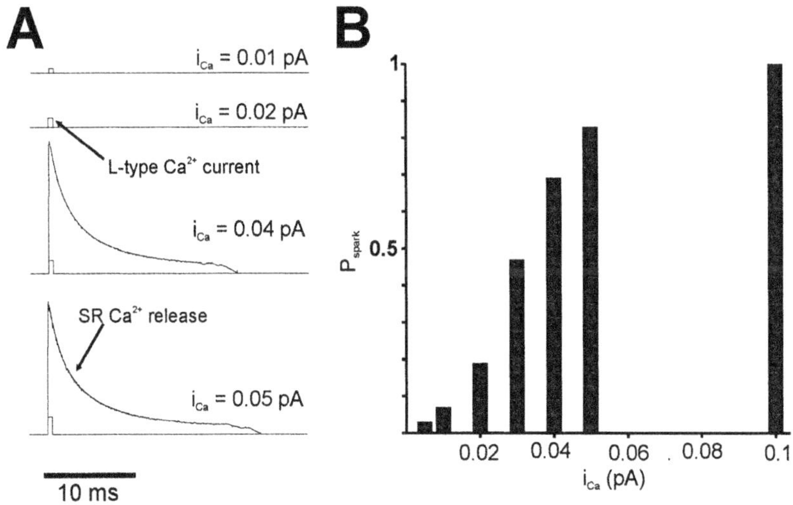

Fig. 5. CICR at the local level is still all-or-none. **A.** Typical simulations performed with our stochastic, Monte-Carlo model of the triggering of Ca^{2+} sparks (52). L-type Ca^{2+} channel openings and the resulting local release fluxes are displayed together as labeled. So that they can be better visualized, L-type currents are scaled up by a factor of 10. Small L-type channel currents generally do not trigger sparks (top two traces) whereas larger L-type currents typically trigger Ca^{2+} release fluxes of relatively constant amplitude (bottom two traces). **B.** Plot of L-type channel current amplitude versus probability of Ca^{2+} spark triggering. Repeated (200) trials with the Monte-Carlo model were simulated to calculate these probabilities.

Rice et al. examined how factors such as "adaptation" of RyRs (a complicated form of inactivation; (14; 20; 60)), SR load, and local SR depletion influenced the time course of release and the "gain" of EC coupling (44). These authors concluded that adaptation by itself was not strong enough to provide robust termination of Ca^{2+} release. The paper by Stern et al. (57) examined several published RyR gating models to determine the impact of various model characteristics on the stability of Ca^{2+} release. Certain RyR gating models derived from experiments performed in planar lipid bilayers (26; 46; 66) were found to produce an unstable CICR system, leading to the hypothesis that interactions between neighboring RyRs in a cluster might influence the cluster behavior.

A paper that we recently published combines these two approaches (52). In this investigation, we used a novel mathematical representation of RyR gating to simulate the triggering of a cluster of a variable number of RyRs (50 in control conditions) by a single L-type Ca^{2+} channel. Ca^{2+} ions passing through either the L-type Ca^{2+} channel or the RyRs enter a small "sub-space" between the SR and SL membranes. We stochastically simulate gating of the channels and also track diffusion of Ca^{2+} from the sub-space to the cytosol and from neighboring SR regions to the junctional SR that contains the RyR cluster. We couple these simulations to a model of the buffering and diffusion that generates the experimentally-measured Ca^{2+} spark signal to examine how changes in RyR gating affect the spatiotemporal properties of Ca^{2+} sparks. Figure 5A displays examples of simulations from the model to illustrate how graded CICR is produced. The sample traces show L-type channel openings of constant duration (0.5 ms) but variable amplitude (square waveforms), and also display the SR release fluxes that result (decaying waveforms). When single channel L-type current is small (0.01 or 0.02 pA, top traces), no SR release is produced in these typical simulations. When current amplitude is increased to 0.03 pA, however, a Ca^{2+} spark is typically triggered, as shown by the large release flux. This trace, which represents the total Ca^{2+} flux through the cluster of 50 RyRs, quickly peaks at approximately 3 pA, decays to around 10% of the peak value, then terminates after about 20 ms. When the current amplitude is increased to 0.04 pA, a Ca^{2+} release flux with nearly identical peak amplitude is triggered. The duration of this release flux is somewhat longer, but this difference is due simply to stochastic nature of the simulation rather than any additional recruitment by the larger trigger. Thus, at the local level, CICR remains all-or-none: a given L-type channel opening either initiates a Ca^{2+} spark or it does not. The gradedness of CICR therefore does not arise from changes in the local Ca^{2+} release flux produced by different trigger amplitudes but rather from the probabilistic nature of Ca^{2+} spark triggering. Figure 5B shows how the probability of triggering varies with current amplitude for these stereotypical 0.5 ms L-type channel openings. At each current amplitude, we ran 200 Monte-Carlo simulations to determine how often a Ca^{2+} release event occurred. As current amplitude increases, the probability of triggering increases until eventually release saturates and nearly every L-type channel opening initiates a Ca^{2+} spark. Thus, an increase in the whole cell Ca^{2+} current can produce more release by recruiting more Ca^{2+} sparks, even though all Ca^{2+} sparks are very similar in amplitude. However, it remains extremely difficult to predict the number of Ca^{2+} sparks that will be triggered by an action potential because the ongoing evolution of time and voltage generates L-type channel openings of diverse magnitude and duration. In addition, the increase in global, cell-wide Ca^{2+} concentration that occurs as the action potential evolves can bias the system so that late L-type openings may be more likely to trigger sparks than those that occur earlier.

6.6. Mechanisms of Ca^{2+} Spark Termination

The primary motivation of our Ca^{2+} spark modeling efforts was to investigate the mechanisms by which Ca^{2+} sparks terminate. Because Ca^{2+} flowing through open RyRs will maintain a high Ca^{2+} concentration in the vicinity of the RyRs in the same cluster, Ca^{2+} sparks can in principle continue indefinitely. However, some mechanism manages to terminate sparks robustly. The nature of this termination mechanism remains largely unknown and is one of the fundamental gaps in our understanding of cardiac EC coupling, as recent review articles have pointed out (7; 37; 62). We proposed the novel hypothesis that two factors, the influence of local SR lumenal Ca^{2+} concentration on RyR gating (9; 19), and coupled gating between RyRs in a cluster (34; 35), are sufficient to terminate Ca^{2+} sparks. By incorporating these features, we developed a model that is broadly consistent with experimental results such as Ca^{2+} spark amplitude, duration, and spontaneous Ca^{2+} spark rate at rest. Moreover, the model exhibits self-correcting behavior such that Ca^{2+} sparks terminate reliably even if parameters are varied over a wide range of values.

Figure 6 illustrates how the dependence of RyR gating on SR lumenal Ca^{2+} and coupled gating between RyRs act together to produce the negative feedback necessary to terminate Ca^{2+} model sparks and the hysteresis to ensure that Ca^{2+} sparks, once terminated, are not immediately re-triggered. This figure plots the steady-state RyR open probability versus sub-space Ca^{2+} concentration under different conditions. In diastole, when the SR load is relatively high, it is relatively easy for an increase in sub-space $[Ca^{2+}]$ to trigger a spark (thin solid line). After the spark is triggered, the local SR load (i.e. free $[Ca^{2+}]$ in the SR lumen) decreases from ~1 mM to ~0.1 mM, shifting RyR open probability to the right (thin dashed line). This shift means that once a Ca^{2+} spark terminates, it is difficult to initiate a new spark until the SR load returns to its control value. The thick lines show how coupling between RyRs can aid in this process. If SR load is high, and 49 out of 50 RyRs are open, the open probability curve shifts from the thin solid to the thick solid curve. Thus, the remaining RyR is likely to open because all of its neighbors are already open. Conversely, if SR load is low, and some RyRs begin to close spontaneously, coupling shifts the curve further to the right (e.g. thin to thick dashed curves). This shift makes the closed RyRs likely to remain closed and helps induce the remaining channels to close as well. Thus, these two factors, working together, provide a robust control mechanism by which Ca^{2+} sparks can terminate. It should be noted that the thick curves (reflecting RyR-RyR interactions) need not have the same slope as the thin curves for individual RyRs. The slopes of these curves and their placement along the x-axis depend on how coupling between channels is modeled.

6.7. Current Unanswered Questions

Despite the advances in our understanding of the control of cardiac CICR that have been made recently, many questions remain unanswered. These remaining issues, which will likely be the focus of experimental and computational research efforts over the next several years, are outlined briefly here. First, questions remain about the mechanisms by which Ca^{2+} sparks are triggered under different conditions. L-type Ca^{2+} channels are clearly the primary source of Ca^{2+} entry to trigger sparks, but some recent work suggests that other sources, such as the Na^+-Ca^{2+} exchanger, can modulate Ca^{2+} release (29; 58). However, the exact role played by the exchanger or other contributing Ca^{2+} sources depends on where these proteins are situated relative to the L-type Ca^{2+} channels and RyR clusters, critical information that has not yet been clarified. Similarly, inositol-triphosphate (IP3) receptors, extremely important for Ca^{2+} signaling in other cell types, are also present in cardiac muscle, but it is unknown whether these proteins have any role

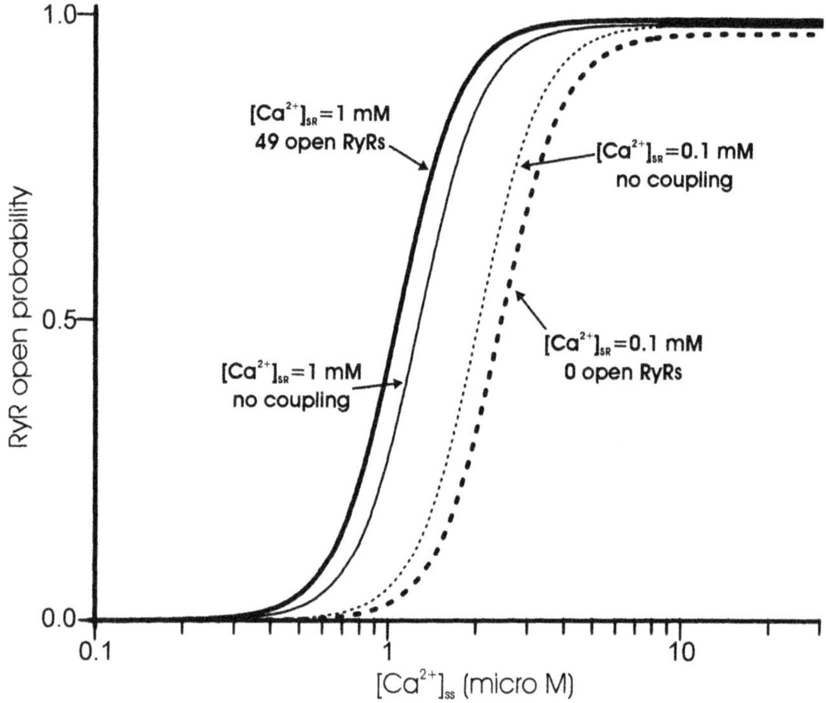

Fig. 6. A potential mechanism for Ca^{2+} spark termination. A plot of RyR open probability versus sub-space Ca^{2+} concentration illustrates how termination arises in the Ca^{2+} spark model presented. A decrease in local SR Ca^{2+} concentration shifts the curve to the right, making RyRs less sensitive (thin solid versus thin dashed curves). Coupling between RyRs adds to this effect. If most RyRs are open, the remaining channels are more likely to be open (thin versus thick solid curves) whereas if most are closed, the remaining channels are more likely to be closed (thin versus thick dashed curves).

in EC coupling. Second, considerable uncertainty still exists regarding how many RyRs open during a Ca^{2+} spark, how much flux passes through each channel, and how many total ions are released during a typical spark. Anatomical evidence suggests that clusters contain many (> 100) RyRs, but experimental spark measurements have provided much smaller estimates of the number of RyRs involved (4; 31; 61).

Third, the proposed mechanisms by which physical contact between RyRs might lead to functional coupling remain largely speculative. More experiments and analysis are necessary to determine the nature of interactions between RyRs, the role played by accessory proteins such as FKBP12.6, and the proper mathematical representation of these effects. Fourth, the factors that control the recovery of the Ca^{2+} transient, or restitution, are still unclear. Knowledge of this process, which is undoubtedly related to the termination of Ca^{2+} release, is necessary to predict how cardiac cells will respond to premature beats or sudden changes in pacing rate, which under certain conditions can lead to alternation of the beat-to-beat Ca^{2+} transient amplitude (10), or arrhythmias (42). Fifth, it is possible that beta-adrenergic stimulation (e.g. epinephrine) leads to phosphorylation of RyRs and altered RyR function (60), but the role that this effect plays in the modulation of cardiac Ca^{2+} sparks remains controversial and unclear (28). Finally, much remains to be learned about how regulation of cardiac CICR is changed in disease states. Heart failure has been shown to be accompanied by a decreased ability to trigger Ca sparks (loss of EC coupling "gain") (17; 18), but it is unclear whether other factors, potentially related to changes in protein phosphorylation, also play a role in the pathophysiology of heart failure.

6.8. Conclusions

Relatively simple modeling of Ca^{2+}-induced Ca^{2+}-release (CICR) and Ca^{2+} sparks reveals that local Ca^{2+} triggering of CICR plays a central role in the modulation of cardiac EC coupling. Because the cellular $[Ca^{2+}]_i$ transient results from the recruitment of individual Ca^{2+} sparks, cardiac CICR is graded despite the fact that the system employs high gain and positive feedback. The triggering of each Ca^{2+} spark depends, in general, on events occurring in the immediate vicinity of the RyR cluster producing the spark, but changes in global $[Ca^{2+}]_i$ can also influence spark initiation by changing the baseline level of Ca^{2+}. Simulation results have suggested that local depletion of SR Ca^{2+} and coupling between adjacent RyRs may play a central role in the termination of Ca^{2+} sparks, but many experimental and theoretical studies must be performed before the regulation of these elementary Ca^{2+} release events is understood fully.

Appendix

The equations that govern the simple, whole-cell CICR model presented above (see Fig. 3) are presented here. Model equations are derived in general from the Phase I Luo-Rudy model (32), but extraneous currents have been eliminated, and a novel formulation of cardiac CICR has been implemented. In this extremely simple version, SR Ca^{2+} is assumed to be constant at 1 mM; thus, beat to beat changes in SR content are not tracked.

L-type Ca^{2+} current

$$I_{Ca} = G_{Ca} \cdot m \cdot h \cdot (V - E_{Ca}) \tag{1}$$

$$E_{Ca} = 7.7 - 13.0287 \ln\left([Ca^{2+}]_i\right) \tag{2}$$

$$\frac{dm}{dt} = \alpha_m (1 - m) - \beta_m m \tag{3}$$

$$\frac{dh}{dt} = \alpha_h (1 - h) - \beta_h h \tag{4}$$

$$\alpha_m = 0.095 \frac{e^{-0.01(V-5)}}{1 + e^{-0.072(V-5)}} \tag{5}$$

$$\beta_m = 0.07 \frac{e^{-0.017(V+44)}}{1 + e^{0.05(V+44)}} \tag{6}$$

$$\alpha_h = 0.012 \frac{e^{-0.08(V+28)}}{1 + e^{0.15(V+28)}} \tag{7}$$

$$\beta_h = 0.0065 \frac{e^{-0.02(V+30)}}{1 + e^{-0.2(V+30)}} \tag{8}$$

SR Ca^{2+} release

$$I_{release} = 30 \cdot m_{release} \cdot h_{release} (1 - [Ca^{2+}]_i) \tag{9}$$

$$\frac{dm_{release}}{dt} = \frac{\dfrac{[Ca^{2+}]_i^3}{[Ca^{2+}]_i^3 + (0.001)^3} - m_{release}}{0.2} \tag{10}$$

$$\frac{dh_{release}}{dt} = \frac{1 - \dfrac{[Ca^{2+}]_i}{[Ca^{2+}]_i + (0.0005)} - h_{release}}{5} \tag{11}$$

$$\frac{d[Ca^{2+}]_i}{dt} = 10^{-4}\left(I_{release} - I_{Ca}\right) - 0.1104[Ca^{2+}]_i \tag{12}$$

Overall cellular Ca^{2+} balance

$$\frac{d[Ca^{2+}]_i}{dt} = 10^{-4}\left(I_{release} - I_{Ca}\right) - 0.1104[Ca^{2+}]_i \tag{13}$$

References

Balke CW and Shorofsky SR. Alterations in calcium handling in cardiac hypertrophy and heart failure. *Cardiovasc Res* 37: 290-299, 1998.

Barcenas-Ruiz L and Wier WG. Voltage dependence of intracellular $[Ca^{2+}]_i$ transients in guinea pig ventricular myocytes. *Circ Res* 61: 148-154, 1987.

Beuckelmann DJ and Wier WG. Mechanism of release of calcium from sarcoplasmic reticulum of guinea- pig cardiac cells. *J Physiol (Lond)* 405: 233-255, 1988.

Bridge JH, Ershler PR and Cannell MB. Properties of Ca^{2+} sparks evoked by action potentials in mouse ventricular myocytes. *J Physiol (Lond)* 518: 469-478, 1999.

Cannell MB, Berlin JR and Lederer WJ. Effect of membrane potential changes on the calcium transient in single rat cardiac muscle cells. *Science* 238: 1419-1423, 1987.

Cannell MB and Soeller C. Numerical analysis of ryanodine receptor activation by L-type channel activity in the cardiac muscle diad. *Biophys J* 73: 112-122, 1997.

Cannell MB and Soeller C. Sparks of interest in cardiac excitation-contraction coupling. *Trends Pharmacol Sci* 19: 16-20, 1998.

Cheng H, Lederer WJ and Cannell MB. Calcium sparks: elementary events underlying excitation-contraction coupling in heart muscle. *Science* 262: 740-744, 1993.

Ching LL, Williams AJ and Sitsapesan R. Evidence for Ca^{2+} activation and inactivation sites on the luminal side of the cardiac ryanodine receptor complex. *Circ Res* 87: 201-206, 2000.

Chudin E, Goldhaber J, Garfinkel A, Weiss J and Kogan B. Intracellular Ca^{2+} dynamics and the stability of ventricular tachycardia. *Biophys J* 77: 2930-2941, 1999.

Cleemann L and Morad M. Role of Ca^{2+} channel in cardiac excitation-contraction coupling in the rat: evidence from Ca^{2+} transients and contraction. *J Physiol (Lond)* 432: 283-312, 1991.

DuBell WH and Houser SR. Voltage and beat dependence of Ca^{2+} transient in feline ventricular myocytes. *Am J Physiol* 257: H746-H759, 1989.

Fabiato A. Calcium-induced release of calcium from the sarcoplasmic reticulum. *J Gen Physiol* 85: 189-320, 1985.

Fill M, Zahradnikova A, Villalba-Galea CA, Zahradnik I, Escobar AL and Gyorke S. Ryanodine receptor adaptation. *J Gen Physiol* 116: 873-882, 2000.

Glukhovsky A, Adam DR, Amitzur G and Sideman S. Mechanism of Ca^{++} release from the sarcoplasmic reticulum: A computer model. *Annals of Biomedical Engineering* 26: 213-229, 1998.

Glukhovsky A, Amitzur G, Adam D and Sideman S. A model of Ca^{2+} release from the sarcoplasmic reticulum. *Adv Exp Med Biol* 382: 97-108, 1995.

Gomez AM, Guatimosim S, Dilly KW, Vassort G and Lederer WJ. Heart failure after myocardial infarction: altered excitation-contraction coupling. *Circulation* 104: 688-693, 2001.

Gomez AM, Valdivia HH, Cheng H, Lederer MR, Santana LF, Cannell MB, McCune SA, Altschuld RA and Lederer WJ. Defective excitation-contraction coupling in experimental cardiac hypertrophy and heart failure. *Science* 276: 800-806, 1997.

Gyorke I and Gyorke S. Regulation of the cardiac ryanodine receptor channel by luminal Ca^{2+} involves luminal Ca^{2+} sensing sites. *Biophys J* 75: 2801-2810, 1998.

Györke S and Fill M. Ryanodine receptor adaptation: Control mechanism of Ca^{2+}- induced Ca^{2+} release in heart. *Science* 260: 807-809, 1993.

Hilgemann DW and Noble D. Excitation-contraction coupling and extracellular calcium transients in rabbit atrium: reconstruction of basic cellular mechanisms. *Proc R Soc Lond [B]* 230: 163-205, 1987.

Houser SR, Piacentino V and Weisser J. Abnormalities of calcium cycling in the hypertrophied and failing heart. *J Mol Cell Cardiol* 32: 1595-1607, 2000.

Izu LT, Mauban JR, Balke CW and Wier WG. Large currents generate cardiac Ca^{2+} sparks. *Biophys J* 80: 88-102, 2001.

Izu LT, Wier WG and Balke CW. Theoretical analysis of the Ca^{2+} spark amplitude distribution. *Biophys J* 75: 1144-1162, 1998.

Jafri MS, Rice JJ and Winslow RL. Cardiac Ca^{2+} dynamics: the roles of ryanodine receptor adaptation and sarcoplasmic reticulum load. *Biophys J* 74: 1149-1168, 1998.

Keizer J and Levine L. Ryanodine receptor adaptation and Ca^{2+}-induced Ca^{2+} release- dependent Ca^{2+} oscillations. *Biophys J* 71: 3477-3487, 1996.

Lederer WJ, Niggli E and Hadley RW. Sodium-calcium exchange in excitable cells: fuzzy space. *Science* 248: 283, 1990.

Li Y, Kranias EG, Mignery GA and Bers DM. Protein kinase A phosphorylation of the ryanodine receptor does not affect calcium sparks in mouse ventricular myocytes. *Circ Res* 90: 309-316, 2002.

Litwin SE, Li J and Bridge JH. Na-Ca exchange and the trigger for sarcoplasmic reticulum Ca release: studies in adult rabbit ventricular myocytes. *Biophys J* 75: 359-371, 1998.

London B and Krueger JW. Contraction in voltage-clamped, internally perfused single heart cells. *J Gen Physiol* 88: 475-505, 1986.

Lukyanenko V, Gyorke I, Subramanian S, Smirnov A, Wiesner TF and Gyorke S. Inhibition of Ca^{2+} sparks by ruthenium red in permeabilized rat ventricular myocytes. *Biophys J* 79: 1273-1284, 2000.

Luo CH and Rudy Y. A model of the ventricular cardiac action potential. Depolarization, repolarization, and their interaction. *Circ Res* 68: 1501-1526, 1991.

Luo CH and Rudy Y. A dynamic model of the cardiac ventricular action potential. I. Simulations of ionic currents and concentration changes. *Circ Res* 74: 1071-1096, 1994.

Marx SO, Gaburjakova J, Gaburjakova M, Henrikson C, Ondrias K and Marks AR. Coupled gating between cardiac calcium release channels (ryanodine receptors). *Circ Res* 88: 1151-1158, 2001.

Marx SO, Ondrias K and Marks AR. Coupled gating between individual skeletal muscle Ca^{2+} release channels (ryanodine receptors). *Science* 281: 818-821, 1998.

Nabauer M, Callewaert G, Cleemann L and Morad M. Regulation of calcium release is gated by calcium current, not gating charge, in cardiac myocytes. *Science* 244: 800-803, 1989.

Niggli E. Localized intracellular calcium signaling in muscle: calcium sparks and calcium quarks. *Annu Rev Physiol* 61:311-35: 311-335, 1999.

Niggli E and Lederer WJ. Voltage-independent calcium release in heart muscle. *Science* 250: 565-568, 1990.

Nordin C. Computer model of membrane current and intracellular Ca^{2+} flux in the isolated guinea pig ventricular myocyte. *Am J Physiol* 265: H2117-H2136, 1993.

O'Rourke B, Kass DA, Tomaselli GF, Kaab S, Tunin R and Marban E. Mechanisms of altered excitation-contraction coupling in canine tachycardia-induced heart failure, I: experimental studies. *Circ Res* 19;84: 562-570, 1999.

Pogwizd SM, Hoyt RH, Saffitz JE, Corr PB, Cox JL and Cain ME. Reentrant and focal mechanisms underlying ventricular tachycardia in the human heart. *Circulation* 86: 1872-1887, 1992.

Pogwizd SM, Schlotthauer K, Li L, Yuan W and Bers DM. Arrhythmogenesis and contractile dysfunction in heart failure: Roles of sodium-calcium exchange, inward rectifier potassium current, and residual beta-adrenergic responsiveness. *Circ Res* 88: 1159-1167, 2001.

Pratusevich VR and Balke CW. Factors shaping the confocal image of the calcium spark in cardiac muscle cells. *Biophys J* 71: 2942-2957, 1996.

Rice JJ, Jafri MS and Winslow RL. Modeling gain and gradedness of Ca^{2+} release in the functional unit of the cardiac diadic space. *Biophys J* 77: 1871-1884, 1999.

Rose WC, Balke CW, Wier WG and Marban E. Macroscopic and unitary properties of physiological ion flux through L- type Ca^{2+} channels in guinea-pig heart cells. *J Physiol (Lond)* 456: 267-284, 1992.

Schiefer A, Meissner G and Isenberg G. Ca^{2+} activation and Ca^{2+} inactivation of canine reconstituted cardiac sarcoplasmic reticulum Ca^{2+}-release channels. *J Physiol (Lond)* 489: 337-348, 1995.

Sham JS, Cleemann L and Morad M. Gating of the cardiac Ca^{2+} release channel: the role of Na^+ current and Na^+-Ca^{2+} exchange. *Science* 255: 850-853, 1992.

Sipido KR, Carmeliet E and Van de WF. T-type Ca^{2+} current as a trigger for Ca^{2+} release from the sarcoplasmic reticulum in guinea-pig ventricular myocytes. *J Physiol (Lond)* 508: 439-451, 1998.

Sipido KR, Maes M and Van de WF. Low efficiency of Ca^{2+} entry through the Na^+-Ca^{2+} exchanger as trigger for Ca^{2+} release from the sarcoplasmic reticulum. A comparison between L-type Ca^{2+} current and reverse-mode Na^+-Ca^{2+} exchange. *Circ Res* 81: 1034-1044, 1997.

Smith GD, Keizer JE, Stern MD, Lederer WJ and Cheng H. A simple numerical model of calcium spark formation and detection in cardiac myocytes. *Biophys J* 75: 15-32, 1998.

Snyder SM, Palmer BM and Moore RL. A mathematical model of cardiocyte Ca^{2+} dynamics with a novel representation of sarcoplasmic reticular Ca^{2+} control. *Biophys J* 79: 94-115, 2000.

Sobie EA, Dilly KW, Dos Santos CJ, Lederer WJ and Jafri MS. Termination of cardiac Ca^{2+} sparks: an investigative mathematical model of calcium-induced calcium release. *Biophys J* 83: 59-78, 2002.

Soeller C and Cannell MB. Numerical simulation of local calcium movements during L-type calcium channel gating in the cardiac diad. *Biophys J* 73: 97-111, 1997.

Soeller C and Cannell MB. Estimation of the sarcoplasmic reticulum Ca^{2+} release flux underlying Ca^{2+} sparks. *Biophys J* 82: 2396-2414, 2002.

Stern MD. Theory of excitation-contraction coupling in cardiac muscle. *Biophys J* 63: 497-517, 1992.

Stern MD, Pizarro G and Rios E. Local control model of excitation-contraction coupling in skeletal muscle. *J Gen Physiol* 110: 415-440, 1997.

Stern MD, Song LS, Cheng H, Sham JS, Yang HT, Boheler KR and os E. Local control models of cardiac excitation-contraction coupling. A possible role for allosteric interactions between ryanodine receptors. *J Gen Physiol* 113: 469-489, 1999.

Su Z, Sugishita K, Ritter M, Li F, Spitzer KW and Barry WH. The sodium pump modulates the influence of I_{Na} on $[Ca^{2+}]_i$ transients in mouse ventricular myocytes. *Biophys J* 80: 1230-1237, 2001.

Tang Y and Othmer HG. A model of calcium dynamics in cardiac myocytes based on the kinetics of ryanodine-sensitive calcium channels. *Biophys J* 67: 2223-2235, 1994.

Valdivia HH, Kaplan JH, Ellis-Davies GC and Lederer WJ. Rapid adaptation of cardiac ryanodine receptors: modulation by Mg^{2+} and phosphorylation. *Science* 267: 1997-2000, 1995.

Wang SQ, Song LS, Lakatta EG and Cheng H. Ca^{2+} signalling between single L-type Ca^{2+} channels and ryanodine receptors in heart cells. *Nature* 410: 592-596, 2001.

Wier WG and Balke CW. Ca^{2+} release mechanisms, Ca^{2+} sparks, and local control of excitation-contraction coupling in normal heart muscle. *Circ Res* 85: 770-776, 1999.

Wier WG, Egan TM, Lopez-Lopez JR and Balke CW. Local control of excitation-contraction coupling in rat heart cells. *J Physiol (Lond)* 474: 463-471, 1994.

Winslow RL, Rice J, Jafri S, Marban E and O'Rourke B. Mechanisms of altered excitation-contraction coupling in canine tachycardia-induced heart failure, II: model studies. *Circ Res* 19;84: 571-586, 1999.

Wong AY, Fabiato A and Bassingthwaighte JB. Model of calcium-induced calcium release in cardiac cells. *Bulletin of Mathematical Biology* 54: 95-116, 1992.

Zahradnikova A and Zahradnik I. A minimal gating model for the cardiac calcium release channel. *Biophys J* 71: 2996-3012, 1996.

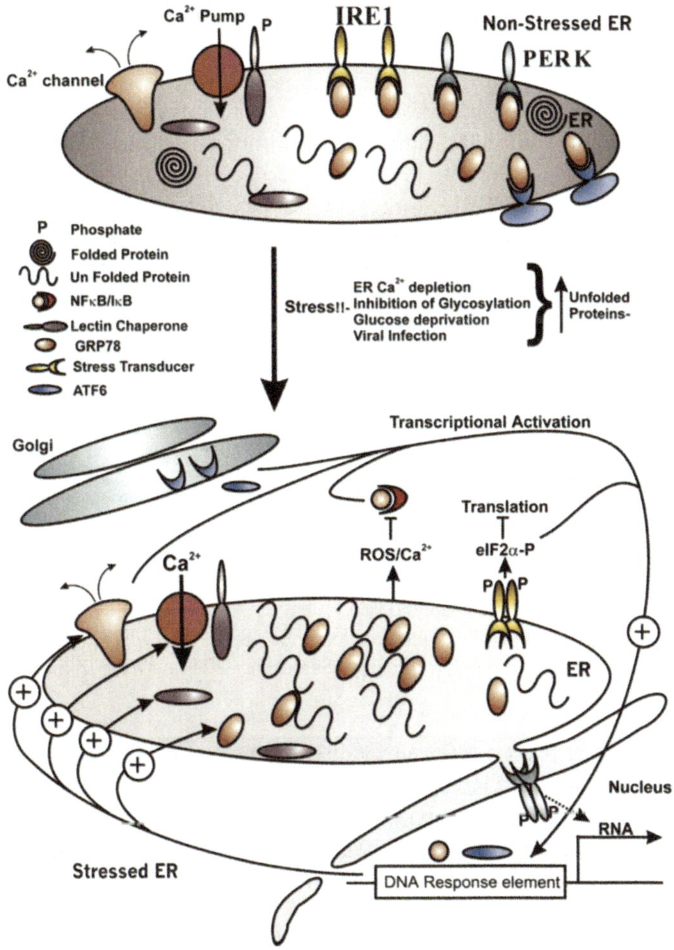

The Endoplasmic Reticulum: a central player in cell signalling and protein synthesis: Fig. 2. ER regulation of the stress response. In the non-stressed ER Ca^{2+} homeostasis is maintained and proteins are undergoing protein folding and oligomerisation prior to transport along the secretory pathway. Induction of ER stress results in accumulation of malfolded proteins. Under these conditions the activity of many ER chaperones is lost. To compensate for this the ER lectin chaperones CRT and CLNX dissociate from SERCA. prompting its activity. This results in increased Ca^{2+} within the lumen; a requirement of efficient protein folding. The increase in malfolded proteins results in GRP78 dissociating from the ER stress sensors/transducers IRE1, PERK and ATF6. Following dissociation IRE1 and PERK dimerise and trans-autophosphorylate. ATF6 traffics to the Golgi where the cytosolic domain is released by golgi resident proteases. A short term consequence of this is that general protein synthesis is inhibited through phosphorylation of eIF2α by PERK thus preventing further overloading of the folding machinery. As a result of ATF6 translocation to the nucleus and the splicing of XBP mRNA by IRE1 the transcription of many stress responsive genes is upregulated including SERCA, CRT, GRP78 and 94. In addition to increasing the abundance of ER chaperones in the ER their activity is increased as a result of the increased Ca^{2+} sequestration and storage capacity of the ER. In addition to or independent of the UPR the ER overload response may be activated. This results in activation of NFκB through an increase in cytosolic Ca^{2+} and reactive oxygen metabolites (ROS). NFκB translocates to the nucleus activating genes involved in inflammation.

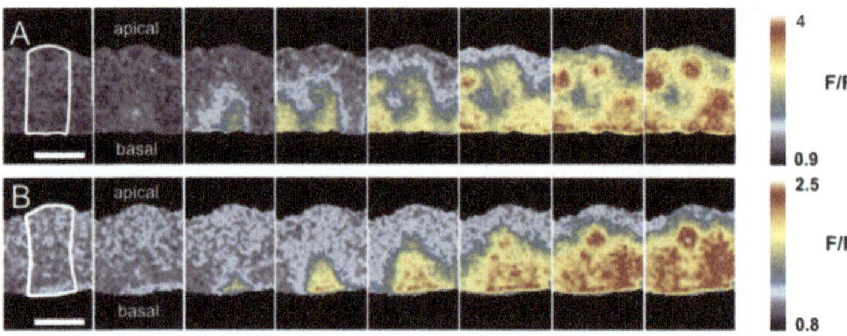

Hormone-mediated intercellular calcium signalling in an insect salivary gland: Fig. 2. Time series of $[Ca^{2+}]$ changes induced by 5-HT (100 nM in **A**, 3 nM in **B**) recorded by confocal imaging of fluo-4-loaded glands. $[Ca^{2+}]_i$ begins to rise at the basal cell pole. Subsequently, a wave of elevated $[Ca^{2+}]_i$ spreads to the cell apex. The white lines outline a single cell in A and B. Image intervals: 400 ms. Scale bar: 20 μm (from Zimmermann 2000, modified).

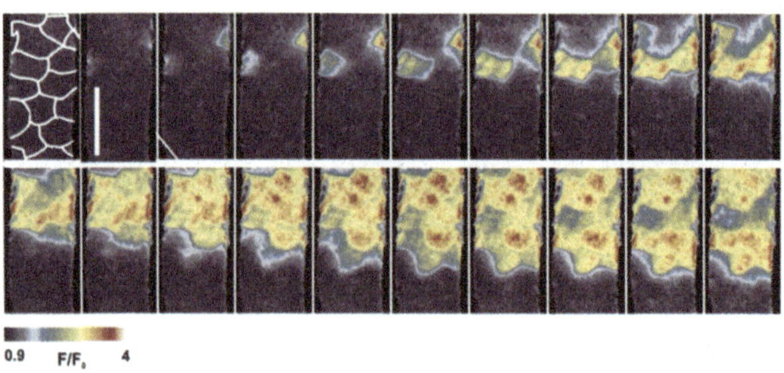

Hormone-mediated intercellular calcium signalling in an insect salivary gland: Fig. 3. Intercellular wave propagation in the basal cell portions (tangential optical section through the epithelium) recorded from a fluo-4-loaded gland by confocal imaging. A Ca^{2+} wave sequentially spreads through several cells. White lines indicate cell boundaries; image intervals: 200 ms; scale bar: 50 μm (from Zimmermann 2000, modified).

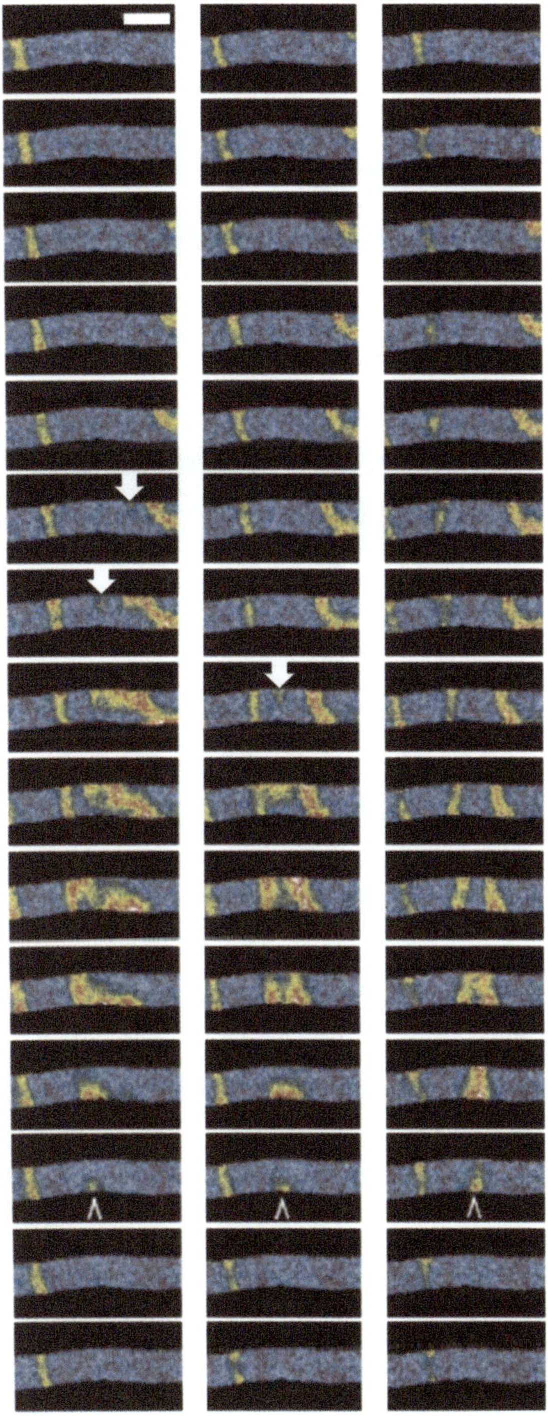

Hormone-mediated intercellular calcium signalling in an insect: Fig. 5. Intercellular Ca^{2+} waves induced by 3 nM 5-HT. Time series of spatial changes in $[Ca^{2+}]_i$ recorded by ratiometric imaging from a fura-2-loaded gland. Ca^{2+} waves are initiated at multiple sites (white arrows) and propagate centrifugally with no decrease in amplitude. Colliding waves annihilate each other (open arrowheads). Image intervals: 1s; scale bar 100 μm (from Zimmermann and Walz 1997, modified).

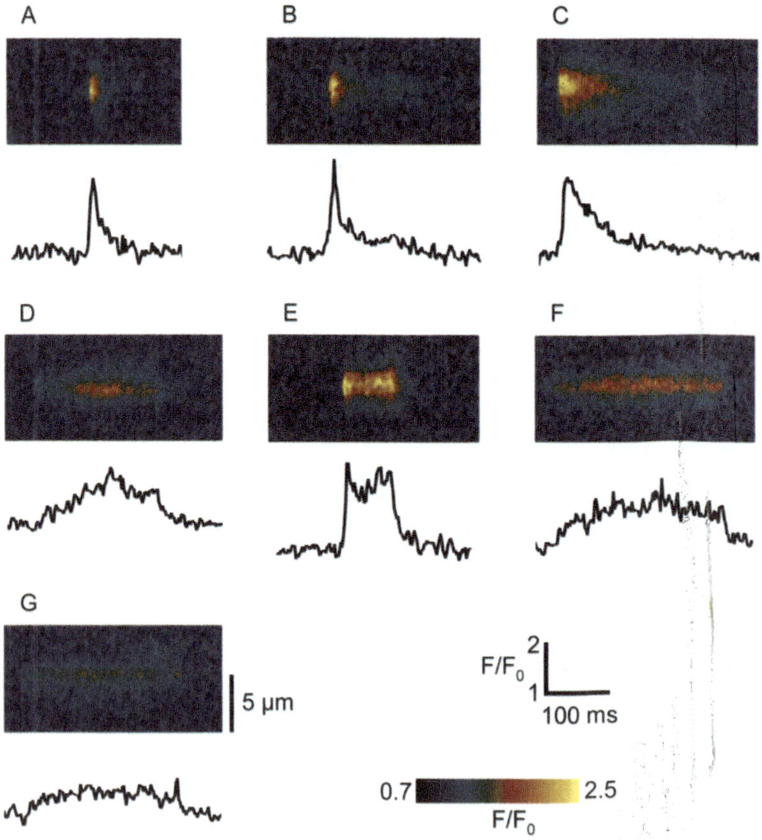

Mathematical modelling and high resolution fluorescence imaging for studying calcium regulation: Fig. 7. Confocal line scan images of elementary Ca^{2+}-release events in permeabilised mammalian skeletal muscle fibres. The events were recorded with 60 µM of the Ca^{2+}-sensitive fluorescence indicator fluo-4 using an 40x water immersion objective. Selected line scan images of normalized fluorescence show the diversity of spontaneous ECRE in chemically or mechanically skinned mammalian fibres. The trace below each picture shows the time course at the centre of the event. Events A-C show typical spark-type ECRE whereas events D-G show ember-type events. From Uttenweiler et al. (2002).

7. Hormone-Mediated Intercellular Calcium Signalling in an Insect Salivary Gland - Pathways and Mechanisms

Bernhard Zimmermann and Bernd Walz

Institute of Biochemistry and Biology, Department of Animal Physiology, University of Potsdam, P.O.B. 60 15 53, 14415 Potsdam, Germany

Abstract. The salivary glands of the blowfly *Calliphora vicina* are a favourable preparation for investigations into spatio-temporal Ca^{2+} dynamics in an intact miniorgan by using Ca^{2+}-sensitive indicator dyes and digital imaging techniques, including confocal microscopy, in combination with pharmacological approaches. The review summarizes the available data on the spatio-temporal patterns of the hormone-induced and IP_3-mediated Ca^{2+} dynamics at both the intracellular and the intercellular level (intra- and intercellular Ca^{2+} waves). The underlying signaling mechanisms are addressed, as well as the pathways of intercellular communication responsible for the complex spatio-temporal Ca^{2+} dynamics. In addition, we review evidence for the exchange of Ca^{2+} between IP_3 sensitive intracellular Ca^{2+} stores and mitochondria including a modulatory effect of mitochondrial Ca^{2+} uptake on the frequency of IP_3-induced Ca^{2+} spiking.

7.1. Introduction

Excitation-secretion coupling in salivary glands of the blowfly, *Calliphora vicina* is stimulated by the neurohormone serotonin (5-HT) and mediated by a 5-HT-induced elevation in intracellular Ca^{2+} and cAMP concentrations. This preparation has been used extensively as a model system for hormone-induced Ca^{2+} and cAMP signaling in the past because of its simplicity and experimental accessability (Berridge et al. 1974). The tubular secretory portion consists of gap-junction-coupled homotypic cells forming a single-layered epithelium around a central lumen (Oschman and Berridge 1970). Moreover, the *Calliphora* salivary gland was the first preparation in which a functional link between hormone-induced phosphoinositide metabolism and the generation of Ca^{2+} signals was established (Berridge and Prince 1972; Fain and Berridge 1979; Berridge et al. 1983). 5-HT elicits, at intermediate concentrations, oscillatory changes in the transepithlial potential (TEP) of the gland; these fluctuations are caused by Ca^{2+}-dependent oscillatory changes of the Cl^- conductance in the epithelial cell membrane (e.g. Rapp and Berridge 1981). The underlying Ca^{2+} oscillations had not previously been characterized by direct intracellular Ca^{2+} measurements, and, a few years ago, we therefore presumed that these glands might be a favourable preparation for studying the mechanisms of intracellular Ca^{2+} oscillations. Indeed, *Calliphora* salivary glands turned out to be ideally suited for investigations into the spatio-temporal Ca^{2+} dynamics in a morphologically and functionally intact mini-organ by using Ca^{2+}-sensitive indicator dyes and digital imaging techniques, including confocal microscopy, in combination with pharmacological approaches. This review summarizes the available data on the spatio-temporal patterns of the hormone-induced and $InsP_3$-mediated Ca^{2+} dynamics at both the intracellular and the intercellular levels. The review also addresses the underlying signaling mechanisms and the pathways of intercellular communication responsible for the complex spatio-temporal Ca^{2+} dynamics.

7.2. The Temporal Pattern of 5-HT-Induced Intracellular Calcium Changes

5-HT concentrations between 1 and 10 nM applied to isolated glands elicit periodic transients in intracellular Ca^{2+} concentration, i.e. a series of Ca^{2+} spikes (Fig. 1). Spike frequency is a function of 5-HT concentration and increases from ~ 0.7 min^{-1} at 1 nM 5-HT to ~ 6.4 min^{-1} at 10 nM 5-HT. At 3 to 10 nM 5-HT, the Ca^{2+}-spike amplitudes range from 100 nM to 1.1 μM, whereby amplitudes decrease from an initial maximum (1st spike) to 36% of the maximal amplitude within the first minute of stimulation, suggesting desensitization of the Ca^{2+}-elevating system. Higher 5-HT concentrations (30 nM to 3 μM) elicit a variable number of initial Ca^{2+} transients that are followed by a tonic Ca^{2+} elevation (Fig.1) (Zimmermann and Walz 1997). Because Ca^{2+}-spike frequency and the frequencies of oscillations in the TEP recorded in the presence of 3 nM 5-HT are equivalent, we concluded that the intracellular free Ca^{2+} concentration ($[Ca^{2+}]_i$) is a key regulator of Cl^- channel opening as suggested previously (Berridge et al. 1975). Since Ca^{2+}-spike frequency, but not amplitude, depends on the agonist concentration, blowfly salivary glands constitute a further example of the frequency modulation of a cellular response to an amplitude-modulated external signal. The finding that the dose-response relationship (spike frequency vs. [5-HT]) compares favourably with the 5-HT dependence of fluid secretion (Berridge 1970; Berridge and Prince 1972) suggests that secretion rates are controlled by the frequency of Ca^{2+} spikes.

With respect to the Ca^{2+} sources for the generation of the 5-HT-induced Ca^{2+} oscillations, we have determined that these are not acutely dependent on extracellular Ca^{2+} and are abolished by the inhibition of Ca^{2+} reuptake into the endoplasmic reticulum (ER) by thapsigargin. We have found no evidence for an involvement of ryanodine receptor Ca^{2+} channels, because neither caffeine nor ryanodine affects 5-HT-induced Ca^{2+} transients. Thus, the upstroke of the Ca^{2+} transients is caused predominantly by Ca^{2+} release from the ER and is mediated by agonist-evoked formation of $InsP_3$ (see below, and Zimmermann and Walz 1997, 1999; Zimmermann 2000).

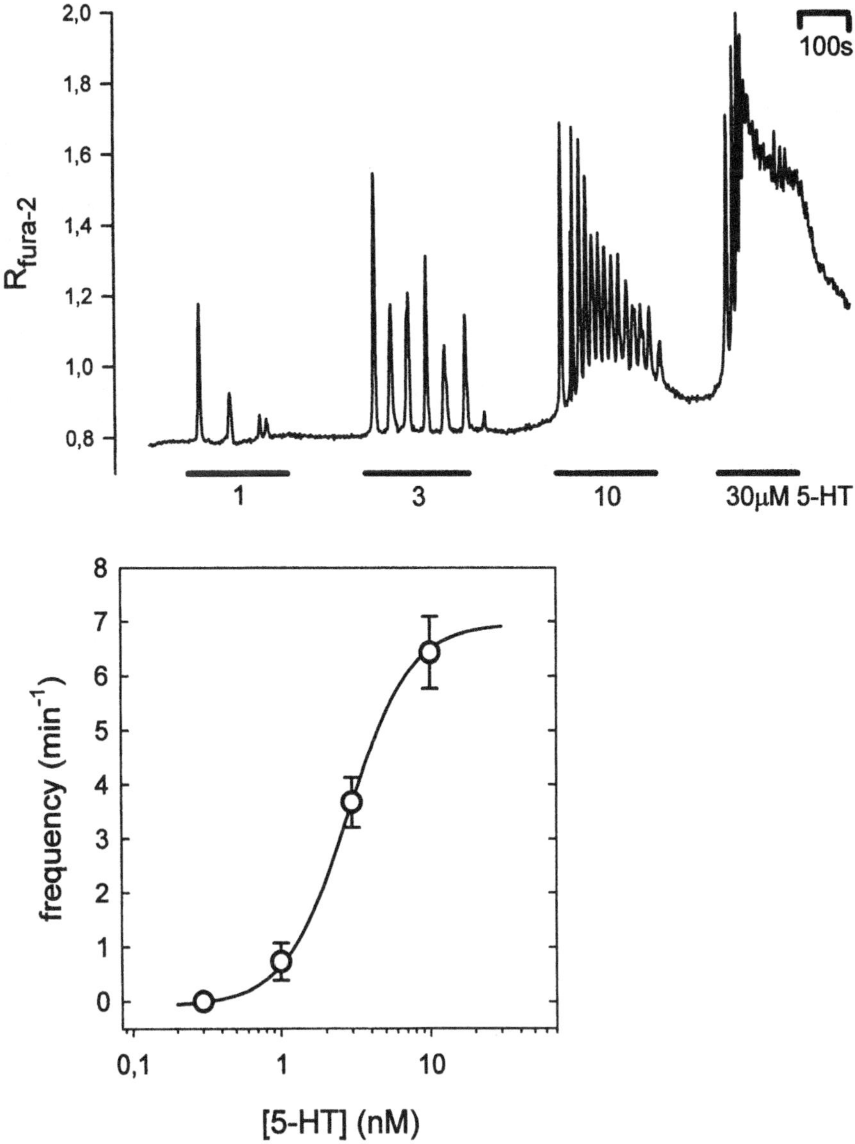

Fig. 1. Temporal $[Ca^{2+}]_i$ dynamics evoked by various concentrations of 5-HT. 5-HT concentrations between 1 and 10 nM elicit periodic Ca^{2+} spikes of variable amplitude. Spike frequency increases with increasing [5-HT]. Higher [5-HT], e.g. 30 μM, induce a variable number of initial spikes followed by a tonic elevation of $[Ca^{2+}]_i$. Original recording from a fura-2 loaded gland. Fluorescence signals were recorded from a region corresponding to the size of a single cell. Horizontal bars: Periods of 5-HT application. Lower panel: Dependence of Ca^{2+}-spike frequency on the concentration of 5-HT.

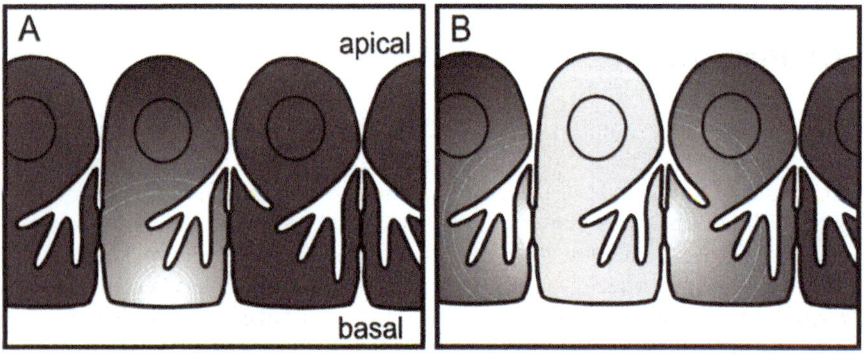

Fig. 4. Schematic representation of the cellular patterns of Ca^{2+} changes associated with Ca^{2+} oscillations and intercellular Ca^{2+} waves. [Ca^{2+}] is encoded in greys, where light and dark grey represent high and low concentrations, respectively. **A**, [Ca^{2+}] elevations are triggered near the basal cell membrane. The radial spread of the Ca^{2+} signal results in a basally to apically propagating Ca^{2+} wave. **B**, Following a temporal delay at the cell boundaries, [Ca^{2+}] begins to rise at the lateral membranes of the adjacent cells. From here, Ca^{2+} waves spread through these and adjacent cells. (From Zimmermann 2000, modified)

However, the 5-HT-induced Ca^{2+} signals are not completely independent of extracellular Ca^{2+}. The baseline Ca^{2+} oscillations elicited by the application of threshold doses of 5-HT (3 nM) in Ca^{2+}-free bathing saline (0 Ca^{2+}, 2 mM EGTA) display unchanged amplitudes but their frequency decreases by ~ 21%. Saturating 5-HT concentrations (e.g. 300 nM) applied in the presence of 2 mM extracellular Ca^{2+} induces biphasic intracellular Ca^{2+} elevations consisting of an initial transient followed by a plateau of elevated Ca^{2+} (Fig. 1). In Ca^{2+}-free saline, the amplitude of the initial transient is almost unchanged, but the subsequent plateau phase is attenuated by more than 90%. These results suggest that the initial upstroke of $[Ca^{2+}]_i$ is attributable to Ca^{2+} release from the ER, whereas the plateau phase arises from Ca^{2+} influx (Zimmermann 1998). Earlier tracer experiments had indirectly demonstrated Ca^{2+} entry as a function of 5-HT concentration, including the activation of Ca^{2+} influx during Ca^{2+} spiking (Prince and Berridge 1973). A detailed study (Zimmermann 1998) of the Ca^{2+} influx mechanisms in blowfly salivary glands has revealed two distinct store-operated mechanisms (Putney 1986, 1990) that support Ca^{2+} influx across the baso-lateral plasma membrane in response to depletion of intracellular Ca^{2+} stores. These two mechanisms are distinguished by their sensitivity to lanthanides and the Ca^{2+} channel blocker SK&F 96365. Essentially, these experiments demonstrate that La^{3+} inhibits the 5-HT-induced Ca^{2+} entry across the baso-lateral membrane completely, thereby mimicking the effects of extracellular Ca^{2+} withdrawal (attenuation of the plateau phase of the Ca^{2+} transient evoked by saturationg 5-HT concentrations). At threshold 5-HT concentrations, La^{3+} has no immediate effects on Ca^{2+} spike amplitudes or frequency. These findings suggest that the Ca^{2+} spike and tonic elevations have different La^{3+} sensitivities, and that store-operated Ca^{2+} influx inhibited by La^{3+} contributes only to

the plateau phase at high 5-HT concentrations and does not support repetitive Ca^{2+} spiking.
In addition, we have found that mitochondria contribute to the control of the temporal pattern of agonist-induced cytoplasmic Ca^{2+} dynamics (see below, and Zimmermann 2000).

7.3. Subcellular Organization of Calcium Waves

In the majority of mammalian secretory epithelial cells, secretagogs stimulate either local Ca^{2+} transients restricted to the cell apex or polarized subcellular Ca^{2+} waves that are initiated at the apical cell pole and spread throughout the cell as an apical-to-basal wave (e.g. Kasai and Augustine 1990; Thorn et al. 1993; for a review see Ashby and Tepikin 2002). We have studied the subcellular organization of 5-HT-induced Ca^{2+} signals in intact *Calliphora* salivary glands by confocal imaging of Fluo-4-loaded cells. We have found that, in this preparation, the spatio-temporal pattern of the 5-HT-induced intracellular Ca^{2+} changes clearly differs from the patterns previously described in almost all studied mammalian systems (Zimmermann 2000).

Saturated 5-HT concentrations (100 nM) induce in all gland cells, an almost simultaneous increase in $[Ca^{2+}]_i$ that consists of an initial Ca^{2+} rise to a peak, a subsequent plateau phase, and, after 5-HT removal, a return to baseline. In individual cells, the increases in $[Ca^{2+}]_i$ are initiated at the basal pole and spread toward the apex at a mean rate of 23 µm/s (Fig. 2, see colorplate). Under Ca^{2+} free conditions, neither the spatio-temporal organization of the initial Ca^{2+} dynamics nor wave velocity is significantly affected, indicating that the initial $[Ca^{2+}]_i$ elevations result from Ca^{2+} release from internal stores.

At subsaturating 5-HT concentrations (3 nM) that elicit intracellular Ca^{2+} oscillations, Ca^{2+} transients are triggered asynchronously in only a subset of cells along the glands. Within these "trigger" cells, $[Ca^{2+}]_i$ initially begins to increase at discrete sites close to the basal plasma membranes; the Ca^{2+} elevations then spread from these sites radially, thereby producing basally to apically propagating intracellular Ca^{2+} waves (Fig. 2, see colorplate). Ca^{2+} wavefronts spread in the lateral cell planes (Fig. 2, see colorplate) at a mean rate of 27 µm/s. Lateral spreading of the Ca^{2+} wavefronts does not stop at the cell boundaries but rather, after a temporal delay of 0.2 to 1.2 s, spread into and propagats within neighboring cells. The Ca^{2+} waves spreads into neighboring cells only after $[Ca^{2+}]_i$ in the first cell has reached its maximum. Overall, this behavior results in a wave-like progression of the Ca^{2+} signal through the epithelium (Fig. 3, see colorplate), whereby the waves are propagated without decrement in amplitude through numerous cells (see next section).

Ca^{2+} elevations in adjacent cells are first detectable exactly at those portions of the lateral plasma membrane that face a previously activated cell. Electron microscopy has revealed that the sites of Ca^{2+} wave transmission between cells are correlated with the distribution of gap junctions that cluster in the basal cell portions (Zimmermann 2000). Indeed, cell-to-cell propagation of Ca^{2+} waves is severely inhibited by the gap junction blocker 1-octanol (Zimmermann and Walz 1997). The schematic representation in Fig. 4 summarizes the subcellular patterns of intracellular Ca^{2+} oscillations and waves and intercellular Ca^{2+} waves.

As mentioned above, intracellular Ca^{2+} oscillations and waves contribute to the stimulation of 5-HT-induced fluid secretion in *Calliphora* salivary glands. Our finding of basal-to-apical intracellular Ca^{2+} waves in this preparation imposes some constraints on the functional significance of intracellular Ca^{2+} waves and potential mechanisms underlying fluid secretion. As in many mammalian fluid-secreting epithelial cells, Ca^{2+} stimulates secretion by increasing Cl^- conductances of the apical and basal cell membranes (e.g. Iwatsuki and Petersen 1977; Petersen and Philpott 1980; Petersen 1992; Berridge et al. 1975, 1976; Prince and Berridge 1972). The observation of apical-to-basal Ca^{2+} waves in pancreatic acinar cells and their correlation with agonist-stimualted ion currents led Kasai and Augustine (1990) to develop the "push-pull" model for Ca^{2+}-induced unidirectional Cl^- secretion. In the so-called "push" phase, apically located Ca^{2+}-dependent Cl^- channels are activated and secrete Cl^- into the gland lumen. When the Ca^{2+} wave reaches the basal membrane, it stimulates ("pulls") Cl^- uptake into the cell. According to this model, the sequential activation of the two phases is necessary, because the "push" phase causes a depolarization of the apical membrane, which is a prerequisite for the "pulling" of Cl^- into the cell basally. Because (1) the direction of the intracellular Ca^{2+} waves is reversed, (2) the finding that 5-HT hyperpolarizes rather than depolarizes the basolateral membrane (Berridge et al. 1975), and (3) the apical membrane forms deep infoldings that spatially approach the basal membrane very closely, this model, or simply its reversal ("pull-push"), cannot account for the secretion in *Calliphora* salivary glands.

7.4. Regenerative Intercellular Calcium Waves - Phenomenology

Ca^{2+}-wave propagation along the tubular *Calliphora* salivary glands over extensive distances has been studied by conventional ratiometric imaging of Fura-2 loaded glands (Zimmermann and Walz 1997, 1999). We have found that the oscillatory activity in individual cells described above is synchronized by regenerative intercellular Ca^{2+} waves that propagate over distances greater than 400 µm. The lowest 5-HT concentrations (1 nM) that elicit Ca^{2+} elevations in individual cells produce only local non-propagating Ca^{2+} transients and/or short-lived Ca^{2+} waves. These waves are propagated over variable distances from their initiation sites, sometimes following spiral pathways along the circumference of the epithelial tube and usually subsiding without collision with waves originating from other initiation sites. 5-HT concentrations (1-10 nM) that evoke periodic Ca^{2+} spikes produce Ca^{2+} waves that are asynchronously initiated at irregularly spaced foci, spread radially, and, finally, progress as regular waves along the longitudinal axis of the gland. Colliding waves annihilate each other (Fig. 5, see colorplate). In the continuous presence of 5-HT, waves are repeatedly initiated at the same sites. Waves travel without decrement in amplitude and velocity at rates between 12.5-17.3 µm/s over ~ 400 µm thereby passing through more than 20 cells.

The described phenomenology of Ca^{2+} wave propagation indicates that it is not the result merely of passive diffusion but rather involves a regenerative process. We have excluded the possibility that wave progression is caused by an extracellular signal because wave progression is not affected by the direction of movement of the surrounding bathing solution. Moreover, local stimulation of the glands by a focal application of 5-HT to a gland segment of 50-80 µm in width elicits $[Ca^{2+}]_i$ elevations only in the stimulated area, but not in adjacent cells. This experiment indicats that the local formation, diffusion, and regeneration of a messenger are not sufficient to promote Ca^{2+}-wave propagation. However, when local 5-HT pulses are given on a global background stimulation with a low (1 nM) 5-HT concentration, propagating Ca^{2+} waves are initiated. These travel without decrement up to 200 µm away from the site of the local 5-HT stimulus. This positive effect of the background 5-HT stimulation cannot be mediated by a Ca^{2+} elevation, because waves can also be triggered by local stimulation when $[Ca^{2+}]_i$ is at resting levels.

How are the Ca^{2+} waves and their non-decremental and spatially organized propagation generated? Activation of 5-HT receptors in *Calliphora* salivary glands leads, via stimulation of phospholipase C and adenylate cyclase activity, to an increase in $InsP_3$, Ca^{2+}, and cAMP concentrations (Heslop and Berridge 1980; Berridge et al. 1983; Zimmermann and Walz 1997). Thus, either of these second messengers might directly or indirectly allow the propagation of Ca^{2+} waves by sensitizing $InsP_3$-induced Ca^{2+} release to the diffusing messenger. What is the nature of this diffusing messenger, the messenger that diffuses through the gap junctions in order to promote and synchronize Ca^{2+} elevations in adjacent cells?

7.5. The Mechanisms of Intra- and Intercellular Calcium Waves

Intra- and intercellular Ca^{2+} waves have now been observed in numerous systems, including cultured cells, small groups of acutely isolated cells, and in intact tissues (for reviews, see, for example, Ashby and Tepikin 2002; Berridge et al. 1998; Jaffee and Creton 1998; Sanderson 1996). Current models of agonist-induced regenerative intra- and intercellular Ca^{2+} waves have led to the postulation of spatio-temporal fluctuations either of $InsP_3$ (Meyer and Stryer 1991) or of Ca^{2+} at constant $InsP_3$ concentrations (e.g. Jaffee 1991; DeLisle and Welsh 1992; Lechleiter and Clapham 1992; Atri et al. 1993). Propagating waves of elevated $InsP_3$ concentration have been suggested to arise from $InsP_3$ diffusion, $InsP_3$-induced Ca^{2+} release, and feedback activation of phospholipase C (PLC) by released Ca^{2+} (Meyer and Stryer 1991; Harootunian et al. 1991) or other products of PLC activity (Cuthbertson and Chay 1991; Sneyd et al. 1994). Indeed, analyses of the spatio-temporal pattern of changes in intracellular $InsP_3$ concentration in Madin-Darby canine kidney epithelial cells by using the green fluorescent protein-tagged pleckstrin homology domain of PLCδ_1 as an $InsP_3$ indicator have revealed (1) $InsP_3$ oscillations accompanying Ca^{2+} oscillations, and (2) intracellular [$InsP_3$] waves accompanying Ca^{2+} waves (Hirose et al. 1999). This study supports models in which regenerative Ca^{2+}-mediated $InsP_3$ production accompanies Ca^{2+} oscillations and waves. Alternative models are based on the finding that the opening of the $InsP_3$ receptor channel is regulated both positively and negatively by Ca^{2+} (for a review, see Taylor and Laude 2002). According to these models, Ca^{2+} waves propagate in the presence of constant sub-threshold $InsP_3$ concentrations by Ca^{2+} diffusion and Ca^{2+}-induced Ca^{2+} release via the $InsP_3$ receptor (e.g. Lechleiter and Clapham 1992; for a review, see Clapham 1995). Cells that co-express ryanodine receptor Ca^{2+} channels have more complex functional interactions between $InsP_3$, Ca^{2+} and Ca^{2+} mobilization.

A key question addressed in many studies is the identity of the chemical signal that mediates intercellular wave transmission. Both $InsP_3$ and Ca^{2+} have been proposed as intercellular messengers.

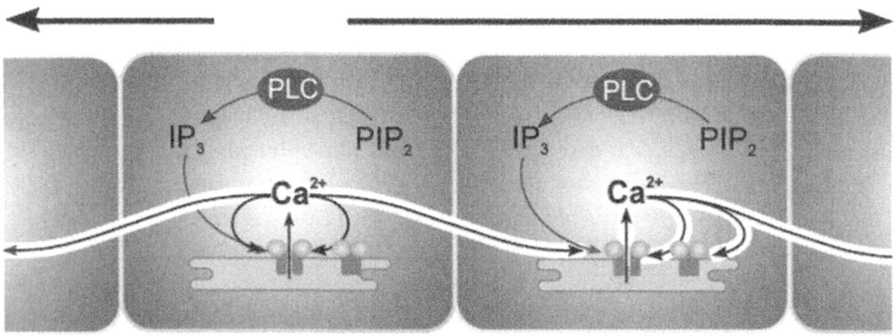

Fig. 6. Schematic representation of the mechanism of Ca^{2+} wave propagation in *Calliphora* salivary glands. Gland stimulation with 5-HT activates the phosphoinositide signalling cascade and the formation of $InsP_3$. $InsP_3$ causes Ca^{2+} release from the ER and an increase in $[Ca^{2+}]_i$. Released Ca^{2+} diffuses through gap junctions into neighboring cells and triggers Ca^{2+}-induced Ca^{2+} release from the ER by way of $InsP_3$ receptor channels, which are sensitised by the presence of supra-basal [$InsP_3$].

Our study of the mechanisms mediating regenerative Ca^{2+} waves in the *Calliphora* salivary glands has revealed a clearcut scenario for this preparation (Zimmermann and Walz 1999). We have shown that iontophoretic injections of $InsP_3$ into salivary gland cells evoke regenerative intercellular Ca^{2+} waves that spread through the impaled cell and several rows of adjacent cells. Ca^{2+} increases induced by microinjection of Ca^{2+}, in contrast, remain confined to the injected cell and its nearest neighbors. However, activation of Ca^{2+} release and intercellular Ca^{2+} waves that consecutively pass through more than 10 cells become clearly evident when Ca^{2+} is injected in the presence of low concentrations of 5-HT (1 or 3 nM). Thus, Ca^{2+} elevations are sufficient to trigger Ca^{2+} waves when $InsP_3$ reaches a critical supra-basal concentration. In order to test for a possible involvement of Ca^{2+}-mediated activation of PLC in Ca^{2+} wave propagation, we have interfered with the production of endogenous $InsP_3$ by inhibiting PLC with U73122. Under these conditions, 5-HT-induced Ca^{2+} changes are completely blocked, whereas the inactive structural analog U73343 is without effect. Most significantly, injections of $InsP_3$ into glands pre-treated with U73122 produce $InsP_3$-induced Ca^{2+} spiking in the activated region accompanied by the propagation of intercellular Ca^{2+} waves.

These results demonstrate that the propagation of agonist-induced Ca^{2+} waves in *Calliphora* salivary glands requires supra-basal $InsP_3$ concentrations but does not depend on feedback activation of PLC. Rather, our results suggest that the generation of Ca^{2+} waves involves the intercellular diffusion of Ca^{2+} and Ca^{2+}-induced Ca^{2+} release from the ER by way of $InsP_3$ receptor channels that are Ca^{2+} sensitized by supra-basal $InsP_3$ concentrations. Intra- and intercellular Ca^{2+} waves are propagated by the same mechanism as outlined schematically in Fig. 6.

This model is further supported by our investigation of the mechanisms that contribute to the control of $InsP_3$-induced Ca^{2+} oscillations. For this purpose, we loaded the Ca^{2+} sensor mag-fura-2 into the ER and monitored $InsP_3$-induced Ca^{2+} dynamics of the intraluminal Ca^{2+} concentrations ($[Ca^{2+}]_L$ in β-escin permeabilized salivary glands. Cytoplasmic Ca^{2+} changes were simultaneously monitored by Calcium Green C_{18}, a Ca^{2+} indicator whose lipophilic alkyl side-chain intercalates into lipid membranes. This dye therefore monitors Ca^{2+} in the immediate vicinity of membranes (Zimmermann 2000). The most important observations were as follows: (1) Application of $InsP_3$ to permeabilized gland cells caused periodic Ca^{2+} release from the ER into the cytosol and thereby reproduced 5-HT-induced cytoplasmic Ca^{2+} oscillations in the intact gland. Remarkably, these oscillatory Ca^{2+} release episodes propagated through several cells as a coordinated wave, as did cytoplasmic Ca^{2+} oscillations in intact cells. (2) Buffering the incubation medium with high EGTA concentrations prevented oscillatory changes in $[Ca^{2+}]_L$. Altogether we found that Ca^{2+} release in permeabilized cells bathed in weakly Ca^{2+}-buffered media consisted of two components: Periodic transient Ca^{2+} release events dependent on fluctuations in cytoplasmic Ca^{2+} concentration, and a steady $InsP_3$-concentration-dependent Ca^{2+} efflux component between Ca^{2+} spikes. We concluded from these experiments that Ca^{2+} oscillations are essentially driven by $InsP_3$-induced Ca^{2+} release periodically triggering Ca^{2+}-induced Ca^{2+} release from the ER. In this scenario, the Ca^{2+} release initiated by subsaturating $InsP_3$ concentrations is boosted because of the Ca^{2+} sensitivity of the $InsP_3$ receptor for $InsP_3$ (Taylor 2002). The resultant local cytoplasmic Ca^{2+} elevation then excites neighboring $InsP_3$ receptors and induces the regenerative and propagating release of additional Ca^{2+}.

7.6. Mitochondria Modulate the Frequency of $InsP_3$-Induced Calcium Oscillations

Mitochondria contribute to the regulation of the dynamics of cytosolic Ca^{2+} signals in many cell types (for reviews, see Pozzan and Rizzuto 2000, Rizzuto et al.

1999). In experiments on permeabilized *Calliphora* salivary glands (see above, and Zimmermann 2000), we have observed that the frequency modulation of Ca^{2+} oscillations in intact glands is completely lost after permeabilization. In contrast, incremental concentrations of $InsP_3$ applied to the permeabilized cells in the presence of the oxidizable mitochondrial substrates citrate, succinate, or pyruvate-malate induce repetitive transients in $[Ca^{2+}]_L$ whose frequency increases with the concentration of $InsP_3$. This $InsP_3$ concentration-dependent modulation of oscillation frequency is abolished after dissipation of the mitochondrial membrane potential by combined treatment with FCCP + oligomycin or after application of Ruthenium Red, an inhibitor of mitochondrial Ca^{2+} uptake. These data indicate that energized mitochondria exert a negative control over the frequency of $InsP_3$-induced Ca^{2+} oscillations. Thus, in *Calliphora* salivary glands, as in many other systems, mitochondria play an important role in determining the duration of inter-spike intervals and, therefore, for the encoding of amplitude-modulated $InsP_3$-mediated cytosolic Ca^{2+} oscillations in the frequency domain.

References

Ashby MC, Tepikin, AV (2002) Polarized calcium and calmodulin signalling in secretory epithelia. Physiol Rev 82: 701-734

Atri AJ, Amundson D, Clapham D, Sneyd J (1993) A single-pool model for intracellular calcium oscillations and waves in the *Xenopus laevis* oocyte. Biophys J 65: 1727-1739

Berridge MJ (1970) The role of 5-hydroxytryptamine and cyclic AMP in the control of fluid secretion by isolated salivary glands. J Exp Biol 53: 171-186

Berridge MJ, Prince WT (1972) Transepithelial potential changes during stimulation of isolated salivary glands with 5-hydroxytryptamine and cyclic AMP. J Exp Biol 56: 139-153

Berridge MJ, Lindley BD, Prince WT (1974) Role of calcium and cAMP in controlling fly salivary gland secretion. In: Thorn NA, Petersen OH (eds) Secretory mechanisms of exocrine glands. Munksgaard, Copenhagen, pp331-343

Berridge MJ, Lindley BD, Prince WT (1975) Membrane permeability changes during stimulation of isolated salivary glands of *Calliphora* by 5-hydroxytryptamine. J Physiol 244: 549-567

Berridge MJ, Lindley BD, Prince WT (1976) Studies on the mechanism of fluid secretion by isolated salivary glands of *Calliphora*. J Exp Biol 64: 311-322

Berridge MJ, Dawson RMC, Downes CP, Heslop JP, Irvine RF (1983) Changes in the levels of inositol phosphates after agonist-dependent hydrolysis of membrane phosphoinositides. Biochem J 212: 473-482

Berridge MJ, Bootman MD, Lipp P (1998) Calcium - a life and death signal. Nature 395: 645-648

Clapham DE (1995) Calcium signalling. Cell 80: 259-268

Cuthbertson KS, Chay TR (1991) Modeling receptor-controlled intracellular calcium oscillators. Cell Calcium 12: 97-109

DeLisle S, Welsh MJ (1992) Inositol trisphosphate is required for the propagation of calcium waves in *Xenopus* oocytes. J Biol Chem 267: 7963-7966

Fain JN, Berridge MJ (1979) Relationship between hormonal activation of phosphatidylinositol hydrolysis, fluid secretion and calcium flux in the blowfly salivary gland. Biochem J 178: 45-58

Harootunian AT, Kao JPY, Paranjape S, Tsien RY (1991) Generation of calcium oscillations in fibroblasts by positive feedback between calcium and IP_3. Science 251: 75-78

Heslop JP, Berridge MJ (1980) Changes in cyclic AMP and cyclic GMP concentrations during the action of 5-hydroxytryptamine on an insect salivary gland. Biochem J 192: 247-255

Hirose K, Kadowaki S, Tanabe M, Takeshima H, Iino M (1999) Spatiotemporal dynamics of inositol 1,4,5-trisphosphate that underlies Ca^{2+} mobilization patterns. Science 284: 1527-1530

Iwatsuki N, Petersen OH (1977) Pancreatic acinar cells: localization of acetylcholine receptors and the importance of chloride and calcium for acetylcholine-induced depolarization. J Physiol 269: 723-733

Jaffe LF (1991) The path of calcium in cytosolic calcium oscillations: a unifying hypothesis. Proc Natl Acad Sci USA 88: 9883-9887

Jaffe LF, Créton R (1998) On the conservation of calcium wave speeds. Cell Calcium 24: 1-8

Kasai H, Augustine GJ (1990) Cytosolic Ca^{2+} gradients triggering unidirectional fluid secretion from exocrine pancreas. Nature 348: 735-738

Lechleiter JD, Clapham DE (1992) Molecular mechanisms of intracellular calcium excitability in *X. laevis* oocytes. Cell 69: 283-294

Meyer T, Stryer L (1991) Calcium spiking. Annu Rev Biophys Biophys Chem 20: 153-174

Oschman JL, Berridge MJ (1970) Structural and functional aspects of salivary fluid secretion. Tissue Cell 2: 281-310

Petersen OH (1992) Stimulus-secretion coupling: cytoplasmic calcium signals and the control of ion channels in exocrine acinar cells. J Physiol 448: 1-51

Petersen OH, Philpott HG (1980) Pancreatic acinar cells: the anion selectivity of the acetylcholine-evoked chloride pathway. J Physiol 306: 481-492

Pozzan T, Rizzuto R (2000) The renaissance of mitochondrial calcium transport. Eur J Biochem 267: 5269-5273

Prince WT, Berridge MJ (1972) The effects of 5-hydroxytryptamine and cyclic AMP on the potential profile across isolated salivary glands. J Exp Biol 56: 323-333

Prince WT, Berridge MJ (1973) The role of calcium in the action of 5-hydroxytryptamine and cyclic AMP on salivary glands. J Exp Biol : 367-384

Putney JW (1986) A model for receptor-regulated calcium entry. Cell Calcium 7: 1-12

Putney JW (1990) Capacitative calcium entry revisited. Cell Calcium 11: 611-624

Rapp PE, Berridge MJ (1981) The control of transepithelial potential oscillations in the salivary gland of *Calliphora erythrocephala*. J Exp Biol 93: 119-132

Rizzuto R, Pinton P, Brini M, Chiesa A, Filippin L, Pozzan T (1999) Mitochondria as biosensors of calcium microdomains. Cell Calcium 26: 193-199

Sanderson MJ (1996) Intercellular waves of communication. News Physiol Sci 11: 262-269

Sneyd J, Charles AC, Sanderson MJ (1994) A model for the propagation of intercellular calcium waves. Am J Physiol 266: C293-C302

Taylor CW, Laude AJ (2002) IP_3 receptors and their regulation by calmodulin and cytosolic Ca^{2+}. Cell Calcium 32: 321-334

Thorn P, Lawrie AM, Smith PM, Gallacher DV, Petersen OH (1993) Local and global cytosolic Ca^{2+} oscillations in exocrine cells evoked by agonists and inositol trisphosphate. Cell 74: 661-668

Zimmermann B (1998) Calcium store depletion activates two distinct calcium entry pathways in secretory cells of the blowfly salivary gland. Cell Calcium 23: 53-63

Zimmermann B (2000) Control of $InsP_3$-induced Ca^{2+} oscillations in permeabilized blowfly salivary gland cells: contribution of mitochondria. J Physiol 525: 707-719

Zimmermann B, Walz, B (1997) Serotonin-induced intercellular calcium waves in salivary glands of the blowfly *Calliphora erythrocephala*. J Physiol 500: 17-28

Zimmermann B, Walz B (1999) The mechanism mediating regenerative intercellular Ca^{2+} waves in the blowfly salivary gland. EMBO J 18: 3222-3231

8. From Simple to Complex Ca^{2+} Oscillations: Regulatory Mechanisms and Theoretical Models.

G. Dupont, G. Houart and A. Goldbeter

Unité de Chronobiologie Théorique, Université Libre de Bruxelles, Faculté des Sciences CP231, Boulevard du Triomphe, B-1050 Brussels, Belgium

Abstract. Intracellular Ca^{2+} oscillations most often appear as regular spikes. However, complex Ca^{2+} oscillations in the form of bursting or chaos have also been observed. These oscillations most probably originate from the interplay between multiple regulatory mechanisms. From an experimental point of view, the mechanism underlying these types of Ca^{2+} oscillations remain poorly understood. In this review, we first briefly review the various types of regulatory mechanism known to regulate Ca^{2+} oscillations, together with the related theoretical models. We show that the interplay between the activation of an $InsP_3$-metabolizing enzyme ($InsP_3$ 3-kinase) by Ca^{2+} and Ca^{2+}-induced Ca^{2+} release is a plausible explanation for at least some complex Ca^{2+} oscillations. This regulation indeed provides a mechanism for the self-modulation of the $InsP_3$ signal, thus leading to a phenomenon analogous to the periodic forcing of an autonomous oscillator. We also briefly assess the role of luminal Ca^{2+} in the onset of Ca^{2+} oscillations, as well as the link between bistability and Ca^{2+} oscillations.

8.1. Introduction

Over the last 15 years, oscillations in intracellular Ca^{2+} have become a major example of oscillatory behavior at the cellular level (Berridge, 1993). Since their first direct observation in hepatocytes (Woods *et al.*, 1986), Ca^{2+} oscillations have been studied in a large number of cell types, such as *Xenopus* oocytes, mammalian eggs, chromaffin, endothelial, or pancreatic acinar cells, etc (Berridge, 1993). These oscillations occur either spontaneously, or when a cell is stimulated by a hormone or neurotransmitter acting on receptors coupled to the phospho-inositide cascade. The period of Ca^{2+} oscillations ranges from seconds to minutes, depending on the cell type. Furthermore, for a given cell type, the frequency increases with the level of stimulation. This allows for the frequency-encoding of Ca^{2+} oscillations. Key phenomena such as secretion (Holl *et al.*, 1988), gene expression (Dolmetsch *et al.*, 1998; Llopis *et al.*, 1998) or neuronal differentiation (Gu and Spitzer, 1999) are thus regulated by the frequency rather than by the amplitude of Ca^{2+} oscillations.

Although Ca^{2+} oscillations generally take the form of regular spikes, complex Ca^{2+} oscillations have also been observed. Such complex oscillations are reminiscent of the bursting-type behavior often displayed by electrically excitable cells. One of the best known examples of Ca^{2+} bursting is the complex oscillatory pattern seen in hepatocytes after stimulation by specific agonists such as cAMP or ATP (Dixon *et al.*, 1993; 1995; Green *et al.*, 1997). The mechanism underlying this type of Ca^{2+} oscillations remains poorly understood from an experimental point of view and has mainly been investigated by several theoretical approaches (Chay *et al.*, 1995; Shen and Larter, 1995; Borghans *et al.*, 1997; Houart *et al.*, 1999; Kummer *et al.*, 2000; Marhl *et al.*, 2000).

Intracellular waves of Ca^{2+} often accompany the oscillations (Berridge, 1993). Spatial propagation of the repetitive Ca^{2+} rises from cell to cell has also been observed, for example in the liver (Robb-Gaspers and Thomas, 1995), in epithelia (Sanderson *et al.*, 1994) or in insect salivary glands (see Chap.7). The significance of such a multicellular organization of the Ca^{2+} signal may be to optimize the physiological response at the organ level (see Chap.10).

In the present chapter, we focus on the temporal organization of intracellular Ca^{2+} signals and present a brief overview of the different types of models that have been proposed for simple and complex Ca^{2+} oscillations. We consider in some more detail the origin of both phenomena in a minimal model based on Ca^{2+}-induced Ca^{2+} release (CICR) incorporating the activation of an $InsP_3$-metabolizing enzyme ($InsP_3$ 3-kinase) by Ca^{2+} (Borghans *et al.*, 1997; Houart *et al.*, 1999).

8.2. Regulatory Mechanisms for Ca^{2+} Oscillations and Related Theoretical Models.

The mechanism of Ca^{2+} oscillations relies on the feedback processes that regulate Ca^{2+} levels within the cell. Upon binding to a specific membrane receptor, the external stimulus (hormone or neurotransmitter) triggers the synthesis of inositol 1,4,5-trisphosphate ($InsP_3$), which binds to the $InsP_3$ receptor ($InsP_3R$). This receptor, located on the membrane of the intracellular Ca^{2+} stores (endoplasmic or sarcoplasmic reticulum), behaves as a Ca^{2+} channel. Upon $InsP_3$ binding, Ca^{2+} is released from the stores into the cytosol. Moreover, as Ca^{2+} stimulates the activity of the $InsP_3$ receptor/channel, this release is progressively enhanced in a process known as $InsP_3$-sensitive Ca^{2+}-induced Ca^{2+} release (CICR). This self-activation process represents a key element in the instability mechanism that generates repetitive Ca^{2+} spiking.

However, Ca^{2+} oscillations are regulated by a variety of other processes that could play an active role in the oscillatory mechanism itself or in the shaping of Ca^{2+} spikes. A non-exhaustive list of such Ca^{2+}-regulatory processes is given in Table 1. On the basis of these regulatory processes, a variety of theoretical models have been developed to account for Ca^{2+} oscillations.

8.2.1. Ca^{2+}-Induced Ca^{2+} Release

This regulation, known as CICR, plays a primary role in the origin of Ca^{2+} oscillations (see above). Activation by cytosolic Ca^{2+} of the release of Ca^{2+} from the intracellular stores into the cytosol has first been proposed for the ryanodine receptor (RyR). This receptor, which shares many structural and functional similarities with the $InsP_3R$ (Berridge, 1993), is responsible for Ca^{2+} release in muscle and cardiac cells, as well as in neurons (Endo *et al.*, 1970; Fabiato and Fabiato, 1975; Kuba and Takeshita, 1981). Following its incorporation into theoretical models for $InsP_3$-induced Ca^{2+} oscillations (Goldbeter *et al.*, 1990; Dupont and Goldbeter, 1993), $InsP_3$-sensitive CICR has been experimentally demonstrated (Wakui *et al.*, 1990). A minimal model based on CICR (Goldbeter *et al.*, 1990; Dupont and Goldbeter, 1993) qualitatively reproduces Ca^{2+} oscillations. The equations governing the minimal model are given in Table 2. As shown in Figs. 1 and 2, this model accounts for (1) the existence of a range of stimulation levels producing Ca^{2+} oscillations, (2) the rise in frequency with both the level of stimulation and the concentration of extracellular Ca^{2+}, and (3) the progressive rise in cytosolic Ca^{2+} preceding the onset of the Ca^{2+} spike (see, e.g. Jacob, 1990).

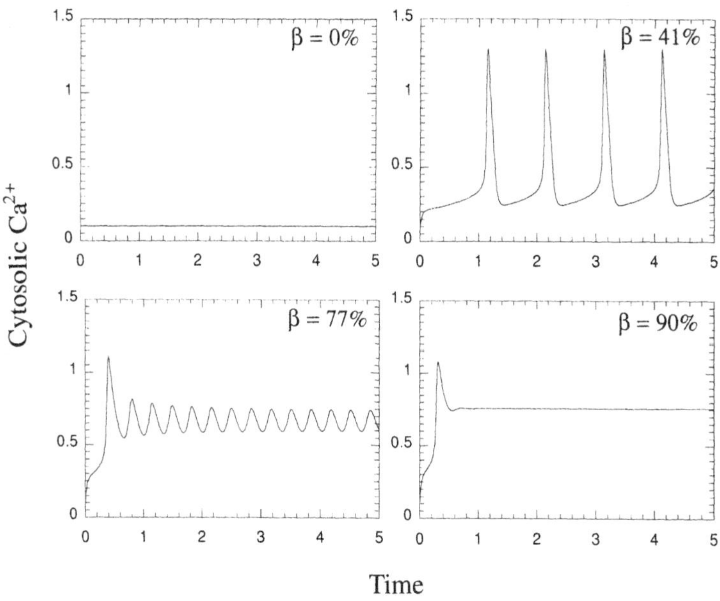

Fig. 1. Different patterns of cytosolic Ca^{2+} oscillations generated by the model based on CICR (Table 2) as a function of the increasing level of external stimulation. The level of $InsP_3$, which rises with the level of external stimulation, is measured by parameter β which gives the saturation function of the $InsP_3R$. Oscillations are obtained in a limited range of β values. A low (high) stable steady state of cytosolic Ca^{2+} concentration is obtained for lower (higher) values of β.

Table 1. Main regulatory processes that bear on the origin and properties of Ca^{2+} oscillations (see text).

Ca^{2+} signalling multiple regulatory mechanisms

- Ca^{2+}-induced Ca^{2+} release
- PLC activation by Ca^{2+}
- Biphasic regulation of $InsP_3R$ by Ca^{2+}
- Capacitative Ca^{2+} entry
- $InsP_3$ 3-kinase activation by Ca^{2+}
- Mitochondrial control of Ca^{2+}
- G-protein regulation
- Phosphorylation of the $InsP_3R$

Table 2. One-pool model for Ca^{2+} oscillations based on CICR (Dupont and Goldbeter, 1993). Formally similar kinetic equations are obtained in the two-pool model based on CICR (Goldbeter et al., 1990).

Minimal model based on CICR

Z : cytosolic Ca^{2+}

Y : Ca^{2+} in the endoplasmic reticulum

$$\frac{dZ}{dt} = v_0 + v_1\beta - V_2 + V_3 + k_f Y - kZ \tag{1}$$

$$\frac{dY}{dt} = V_2 - V_3 - k_f Y \tag{2}$$

with

v_0: basal Ca^{2+} influx from the extracellular medium

β : level of external stimulation

$v_1\beta$: agonist-stimulated Ca^{2+} entry

Ca^{2+} pumping from the cytosol into the ER:

$$V_2 = V_{M2}\frac{Z^2}{K_2^2 + Z^2} \tag{3}$$

Ca^{2+} release from the ER into the cytosol activated by cytosolic Ca^{2+} (CICR):

$$V_3 = \beta V_{M3}\frac{Y^2}{K_R^2 + Y^2}\frac{Z^4}{K_A^4 + Z^4} \tag{4}$$

$k_f Y$: passive efflux of Ca^{2+} from the stores into the cytosol

$k Z$: passive efflux of Ca^{2+} from the cytosol into the extracellular medium

The CICR regulation, first established for type I $InsP_3Rs$ (Bezprozvanny and Ehrlich, 1995), extends to the other sub-types, although in different Ca^{2+} ranges. It is by now clear that this variety in CICR regulation contributes to the diversity of temporal patterns of Ca^{2+} signals observed in different cell types (Taylor *et al.*, 1999; Miyakawa *et al.*, 1999; Thrower *et al.*, 2001; Haberichter *et al.*, 2002). Nearly all models for Ca^{2+} oscillations rely on $InsP_3$-sensitive CICR. These models mainly differ by the degree of detail in the description of $InsP_3R$ kinetics and by the incorporation of one or the other regulatory process considered below (for review, see Sneyd *et al.*, 1995; Tang *et al.*, 1996; Dupont, 1999; Schuster *et al.*, 2002).

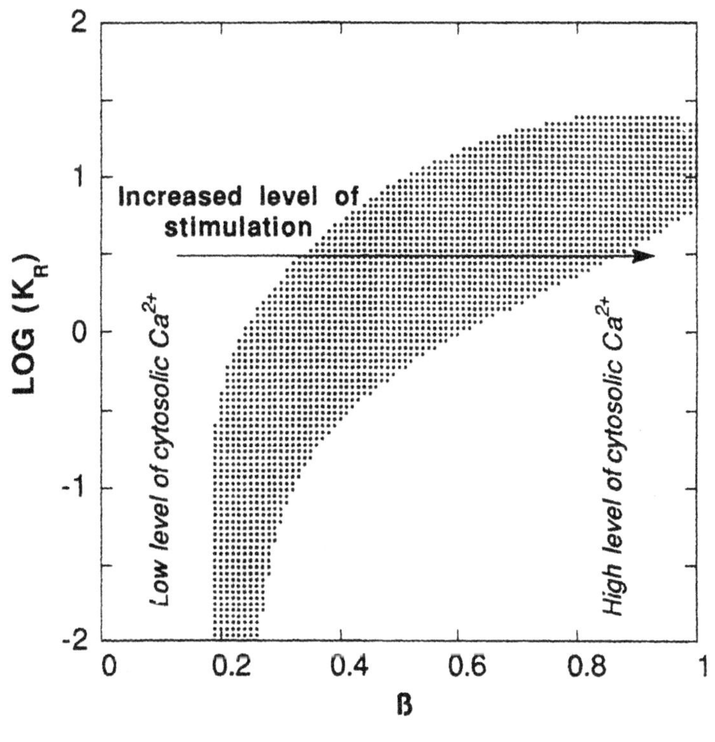

Fig. 2. Diagram showing the stability of the steady state of the model based on CICR (Table 2) as a function of the values of the level of external stimulation (β) and of the constant characterizing the regulation by intraluminal Ca^{2+} of Ca^{2+} release into the cytosol (K_R). The steady state is unstable in the region filled with dots. In this region, sustained oscillations of the levels of cytosolic and intravesicular Ca^{2+} occur. Thus, when increasing the level of stimulation (arrow), one goes through a sequence of behaviors similar to the sequence represented in Fig. 1. Also visible is the fact that Ca^{2+} oscillations can be induced by decreasing the constant K_R. Such a decrease of the threshold level of intraluminal Ca^{2+} inducing Ca^{2+} release into the cytosol may represent the effect of caffeine.

8.2.2. Assessing the Role of Luminal Ca^{2+} in the Onset of Ca^{2+} Oscillations

An alternative view of CICR attributes the observed activation of Ca^{2+} release to the progressive filling of the intracellular stores, which are triggered to release Ca^{2+} when a threshold level of luminal Ca^{2+} is reached. Although this hypothesis (Berridge and Galione, 1988) has received some experimental support (Missiaen *et al.*, 1996; Combettes *et al.*, 1996), mathematical modeling predicts that, at least in the absence of time delays, such a process does not suffice to induce a sustained oscillatory response (Dupont and Goldbeter, 1989; Goldbeter *et al.*, 1990). Application of the Bendixson criterion (Minorsky, 1967) to the equations of the minimal model indeed shows that, in the absence of CICR from the cytosol, activation of Ca^{2+} release by luminal Ca^{2+} cannot generate sustained oscillations. This theoretical prediction has been recently corroborated in DT-40 cells in which the sensitivity of type 1 $InsP_3R$ to cytosolic Ca^{2+} was decreased by a factor of 10 by substitution of the appropriate residue; such cells show greatly diminished Ca^{2+} responsiveness and, most importantly, Ca^{2+} oscillations are abolished (Miyakawa *et al.*, 2001).

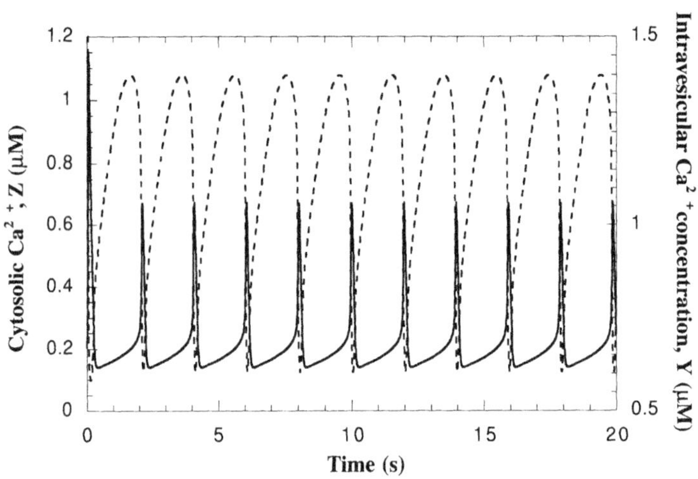

Fig. 3. Oscillations in cytosolic Ca^{2+} (Z, solid line) and intravesicular Ca^{2+} (Y, dashed line) in the extended version of the model based on CICR aimed at investigating the role of intravesicular Ca^{2+} in the onset of Ca^{2+} spikes. Curves have been obtained by numerical integration of (3)-(6).

However, it is clear that the content of the stores has a profound effect on the dynamics of intracellular Ca^{2+}. The activation of Ca^{2+} release by luminal Ca^{2+} was already incorporated into the minimal model for Ca^{2+} oscillations (Goldbeter *et al.*, 1990; see (4) in Table 2). To quantitatively assess the contribution of luminal Ca^{2+} to CICR, we slightly modified the minimal model by considering that the stimulus-activated Ca^{2+} entry ($v_1\beta$) directly feeds into the ER, rather than into the cytosol as previously assumed. The kinetic equations governing the time evolution of cytosolic Ca^{2+} (Z) and intravesicular Ca^{2+} (Y) now take the form of (5) and (6) in which V_2 and V_3 remain given by (3) and (4) in Table 2:

$$\frac{dZ}{dt} = v_0 - V_2 + V_3 + k_f Y - kZ \tag{5}$$

$$\frac{dY}{dt} = v_1\beta + V_2 - V_3 - k_f Y \tag{6}$$

As shown in Fig. 3, the model still displays regular Ca^{2+} oscillations in the cytosol (solid line) and in the ER (dashed line). Moreover, in such conditions, we can directly evaluate the importance of the Ca^{2+} level in the stores by changing parameter v_1. Figure 4 clearly shows that Ca^{2+} oscillations can only be obtained in a limited range of v_1 values. If the rate of stimulated Ca^{2+} entry in the cytosol, v_1, is too low, there is only a stable steady state characterized by a low level of Ca^{2+} in the cytosol and a high level of Ca^{2+} in the ER. When v_1 is too large, oscillations are also prevented and a stable steady-state is reached, in which cytosolic Ca^{2+} is high and intravesicular Ca^{2+} is low. This situation can be ascribed to enhanced Ca^{2+} release through CICR. Thus, although the primary role in generating Ca^{2+} oscillations must be ascribed to the activation of Ca^{2+} release by cytosolic Ca^{2+} (CICR), such oscillations can only be obtained for appropriate levels of Ca^{2+} loading of the stores. Moreover, the spatial distribution of Ca^{2+} inside the ER also affects the dynamics of cytosolic Ca^{2+}. In pancreatic acinar cells, Ca^{2+} movements have indeed been observed inside the ER, suggesting that this organelle can function as a " Ca^{2+} tunnel" (Petersen *et al.*, 2001).

8.2.3. PLC Activation by Ca^{2+}

The first mathematical model for $InsP_3$-induced Ca^{2+} oscillations assumed that this phenomenon relies on an oscillating level of $InsP_3$. The periodic variation of the $InsP_3$ level was based on a positive feedback loop arising from the assumed activation by Ca^{2+} of $InsP_3$ synthesis catalyzed by phospholipase C (PLC) (Meyer and Stryer, 1988 and 1991; De Young and Keizer, 1992). That such positive feedback occurs for Ca^{2+} concentrations in the physiological range of 0.1-1 μM has been demonstrated solely for the γ isoform of the enzyme (Renard *et al.*, 1987), which has only been detected in a few cell types.

Although many early experimental studies indicated that Ca^{2+} oscillations can be observed in the presence of a constant level of $InsP_3$ (Wakui *et al.*, 1989; Hajnoczky and Thomas, 1997), some recent experiments suggest that Ca^{2+} oscillations are accompanied by $InsP_3$ variations (Hirose *et al.*, 1999; Nash *et al.*, 2001a and 2001b). However, as will be emphasized below, $InsP_3$ oscillations could result from the well-characterized stimulation of $InsP_3$ degradation by Ca^{2+} (Takazawa *et al.*, 1990; Sims and Allbritton, 1998), in which case they do not really participate in the core oscillatory mechanism. Moreover, the above-mentioned suppression of Ca^{2+} oscillations in DT40 cells expressing $InsP_3Rs$ with a decreased sensitivity to cytosolic Ca^{2+} (Miyakawa *et al.*, 2001) also argue against a primary role of the stimulation of PLC activity by Ca^{2+} in the generation of Ca^{2+} oscillations.

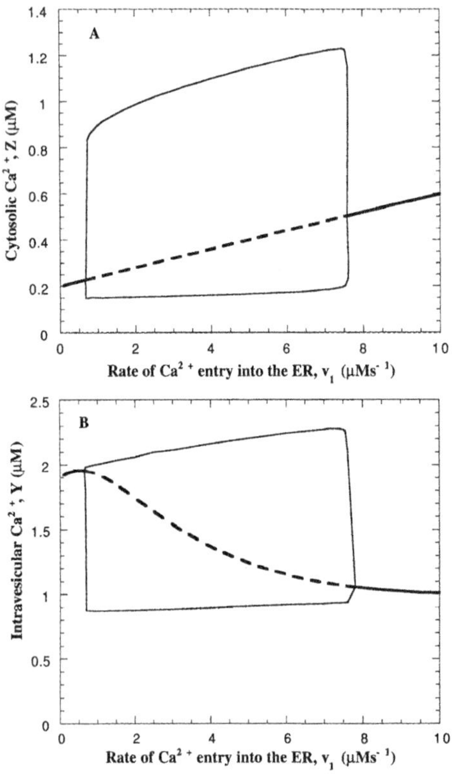

Fig. 4. Bifurcation diagrams in the extended version of the model based on CICR aimed at investigating the role of intravesicular Ca^{2+} in the onset of Ca^{2+} spikes. Shown are the steady-state values of cytosolic Ca^{2+} (A) or of intravesicular Ca^{2+} (B) as a function of the rate of Ca^{2+} entry into the endoplasmic reticulum. Stable steady states are represented by solid lines, and unstable steady states by dashed lines. Moreover, when the steady state is unstable, the minimum and maximum values of the variable reached during oscillations are also shown. Thus, oscillations can only be obtained in a domain of values of the rate of Ca^{2+} entry into the stores, indicating that the level of Ca^{2+} inside the ER plays a crucial role in the occurrence of Ca^{2+} oscillations.

8.2.4. Biphasic Regulation of $InsP_3R$ by Ca^{2+}

The steady-state open probability of the $InsP_3R/Ca^{2+}$ channel presents a bell-shaped dependence on cytosolic Ca^{2+} concentration (Bezprozvanny and Ehrlich, 1995). Thus, besides CICR, slower, Ca^{2+}-induced inhibition of Ca^{2+} release does also take place. From a theoretical point of view, fast activation and slower inhibition are well known to be sufficient to generate an oscillatory behavior (Goldbeter, 1996). Comprehensive overviews of models based on a detailed description of the transitions between the different states of the $InsP_3R$ has been given elsewhere (see, for example, Sneyd *et al.*, 1995; Tang *et al.*, 1996; Schuster *et al.*, 2002). These models reproduce both the oscillations and the equilibrium bell-shaped Ca^{2+} dependence of the $InsP_3R$. A recent variant of this type of models is based on the intriguing possibilities that Ca^{2+} binds to the different states of the receptor with nonlinear kinetics and that Ca^{2+} binding to the receptor both decreases the rate of

$InsP_3$ binding and increases the sensitivity of the receptor to $InsP_3$ (Sneyd and Dufour, 2002). These assumptions greatly enhance the agreement between the observed dynamical behavior of the channel and the predictions of the model.

The main difference between the model based on a detailed description of the transitions between the different states of the $InsP_3R$ and those simply based on $InsP_3$-sensitive CICR pertains to the mechanism responsible for the termination of the Ca^{2+} spike. In the model incorporating the inhibition of the $InsP_3R$ by Ca^{2+}, the latter regulation shuts the channel. In contrast, in models based on simple CICR, the decrease of the Ca^{2+} content of the ER leads to a decrease in cytosolic Ca^{2+} as soon as the rate of release from the ER becomes smaller than the rate of pumping into the ER.

8.2.5. Capacitative Ca^{2+} Entry

During Ca^{2+} oscillations, entry of external Ca^{2+} is somehow controlled by the state of filling of the ER. Indeed, Ca^{2+} entry is prevented when Ca^{2+} stores are charged up, but store depletion immediately promotes Ca^{2+} entry into the cytosol. The mechanism of Ca^{2+} entry into the cytosol is still not totally understood and probably differs from one cell type to the other (Berridge, 1995; Lewis, 2001; Putney *et al.*, 2001). By regulating the level of Ca^{2+} inside the cells, capacitative Ca^{2+} entry (CCE) has a profound effect on the overall intracellular Ca^{2+} dynamics. Incorporated into theoretical models, this regulation has been shown to mitigate the depletion of intracellular stores and favor a stable, high steady-state level of cytosolic Ca^{2+} (Wiesner *et al.*, 1996). The possible role of CCE in the onset of bistable behavior in Ca^{2+} signaling is considered in Sect. 4 below.

8.2.6. $InsP_3$ 3-Kinase Activation by Ca^{2+}

The metabolism of $InsP_3$ by Ins-1,4,5-P_3 3-kinase and 5-phosphatase has been well characterized. It has been shown that the Ca^{2+}/calmodulin complex stimulates 3-kinase activity (Takazawa *et al.*, 1990; Sims and Allbritton, 1998). Thus, each Ca^{2+} spike provokes a temporary increase in $InsP_3$ degradation, a phenomenon that leads to $InsP_3$ oscillations (Borghans *et al.*, 1997; Houart *et al.*, 1999). Thus, it is not necessary to invoke the stimulation of $InsP_3$ synthesis by Ca^{2+} to explain the experimentally observed oscillations in $InsP_3$ (Hirose *et al.*, 1999; Nash *et al.*, 2001a and 2001b). If PLC activation were the driving force of $InsP_3$ and Ca^{2+} oscillations, the peak in $InsP_3$ should precede the peak of Ca^{2+}. The opposite should be true if $InsP_3$ oscillations were passively following Ca^{2+} oscillations through stimulation of $InsP_3$ 3-kinase activity by Ca^{2+}. Current experimental techniques do not allow to discriminate between the two situations. The observation that Ca^{2+} oscillations can be induced by large amounts of $InsP_3$ analogs (Wakui *et al.*, 1990; Dumollard et al., 2002) favors the view that $InsP_3$ oscillations originate from the regulation of $InsP_3$ degradation by Ca^{2+}.

8.2.7. Mitochondrial Control of Ca^{2+}

Mitochondria play an important role in controlling intracellular Ca^{2+} dynamics. Mitochondrial membranes contain many Ca^{2+} transporters, such as the permeability transition pore (PTP), the Ca^{2+} uniporter and the Na^+/Ca^{2+} and H^+/Ca^{2+} exchangers. Through these channels, mitochondria behave as high capacity, low affinity Ca^{2+} buffers (Jouaville *et al.*, 1995) and could also play a more active role in releasing their Ca^{2+} content when overloaded, in a process called mitochondrial Ca^{2+}-induced Ca^{2+} release (Ichas *et al.*, 1994). Also, as a consequence of cytosolic Ca^{2+} oscillations, the level of Ca^{2+} inside mitochondria varies, allowing for the frequency-encoding of intramitochondrial reactions such as those based on pyruvate dehydrogenase activity (Robb-Gaspers *et al.*, 1998).

Mitochondria have been taken into account in several models for Ca^{2+} dynamics (Marhl *et al.*, 2000; Falcke *et al.*, 1999). These models predict that mitochondria can regulate the amplitude and the frequency of cytosolic Ca^{2+} spikes, and can influence the transition from simple to complex Ca^{2+} oscillations.

8.2.8. G-Protein Regulation

G-proteins are involved in the transduction signal cascade between the external hormonal receptor and $InsP_3$ synthesis through the PLC isoform, PLCß. Possible regulations of G-protein by $InsP_3$, Ca^{2+} or diacyl-glycerol (DAG) have been incorporated into various theoretical models (Cuthbertson and Chay, 1991; Kummer *et al.*, 2000; Läer *et al.*, 2001). In these models, which often display complex Ca^{2+} oscillations, the level of $InsP_3$ oscillates together with that of Ca^{2+}. However, solid experimental evidence for the regulation of G-protein by Ca^{2+} or $InsP_3$ is still lacking.

8.2.9. Phosphorylation of the $InsP_3$ Receptor

Phosphorylation of the $InsP_3$ receptor by protein kinases A, C and calmodulin-dependent kinase II has been demonstrated (see Bezprozvanny and Ehrlich, 1995 for review). Type III $InsP_3R$, most abundant in pancreatic acinar cells, has been shown to be largely phosphorylated in response to stimulation by cholecystokinin (LeBeau *et al.*, 1999). Phosphorylation is not observed in response to other agonists such as acetylcholine. If it is assumed that the phosphorylated receptor does not pass Ca^{2+} current, the rate of $InsP_3R$ recovery from inactivation crucially depends on the nature of the agonist. This mechanism might explain the long-period Ca^{2+} spiking observed in these cells after stimulation by cholecystokinin (LeBeau *et al.*, 1999).

8.3. Complex Ca^{2+} Oscillations

8.3.1. Interplay Between Different Oscillatory Mechanisms

Complex Ca^{2+} oscillations of the bursting type, such as those observed in hepatocytes in response to stimulation by specific agonists (Dixon *et al.*, 1993; 1995; Rooney *et al.*, 1996; Green *et al.*, 1997), bring additional information about the regulatory mechanism of Ca^{2+} oscillations. Models indicate that a variety of regulatory processes might lead to complex Ca^{2+} oscillations. In all these models, complex Ca^{2+} oscillations originate from the interplay between at least two instability-generating mechanisms.

The first model, proposed by Shen and Larter (1995), is based on the coupling between CICR and stimulation of PLC activity by Ca^{2+}. However, in this model, complex Ca^{2+} oscillations only arise in a very small region of the parameter space. Activation of PLC by Ca^{2+} was also included in another model for complex Ca^{2+} oscillations, together with the putative self-activation of the G_α subunit of the receptor complex (Kummer *et al.*, 2000). In the latter model, a change in the parameters characterizing the receptor results in the switch from simple to complex oscillations.

On the basis of their experimental observations, an inhibitory role of PKC in the origin of bursting in hepatocytes has been stressed by Dixon *et al.* (1995). Such a regulation was incorporated into the model proposed by Chay (1995), together with CICR, the stimulation of PLC activity by Ca^{2+}, and the indirect inhibition of PLC by PKC, which is itself activated by Ca^{2+}. Although the latter model shows a great richness in dynamical behavior, it includes so many regulations that it is difficult to obtain from it a clear understanding of the mechanisms leading to Ca^{2+} bursting and chaos.

Complex Ca^{2+} oscillations have also been obtained in models in which CICR is coupled to Ca^{2+} sequestration in mitochondria (Marhl *et al.*, 2000) or to modulation of the $InsP_3$ stimulation by oscillating Ca^{2+} levels (see Sect. 3.2).

8.3.2. Model Based on the Self-Modulation of the $InsP_3$ Signal via Ca^{2+} Oscillations

Here, we focus on a model based on the interplay between CICR at the level of the $InsP_3R$ and the stimulation by Ca^{2+} of the activity of the inositol 1,4,5-trisphosphate 3-kinase, one of the $InsP_3$ metabolizing enzymes (Takazawa *et al.*, 1990; Sims and Allbritton, 1998). From a dynamical point of view, this model is of particular interest as it can generate a large variety of dynamical behaviors, including chaos, bursting, quasi-periodicity and birhythmicity (Borghans *et al.*, 1997; Houart *et al.*, 1999). We further show that this model displays chaotic bursting that much resembles the Ca^{2+} oscillatory pattern sometimes observed in hepatocytes.

The model, schematized in Fig. 5, extends the minimal model based on CICR (see Sect. 2.1) by incoporating the dynamics of $InsP_3$. Thus, $InsP_3$ is synthesized at a rate proportional to the level of stimulation (β) and hydrolyzed both into $InsP_2$ by a 5-phosphatase (εA) and into $InsP_4$ by a 3-kinase (V_5). Thus, the evolution equation for $InsP_3$ (A) reads:

$$\frac{dA}{dt} = \beta V_4 - V_5 - \varepsilon A \tag{7}$$

where V_4 represents the maximum rate of $InsP_3$ synthesis by PLC. The rate of $InsP_3$ hydrolysis by the 3-kinase is given by:

$$V_5 = V_{M5} \frac{A}{K_5 + A} \frac{Z^n}{K_d^n + Z^n} \tag{8}$$

This expression derives from the observations that 3-kinase behaves as a Michaelian enzyme with respect to its substrate $InsP_3$, and that stimulation by Ca^{2+} is a cooperative phenomenon (Takazawa *et al.*, 1990; Sims and Allbritton, 1998). As the level of $InsP_3$ is now a variable of the model, the rate of $InsP_3$- and Ca^{2+}-

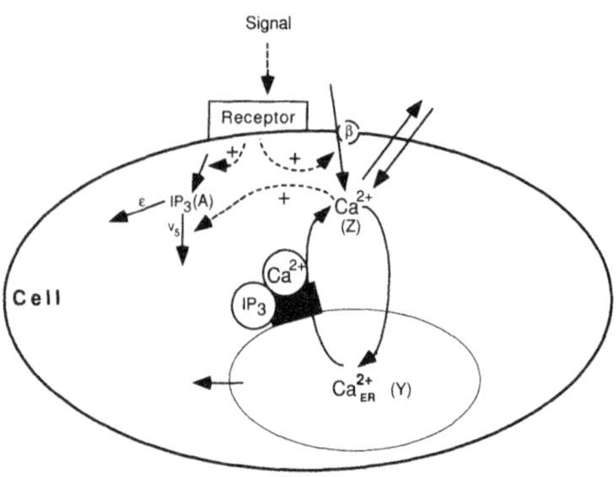

Fig. 5. Schematic representation of the model based on the interplay between CICR and the Ca^{2+}-stimulated degradation of $InsP_3$ (Borghans *et al.*, 1997; Houart *et al.*, 1999). Besides simple periodic behavior, this model can produce complex Ca^{2+} oscillations including burs-ting, chaos, quasiperiodic be-havior, as well as birhyth-micity.

activated Ca^{2+} release from the ER into the cytosol (given by (4) in the minimal model) is now given by:

$$V_3 = V_{M3} \frac{Z^m}{K_Z^m + Z^m} \frac{Y^2}{K_Y^2 + Y^2} \frac{A^4}{K_A^4 + A^4} . \tag{9}$$

The complete model is thus governed by (1)-(3) and (7)-(9).

As expected intuitively, oscillations in the level of $InsP_3$ accompany Ca^{2+} oscillations in this extended model (Fig. 6). Oscillations in $InsP_3$ simply follow the Ca^{2+} spikes, as each increase in Ca^{2+} elicits, after a delay due to 3-kinase, the degradation of $InsP_3$.

8.3.3. Complex Ca^{2+} Oscillations Arising from the Self-Modulation of the $InsP_3$ Signal

If $InsP_3$ oscillations can be entrained by Ca^{2+} oscillations, they can, however, strongly influence the nature of the oscillatory pattern. In the present model indeed, an increase in cytosolic Ca^{2+} has two opposite effects. On one hand, due to CICR, it enhances the release of Ca^{2+} from internal stores. On the other hand, due to 3-kinase stimulation by Ca^{2+}, it brings about a decrease in $InsP_3$, which in turn reduces the rate of Ca^{2+} release into the cytosol. These counteracting effects of Ca^{2+} are the source of bursting and chaos in the present model, because the system somehow behaves as a periodically forced oscillator, for which complex oscillations are known to occur (Goldbeter, 1996). Indeed, CICR, which can proceed in the presence of a constant level of $InsP_3$, provides a mechanism for autonomous oscillations, while the signal ($InsP_3$) that triggers Ca^{2+} oscillations is modulated by them: $InsP_3$ raises the level of cytosolic Ca^{2+}, which in turn decreases the level of $InsP_3$ through the action of the Ca^{2+}-activated 3-kinase.

Some types of complex oscillations that can be obtained with the model are illustrated in Fig. 7 (top row: time series, bottom row: phase space). The leftmost column shows oscillations of the bursting type in which a large amplitude Ca^{2+} spike is followed by smaller variations around a plateau level. The decreased amplitude of Ca^{2+} oscillations on the plateau is due to the reduced level of $InsP_3$ after the massive degradation that follows the first, large Ca^{2+} spike.

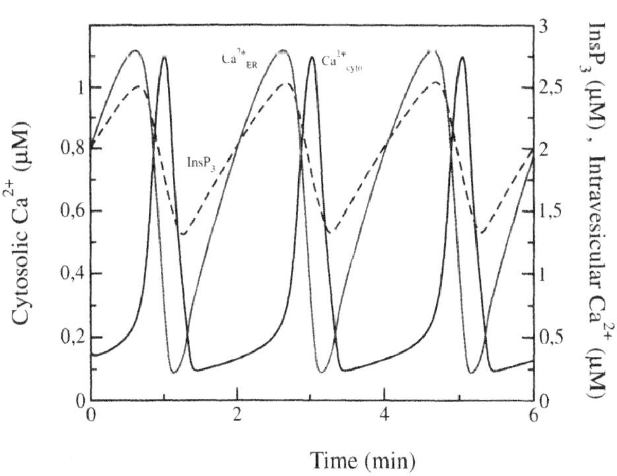

Fig. 6. Simple oscillations of Ca^{2+} and $InsP_3$ in the model model based on the interplay between CICR and the Ca^{2+} stimulated degradation of $InsP_3$. The curves have been obtained by numerical integration of the model defined by (1)-(3), (7)-(9).

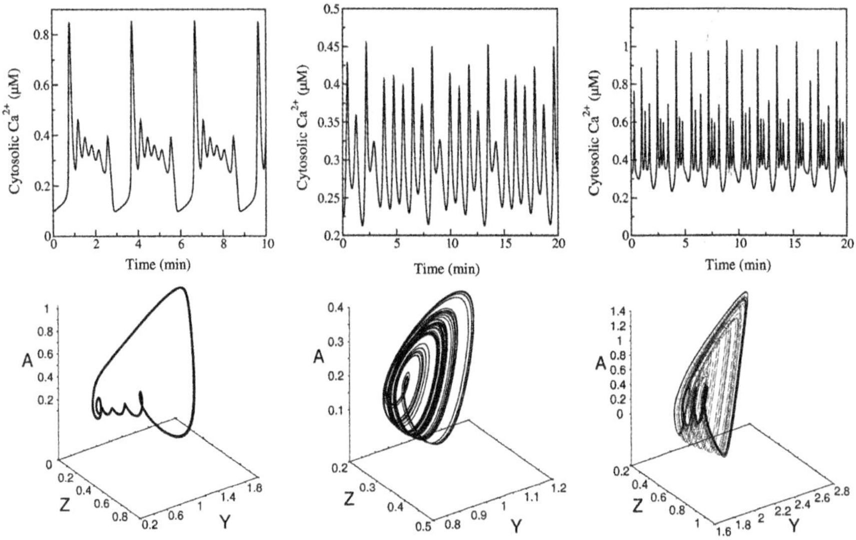

Fig. 7. Various types of complex Ca^{2+} oscillations that can be obtained in the model based on the interplay between CICR and the Ca^{2+}-stimulated degradation of $InsP_3$. From left to right, these complex bahaviors correspond to bursting, chaos and chaotic bursting. The panels in the top row show the time evolution of cytosolic Ca^{2+} concentration (Z) while the panels in the bottom row show the corresponding attractors in the phase space. Results have been obtained by numerical integration of the model defined by (1)-(3), (7)-(9) for three different sets of parameter values. The left and median curves are redrawn from Fig. 4 (a) and (b) of Houart *et al.* (1999). For the rightmost figures, parameter values are the same as for Fig. 4 (b) of Houart *et al.* (1999), except for $\beta = 0.5$, $K_A = 0.12$ μM, $K_d = 0.8$ μM, $k_f = 1.9$ min^{-1}, $\varepsilon = 0.1$ min^{-1}, m=4, $v_1 = 4$ μM min^{-1}, $V_{M2} = 8$ μM min^{-1}, $V_4 = 6$ μM min^{-1}, $V_{M5} = 100$ μM min^{-1}

As illustrated in the middle column of Fig. 7, aperiodic, chaotic oscillations are of reduced amplitude, as compared to the case of bursting. The irregularity of the oscillations shows up both in the amplitude and in the time interval between successive Ca^{2+} spikes.

The last rightmost column of Fig. 7 illustrates another type of dynamical behavior known as "chaotic bursting". This behavior is characterized by an irregular number of secondary spikes, as well as by an irregular spacing between the main spikes. These oscillations much resemble Ca^{2+} oscillations sometimes observed in hepatocytes (Dixon *et al.*, 1993 and 1995; Rooney *et al.*, 1996; Green *et al.*,1997). Similar chaotic bursting was first obtained theoretically in another model for complex Ca^{2+} oscillations based on the regulation of G_α subunits in the signal transduction cascade (Kummer *et al.*, 2000).

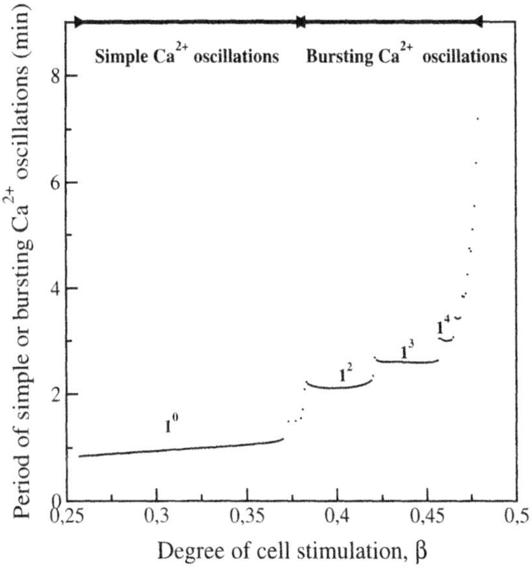

Fig. 8. Relationship between the period of Ca^{2+} oscillations and the degree of stimulation for periodic oscillations of the bursting type. The number of spikes preceding the plateau is always equal to 1 and the number of smaller amplitude spikes in the plateau phase is indicated as an exponent. The period increases in a stepwise manner with β, because the number of spikes in the plateau phase increases at each step. For high values of β, complex oscillations disappear abruptly, and the steady state becomes stable again. When complex oscillations disappear, the period of oscillations tends to infinity, a feature characterisic of a homoclinic bifurcation. Data have been obtained with the model defined by (1)-(3), (7)-(9) (redrawn form Fig. 7 in Houart *et al.*, 1999).

Other types of dynamical behavior can also be obtained in the model of Fig. 5. Among these are birhythmicity, i.e. the coexistence of two stable limit cycles for the same values of the parameters, and quasi-periodicity, i.e. an oscillatory behavior characterized by the existence of multiple frequencies. In the latter situation, the time series greatly resembles those obtained for chaos.

As to the transition between the various modes of oscillatory behavior, the model predicts that, for appropriate parameter values, bursting can be observed when increasing the level of stimulation of a cell exhibiting simple Ca^{2+} oscillations (Houart *et al.*, 1999). As shown in Fig. 8, the stable limit cycle corresponding to simple Ca^{2+} oscillations loses its stability and a sequence of bursting states occurs as the stimulation level increases. Noticeable is the fact that the period increases in a stepwise manner with ß, because the number of spikes in the plateau increases at each step. Such bursting Ca^{2+} oscillations might thus allow the cell to maintain relatively high concentrations of free Ca^{2+} longer during spiking, and thereby to activate slower Ca^{2+}-dependent processes. The possible physiological significance of oscillations of the bursting type is supported by the fact that such oscillations are reminiscent of those seen in electrically excitable cells, in which complex Ca^{2+} oscillations arise from the interplay between a plasma-membrane oscillator and the $InsP_3R$ (Li *et al.*, 1994, Falcke *et al.*, 2000a): in such cells, it is known that cellular processes are activated differently by Ca^{2+} spiking or bursting. Thus, in pancreatic ß cells, granular exocytosis is optimized by long-duration Ca^{2+} bursting (Rorsman and Trube, 1986; Pipeleers, 1987).

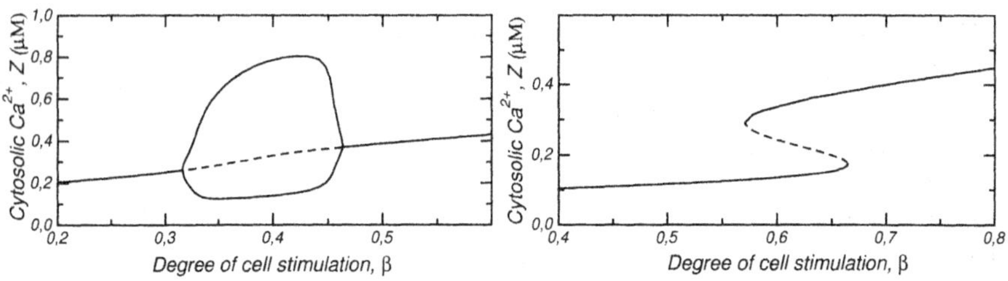

Fig. 9. Bifurcation diagram showing the steady-state values of cytosolic Ca^{2+} as a function of the degree of stimulation (β), for two different values of the constant characterizing capacitative Ca^{2+} entry. Solid lines represent the stable steady state, or the extrema reached during the oscillations when the steady state is unstable. Dashed lines represent unstable steady states. If K_{CCE} is large, e.g. K_{CCE} = 10 µM (A), capacitative calcium entry plays a minor role and oscillations are observed in a limited domain of β values. For smaller values of K_{CCE}, e.g. K_{CCE} = 1 µM (B), the effect of capacitative calcium entry becomes significant and bistability may occur. Then, over a limited range of β values, two stable steady states of cytosolic Ca^{2+} concentration can be obtained for one value of β. The stable steady states are separated by an unstable state. Results have been obtained for the model defined by (3), (4), (9) and (10) and the following parameter values: $K_2 = 0.1$ µM, $K_Y = 0.2$ µM, $K_Z = 0.5$ µM, $k = 10$ min^{-1}, $k_f = 1$ min^{-1}, $v_0 = 2$ µM min^{-1}, $v_1 = 4$ µM min^{-1}, $V_{M2} = 6$ µM min^{-1}, $V_{M3} = 40$ µM min^{-1}.

8.4. Link Between Ca^{2+} Oscillations and Bistability

Besides sustained oscillations, bistability represents another, closely related mode of non-linear behavior. Here we wish to illustrate the link between oscillations and bistability in one particular model for Ca^{2+} signaling, incorporating capacitative Ca^{2+} entry (CCE). One way to represent CCE is to include a dependence of cytosolic Ca^{2+} entry on the content of intracellular stores. This view holds with the fact that the depletion of intracellular stores triggers their refilling directly from the extracellular medium (Berridge, 1995; Putney, 2001). The model incorporating CCE is governed by (10) and (11), in which V_2 and V_3 remain given by (3) and (4):

$$\frac{dZ}{dt} = (v_0 + v_1\beta)\frac{K_{CCE}}{K_{CCE} + Y} - V_2 + V_3 + k_f Y - kZ \qquad (10)$$

$$\frac{dY}{dt} = V_2 - V_3 - k_f Y \quad (11)$$

The first term in the kinetic equation for Z (10) incorporates the effect of CCE: the influx of Ca^{2+} into the cytosol increases as the Ca^{2+} content of intracellular stores (Y) decreases. The constant K_{CCE} measures the importance of CCE: the lower this constant, the more significant the activation of Ca^{2+} entry by the depletion of the stores.

This model predicts the occurrence of either oscillations (Fig. 9A) or bistability (Fig. 9B), depending on the value of K_{CCE}. The largest value (panel A) corresponds to the situation analyzed previously (see Sect. 2.1). When the effect of CCE becomes significant, bistability may occur. Two different, stable steady-state levels of cytosolic (and intravesicular) Ca^{2+} may then coexist, in a range of the degree of stimulation, β (Fig. 9B). In such conditions, as illustrated in Fig. 10, the system evolves towards either one of the two stable steady states, depending on the initial value of Ca^{2+} concentration in the cytosol and in the pool.

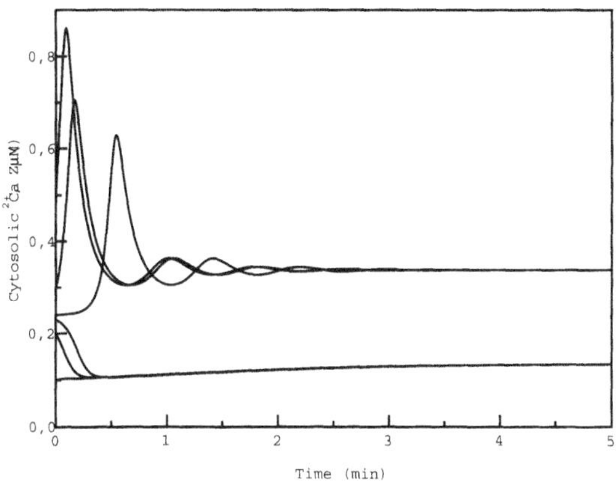

Fig. 10. Evolution towards either one of the two stable steady states in the model incorporating capacitative Ca^{2+} entry. The curves represent the evolutions from different initial values of the cytosolic Ca^{2+} concentration, Z. When the initial value of Z is above the threshold value Z = 0.239569 μM, the system evolves to a high steady-state level of cytosolic Ca^{2+}. When the initial value of Z is below this threshold, a low steady-state level of Z is reached. The initial value for intravesicular Ca^{2+}, Y, is the same for all curves and equal to 1 μM. Results have been obtained by numerical integration of (3), (4), (10) and (11) for the same parameter values as in Fig.9B, with β=0.6.

Bistability is not confined to the model for Ca^{2+} signaling incorporating CCE. In fact, the phenomenon has been observed, in an appropriate parameter range, in nearly all models proposed for Ca^{2+} signaling (Houart, 2001), including the models proposed by Meyer and Stryer (1988) , De Young and Keizer (1992), Atri *et al.* (1993) and Dupont and Swillens (1996). Bistability also plays a key role in a model for Ca^{2+} wave propagation triggered by fertilization in *Xenopus* eggs (Wagner *et al.*, 1998). Furthermore, bistability may have an important role to explain the wave instabilities observed in *Xenopus* eggs, as well as the various characteristics of the Ca^{2+} wave front (Falcke *et al.*, 1999; 2000b).

8.5. Concluding Remarks

In this chapter we presented a brief overview of the regulatory mechanisms involved in the control of Ca^{2+} signaling within the cell. We showed how models based on these mechanisms are used to account for the origin of simple or complex Ca^{2+} oscillations. Spiking relies on the activation of Ca^{2+} release by the signal-induced rise in $InsP_3$ and on Ca^{2+}-induced Ca^{2+} release (CICR). These models mainly differ by the degree of detail used for describing the kinetics of the $InsP_3$ receptor.

Models indicate that complex oscillatory phenomena such as bursting, chaos, quasi-periodicity or birhythmicity may originate from the interplay between several of the regulatory mechanisms that control intracellular Ca^{2+} dynamics. Theoretical models further predict that bistability may also occur in conditions closely related to those that give rise to Ca^{2+} oscillations.

While further experimental studies are needed to assess the physiological significance of bistability in Ca^{2+} signaling, the significance of Ca^{2+} oscillations is related to their frequency encoding, as exemplified by the link between the frequency of Ca^{2+} oscillations and responses such as secretion (Holl *et al.*, 1988), gene expression (Dolmetsch *et al.*, 1998), differentiation (Gu and Spitzer, 1995), and CaM kinase II activation (De Koninck and Schulman, 1998). Here also models prove useful in quantifying the possibility of encoding Ca^{2+} oscillations through their effect on CaM kinase II activation (Dupont *et al.*, submitted for publication) or, more generally, on protein phosphorylation by a Ca^{2+}-activated protein kinase (Goldbeter *et al.*, 1990; Dupont and Goldbeter, 1992; Tang and Othmer, 1995; Gall *et al.*, 2000).

In regard to the spatial propagation of Ca^{2+} signals within the cell or between cells, models indicate that Ca^{2+} oscillations and waves are closely related phenomena. The shape and initiation sites of Ca^{2+} waves are modulated by the many processes (Table 1) that regulate Ca^{2+} oscillations (Wagner *et al.*, 1998; Dumollard *et al.*, 2002). Thus, nearly two decades after the first direct observation of Ca^{2+} oscillations, the spatio-temporal organization of Ca^{2+} signals remains a topic of choice for a joint experimental and a theoretical approach.

Acknowledgement

G.D. is Chercheur Qualifié du Fonds Nationalde la Recherche Scientifique (Belgium).

References

Atri A., Amundson J., Clapham D., Sneyd J. (1993) A single-pool model for intracellular calcium oscillations and waves in the Xenopus laevis oocyte. Biophys. J. 65: 1727-1739.

Berridge M.J., Galione A. (1988) Cytosolic calcium oscillators. FASEB J. 2: 3074-3082.

Berridge M.J. (1993) Inositol trisphosphate and calcium signalling. Nature 361: 315-325.

Berridge M. (1995) Capacitative calcium entry. Biochem. J. 312: 1-11.

Bezprozvanny I. , Ehrlich B. (1995) The inositol 1,4,5-trisphosphate receptor. J. Membr. Biol. 145: 205-216.

Borghans J., Dupont G. , Goldbeter A. (1997) Complex intracellular calcium oscillations. A theoretical exploration of possible mechanisms. Biophys. Chem. 66: 25-41.

Chay T., Fan Y., Lee Y. (1995) Bursting, spiking, chaos, fractals and universality in biological rhythms. Int. J. Bifurcation Chaos 5: 595-635.

Combettes L., Cheek T., Taylor C. (1996) Regulation of inositol trisphosphate receptors by luminal Ca^{2+} contributes to quantal Ca^{2+} mobilization. EMBO J. 15: 2086-2093.

Cuthbertson K., Chay T. (1991) Modelling receptor-controlled intracellular calcium oscillators. Cell Calcium 12: 97-109.

De Koninck P. , Schulman H. (1998) Sensitivity of CaM kinase II to the frequency of Ca^{2+} oscillations. Science 279: 227-230.

De Young G., Keizer J. (1992) A single-pool inositol 1,4,5-trisphosphate-receptor-based model for agonist-stimulated oscillations in Ca^{2+} concentration. Proc. Natl. Acad. Sci. USA 89, 9895-9899.

Dixon C., Cobbold P., Green A. (1993) Adenosine 5'-[αβ-methylene]triphosphate potentiates the oscillatory cytosolic Ca^{2+} responses of hepatocytes to ATP, but not to ADP. Biochem. J. 293: 757-760.

Dixon C., Cobbold P., Green A. (1995) Oscillations in cytosolic free Ca^{2+} induced by ADP and ATP in single rat hepatocytes display differential sensitivity to application of phorbol ester. Biochem. J. 309: 145-149.

Dolmetsch R., Xu K., Lewis R. (1998) Calcium oscillations increase the efficiency and specificity of gene expression. Nature 392: 933-938.

Dumollard R., Carroll J., Dupont G., Sardet C. (2002) Calcium wave pacemakers in eggs. J. Cell Science 115: 3557-3564.

Dupont G., Goldbeter A. (1989) Theoretical insights into the origin of signal-induced Ca^{2+} oscillations. In: Goldbeter A. (ed) Cell to Cell Signaling: From Experiments to Theoretical Models. Academic Press, London, pp. 461-474.

Dupont G., Goldbeter A. (1992) Protein phosphorylation driven by intracellular Ca^{2+} oscillations: A kinetic analysis. Biophys. Chem. 42: 257-270.

Dupont G., Goldbeter A. (1993) One-pool model for Ca^{2+} oscillations involving Ca^{2+} and inositol 1,4,5-trisphosphate as co-agonists for Ca^{2+} release. Cell Calcium 14: 311-322.

Dupont G., Swillens S. (1996) Quantal release, incremental detection, and long-period Ca^{2+} oscillations in a model based on regulatory Ca^{2+}-binding sites along the permeation pathway. Biophys. J. 71: 1714-1722.

Dupont G. (1999) Spatio-temporal organization of cytosolic Ca^{2+} signals: from experimental to theoretical aspects. Comments Theor. Biol. : 305-340.

Endo M., Tanaka M., Ogawa Y. (1970) Calcium-induced release of calcium from the sarcoplasmic reticulum of skinned skeletal muscle cells. Nature 228: 34-36.

Fabiato A., Fabiato F. (1975) Contractions induced by a calcium-triggered release of calcium from the sarcoplasmic reticulum of single skinned cardiac cells. J. Physiol. (London) 249: 469-495.

Falcke M., Hudson J., Camacho P., Lechleiter J. (1999) Impact of mitochondrial Ca^{2+} cycling on pattern formation and stability. Biophys. J. 77: 37-44.

Falcke M., Huerta R., Rabinovitch M., Abarbanel H., Elson R., Selverston A. (2000a) Modeling observed chaotic oscillations in bursting neurons: the role of calcium dynamics and IP_3. Biol. Cybern. 82: 517-527.

Falcke M., Or-Guil M., Markus B. (2000b) Dispersion gap and localized spiral waves in a model for intracellular Ca^{2+} dynamics. Phys. Rev. Lett. 84: 4753-4756.

Gall D., Baus E., Dupont G. (2000) Activation of the liver glycogen phosphorylase by Ca^{2+} oscillations: a theoretical study. J. Theor. Biol. 207: 445-454.

Gu X., Spitzer N. (1995) Distinct aspects of neuronal differentiation encoded by the frequency of spontaneous Ca^{2+} transients. Nature 375: 784-787.

Goldbeter A., Dupont G., Berridge M.J. (1990) Minimal model for signal-induced Ca^{2+} oscillations and for their frequency encoding through protein phosphorylation. Proc. Natl. Acad. Sci. USA 87: 1461-1465.

Goldbeter A. (1996). Biochemical Oscillations and Cellular rhythms. The Molecular Bases of Periodic and Chaotic Behaviour, Cambridge University Press, Cambridge.

Green A., Cobbold P., Dixon C. (1997) Effects on the heaptocyte $[Ca^{2+}]_i$ oscillator of inhibition of the plasma membrane Ca^{2+} pump by carboxyeosin or glucagon-(19-29) Cell Calcium 22: 99-109.

Gu, X. & Spitzer, N.C. (1995) Distinct aspects of neuronal differentiation encoded by frequency of spontaneous Ca^{2+} transients. Nature 375: 784-787.

Haberichter T., Roux E., Marhl M., Mazat J.-P. (2002) The influence of different $InsP_3$ receptor isoforms on Ca^{2+} signaling in tracheal smooth muscle cells. Bioelectrochemistry 57: 129-138.

Hajnoczky G., Robb-Gaspers L., Seitz M., Thomas A. (1995) Decoding of cytosolic calcium oscillations in the mitochondria. Cell 82: 415-424.

Hajnoczky G., Thomas A. (1997) Minimal requirements for calcium oscillations driven by the $InsP_3$ receptor. EMBO J. 16: 3533-3543.

Hirose K., Kadowski S., Tanabe M., Takeshima H., Iino M. (1999) Spatiotemporal dynamics of inositol 1,4,5-trisphosphate that underlies complex Ca^{2+} mobilization patterns. Science 284: 1527-1530.

Holl R., Thorner M., Mandell G., Sullivan J., Sinha Y. & Leong D. (1988) Spontaneous oscillations of intracellular calcium, growth hormone secretion. J. Biol. Chem. 263: 9682-9685.

Houart G., Dupont G., Goldbeter A. (1999) Bursting, chaos and birhythmicity originating from self-modulation of the inositol 1,4,5-trisphosphate signal in a model for intracellular Ca^{2+} oscillations. Bull. Math. Biol. 61: 507-530.

Houart G. (2001) Modélisation de la dynamique du Ca^{2+} intracellulaire: oscillations complexes, bistabilité, et codage par fréquence des signaux calciques par la CaM kinase. Ph.D. thesis, Brussels.

Ichas F., Jouaville L., Mazat J.-P. (1997) Mitochondria are excitable organelles capable of generating and conveying electrical and calcium signals. Cell 89: 1145-1153.

Jacob R. (1990) Imaging cytoplasmic free calcium in histamine stimulated endothelial cells and in fMet-Leu-Phe stimulated neutrophils. Cell Calcium 11: 241-249.

Jouaville L., Ichas F., Holmuhamedov E., Camacho P., Lechleiter J. (1995) Synchronisation of calcium waves by mitochondrial substrates in *Xenopus* laevis oocytes. Nature 377: 438-441.

Kawabata S., Tsutsumi R., Kohara A., Yamaguchi T., Nakanishi S., Okada M. (1996) Control of calcium oscillations by phosphorylation of metabotropic glutamate receptors. Nature 383: 89-92.

Kuba K., Takeshita S. (1981) Simulation of intracellular Ca^{2+} oscillation in a sympathetic neurone. J. Theor. Biol. 93: 1009-1031.
Kummer U., Olsen L., Dixon C., Green A., Bornberg-Bauer E., Baier G. (2000) Switching from simple to complex oscillations in calcium signalling. Biophys. J. 79: 1188-1195.
Läer L., Kloppstech M., Schöfl C., Sejnowski T., Brabant G., Prank K. (2001) Noise enhanced hormonal signal transduction through intracellular calcium oscillations. Biophys. Chem. 91: 157-166.
LeBeau A., Yule D., Groblewski G., Sneyd J. (1999) Agonist-dependent phosphorylation of the inositol 1,4,5-trisphosphate receptor: a possible mechanism for agonist-specific calcium oscillations in pancreatic acinar cells. J. Gen. Physiol. 113: 851-872.
Lewis R. (2001) Calcium signaling mechanisms in T lymphocytes. Annu. Rev. Immunol. 19: 497-521.
Li Y.X., Rinzel J., Keizer J., Stojilkovic S. (1994) Calcium oscillations in pituitary gonadotrophs: comparison of experiments and theory. Proc. Natl. Acad. Sci. USA 91: 58-62.
Llopis W.-H., Whitney M., Zlokarnik G., Tsien R. (1998) Cell-permeant caged $InsP_3$ ester shows that Ca^{2+} spike frequency can optimize gene expression. Nature 392: 936-941.
Marhl M., Haberichter T., Brumen M., Heinrich R. (2000) Complex calcium oscillations and the role of mitochondria and cytosolic proteins. Biosystems 57: 75-86.
Meyer T., Stryer L. (1988) Molecular model for receptor-stimulated calcium spiking. Proc. Natl. Acad. Sci. USA 85: 5051-5055.
Meyer T., Stryer L. (1991) Calcium spiking. Ann. Rev. Biophys. Chem. 20: 153-174.
Minorsky N. (1967) Nonlinear Oscillations ,Van Nostrand Reinhold, Princeton, N.J.
Missiaen L., De Smedt H., Parys J., Sienaert I., Vanlingen S., Casteels R. (1996) Characterization of a cytosolic and a luminal Ca^{2+} binding site in the type I inositol 1,4,5-trisphosphate receptor. J. Biol. Chem. 271: 27005-27012.
Miyakawa T., Mizushima A., Hirose K., Yamazawa T., Bezprozvanny I., Kurosaki T., Iino M. (2001) Ca^{2+}-sensor region of $InsP_3$ receptor controls intracellular Ca^{2+} signaling. EMBO J. 20: 1674-1680.
Nash M., Young K., Challiss R., Nahorski S. (2001a) Single-cell imaging of graded Ins(1,4,5)P_3 production following G-protein-coupled-receptor activation. Biochem. J. 356: 137-142.
Nash M., Young K., Challiss R., Nahorski S. (2001b) Receptor-specific messenger oscillations. Nature 413: 381-382.
Petersen O., Tepikin A., Park M. (2001) The endoplasmic reticulum: one continuum or several separate Ca^{2+} stores? Trends Neurosci. 24: 271-276.
Pipeleers D. (1987) The biosociology of pancreatic β cells. Diabetologia 30: 277-291.
Putney J., Broad L., Braun F., Lievremont J., Bird G. (2001) Mechanisms of capacitative calcium entry. J. Cell Science 114: 2223-2229.
Robb-Gaspers L., Thomas A. (1995) Coordination of Ca^{2+} signaling by intercellular propagation of Ca^{2+} waves in the intact liver. J. Biol. Chem. 270: 8102-8107.
Robb-Gaspers L., Rutter G., Burnett P., Hajnoczky G., Druton R., Thomas A. (1998) Coupling between cytosolic and mitochondrial calcium oscillations: role in the regulation of hepatic metabolism. Biochim. Biophys. Acta 1366: 17-32.
Rooney T., Joseph S., Queen C., Thomas A. (1996) Cyclic GMP induces oscillatory calcium signals in rat hepatocytes. J. Biol. Chem. 271: 19817-19825.
Rorsman P., Trube G. (1986) Calcium and delayed potassium currents in mouse pancreatic β cells under voltage-clamp conditions. J. Physiol. (London) 374: 531-550.
Schuster S., Marhl M., Höfer T. (2002) Modelling of simple and complex calcium oscillations. From single cell responses to intercellular signalling. Eur. J. Biochem. 269: 1333-1355.

Sanderson M., Charles A., Boitano S., Dirksen E. (1994) Mechanisms and function of intercellular calcium signaling. Mol. and Cell. Endocrinol. 98: 173-187.

Shen P., Larter R. (1995) Chaos in intracellular Ca^{2+} oscillations in a new model for non-excitable cells. Cell Calcium 17: 225-232.

Sims C., Allbritton N. (1998) Metabolism of inositol 1,4,5-trisphosphate and inositol 1,3,4,5-tetrakisphosphate by the oocytes of *Xenopus laevis*. J. Biol. Chem. 273: 4052-4058.

Sneyd J., Dufour J.-F. (2002) A dynamic model of the type-2 inositol trisphosphate receptor. Proc. Natl. Acad. Sci. 19: 2398-2403.

Sneyd J., Keizer J., Sanderson M. (1995) Mechanisms of calcium oscillations and waves: a quantitative analysis. FASEB J. 9: 1463-1472.

Takazawa K., Lemos M., Delvaux A., Lejeune C., Dumont J.E., Erneux C. (1990) Rat brain inositol 1,4,5-trisphosphate 3-kinase. Ca^{2+}-sensitivity, purification and antibody production. Biochem. J. 268: 213-217.

Tang Y., Othmer H. (1995) Frequency encoding in excitable systems with applications to calcium oscillations. Proc. Natl. Acad. Sci. 92: 7869-7873.

Tang Y., Stephenson J., Othmer H. (1996) Simplification and analysis of models of calcium dynamics based on IP_3-sensitive calcium channel kinetics. Biophys. J. 70: 246-263.

Taylor C., Genazzani A., Morris S. (1999) Expression of inositol trisphosphate receptors. Cell Calcium 26: 237-251.

Taylor C., Thorn P. (2001) Calcium signalling: $InsP_3$ rises again…and again. Current Biol. 11: R352-R355.

Thrower E., Hagar R., Ehrlich B. (2001) Regulation of $Ins(1,4,5)P_3$ receptor isoforms by endogenous modulators. TiPS 22: 580-586.

Wagner J., Li Y-X, Pearson J., Keizer J. (1998). Simulation of the fertilization Ca^{2+} wave in *Xenopus laevis* eggs. Biophys. J. 75: 2088-2097.

Wakui M., Osipchuk Y., Petersen O. (1990) Receptor-activated cytoplasmic Ca^{2+} spiking mediated by inositol trisphosphate is due to Ca^{2+}-induced Ca^{2+} release. Cell 63: 1025-1032.

Wiesner T., Berk B., Nerem R. (1996) A mathematical model of cytosolic calcium dynamics in human umbilical vein endothelial cells. Am. J. Physiol. 270: C1556-C1569.

Woods N., Cuthbertson K., Cobbold P. (1986) Agonist-induced oscillations in cytoplasmic free calcium concentration in single rat hepatocytes. Cell Calcium 8: 79-100.

9. Information Processing in Calcium Signal Transduction

Ann Zahle Larsen[1] and Ursula Kummer[2]

[1]Celcom, Department of Biochemistry and Molecular Biology, University of Southern Denmark, Campusvej 55, DK-5230 Odense M, Denmark. [2]European Media Laboratory, Schloss-Wolfsbrunnenweg 33, D-69118 Heidelberg, Germany.

Abstract. Calcium signalling often employs calcium oscillations. It is by now a well-known fact that information, e.g. about the amount of stimulus that evoked a calcium response is encoded in the oscillations. Information can be stored in the frequency, amplitude and shape of the oscillations. How this information is decoded by the many enzymes which are regulated by calcium is not thoroughly understood and so far only few experiments and theoretical studies have been made in order to elucidate the responsible mechanisms. We present both experimental and theoretical studies of calcium signal encoding and decoding in non-excitable cells, and try to illuminate the mechanisms by which effective and versatile information processing is made possible.

9.1. Introduction

Calcium signalling occurs in many different cell types and carries information from agonists or action potentials from the extracellular side of the cell membrane to targets in the interior of the cell. The information carried serves many different purposes from triggering the developmental program of fertilized mammalian eggs to mobilizing antigenic response of immune cells and mediating cell death (Berridge et al 1998; Clapham 1995).

The focus of this chapter is the processing of signals induced by agonist binding to specific membrane spanning receptor enzymes in the membrane of non-excitable cells such as the liver cell (or hepatocyte). The binding of an agonist induces a dynamic calcium signal, which carries information to target proteins in the cell. This transfer of information from agonist to signal is referred to as encoding whereas decoding refers to the transfer of encoded information from the calcium signal to a target response. Typical of calcium signalling in non-excitable cells is that the calcium concentration in the cytosol displays sustained oscillations upon non-oscillatory agonist influence. The oscillations show different frequencies, amplitudes and waveforms, depending on type and amount of agonist bound. It is generally believed that the information carried by a specific agonist is stored in the oscillation patterns (Nash et al. 2001). How this information is decoded by the many enzymes which are regulated by calcium is not thoroughly understood and so far only few experiments and theoretical studies have been made in order to elucidate the responsible mechanisms. We present both experimental and theoretical studies of calcium signal encoding and decoding in non-excitable cells, and try to illuminate the mechanisms by which effective and versatile information processing is made possible.

9.2 Dynamic Calcium Signal Encoding; Mechanisms and Models

9.2.1 From Receptors to Signal Targets – the Physiology of Signal Encoding

In the following, we will restrict our discussion to the cell type hepatocyte, since this is one of the best-studied non-excitable cell systems. Protein and peptide hormones, catecholamines like epinephrine, and eicosanoids such as prostaglandines find their receptors on the plasma membrane of target cells. Only a few hormone types are able to enter the cell and thus carrying the information by themselves to molecular targets inside the cell. All other hormones and information carriers in the blood stream need a second messenger within the cell. A surprising variety of these use calcium as their second messenger. The agonists evoking a calcium response in hepatocytes range from hormones such as vasopressin and glucagon (for a review see Clapham 1995) to nucleotides like ATP or UTP (for a review see Dubyak et al. 1993).

The whole process, from agonist binding to target response, is very complex and any existing model only represents a highly simplified picture of what it really going on during agonist stimulation. To give a feeling for this, we would like to give some insight into the complexity of these processes (presented schematically in Fig. 1).

All agonists using calcium as their second messenger bind to their respective specific plasma membrane receptors. These receptors share some common features: They possess an extracellular domain, which binds to their agonist. Transmembrane domains anchor the receptor safely in the membrane and the intracellular domain couples the receptor commonly to G-Proteins (for a review see Bockaert et al. 2002) or display tyrosine kinase activity by itself (Clapham 1995). The receptor molecules are neither isolated by themselves, nor do they possess a fixed location in the membrane. In many cases, other membrane proteins (incl. receptors) or molecules are able to interact with the receptor modulating its activity. Furthermore, some receptors cluster together in the membrane after binding hormone thus facilitating further binding. The impact of agonist binding to the receptor is further complicated by the influence of the state of its own effector molecules (Bockaert et al. 2002).

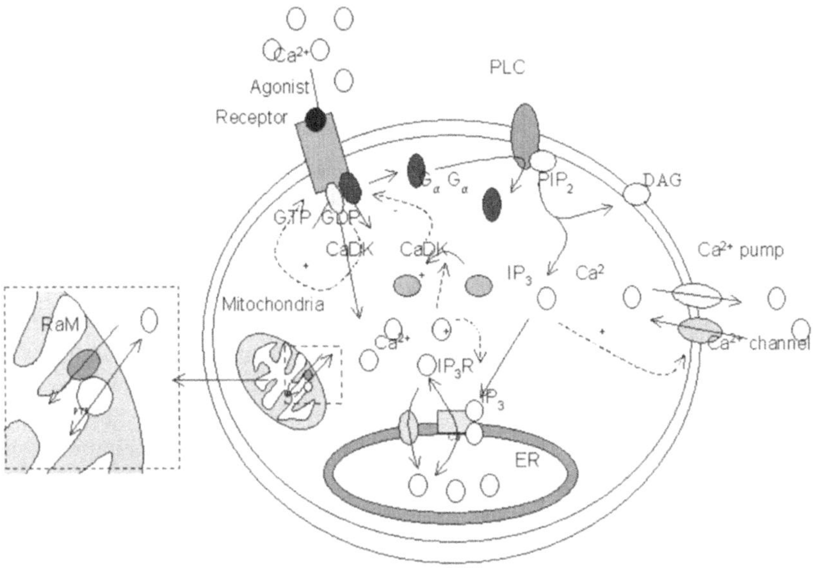

Fig. 1. Schematic diagram of agonist induced calcium signalling via the phosphoinositide pathway in hepatocytes. Only reactions included in the model (Sect. 9.2.3) are presented.

As mentioned above, most receptors transferring agonist signals to the inside of the cell by using calcium are G-protein-coupled. Upon binding of the agonist, the receptor changes conformation such that it is able to bind to the G-protein. G-proteins consist of two different subunits, a G_α-subunit that binds guanine nucleotide and is usually considered to be responsible for activating effector molecules and a $\beta\gamma$-subunit which dissociates from the α-subunit after activation of the G-protein. Again, all of these processes can be regulated at different levels and are therefore highly complex. There are multiple different G_α- and $G_{\beta\gamma}$-subunits in the cell, and those can combine to a huge variety of G-proteins. In the inactive form, the G_α-subunit contains a bound GDP. After binding of the agonist to the receptor and of the receptor to the G-protein, this GDP is released and GTP may bind. This then leads to the dissociation of the two subunits and the thus activated G_α-subunit may do whatever it should do to the respective effector molecule (Clapham 1995). However, this picture is oversimplified. Apart from mediating receptor binding and facilitating the inactivation of the G_α-subunit, the $G_{\beta\gamma}$-subunit is also able to influence and activate its own targets. There exists a family of phospholipase C, which is switched on by these molecules (Sternweis 1994). Similar to the receptor activation, there are also multiple ways in which the cell can interact and fine-tune the activity of the G-proteins. One prominent example here is the existence of a big family of so-called RGS proteins, which stands for „regulators of G protein signalling". These proteins can help shut off the activation of G_α by GTPase action, but can also speed up the activation of signalling (Zerangue et al. 1998). Agonist receptors that do not couple G-proteins, but rather act as tyrosine kinases, activate the effector molecules directly.

A key effector molecule in calcium signalling is phospholipase C (PLC). There are different isoforms of this enzyme, which might be activated by either the G_α-subunit, the $G_{\beta\gamma}$-subunit of the activated G-protein or the tyrosine kinase. PLC also controls its own activation by facilitating the GTP-hydrolysis in the G_α-subunit (Bourne et al. 1992). PLC is responsible for splitting inositole-3-phosphate (IP_3) from phospholipids in the membrane and thereby liberating the diffusive messenger. Different compounds regulate the activity of the different PLC isoforms, e.g. phosphokinase C (Fain 1989).

IP_3 diffuses through the cell and binds to IP_3-receptors at the membrane of the ER. This binding leads to the opening of the associated calcium channels and massive influx of calcium from the ER into the cytosol (Tsien et al. 1990). However, IP_3 is supposed also to be directly or indirectly facilitating calcium influx from extracellular space (Hansen et al. 1991; Striggow et al. 1994). In addition to the action of IP_3, agonist binding to receptors alone can also lead directly to the opening of calcium channels in the plasma membrane. These are called receptor-operated channels (Llopis et al. 1992).

The dominant calcium channel in the membrane of the ER (the IP_3-receptor channel) is highly regulated. Not only does IP_3-binding open it, but additional calcium binding activates the channel further (for details see Chap.1). This fact – calcium facilitating its own liberation – has been called calcium-induced-calcium release (CICR) and represents a classical example for biochemical autocatalysis.

Calcium in the cell will influence diverse enzymes in their activity as well as gene expression, exocytosis, cell development etc. Some dominant calcium binding targets are protein kinases and ATPases (Celio et al. 1996). In many cases, the impact of calcium is mediated by calmodulin. This protein is activated by binding up to four calcium ions in a cooperative way and then activates e.g. protein kinases, phosphatases, ion transporters and cytoskeletal proteins (Celio et al. 1996).

9.2.2. Experimentally Observed Signals and Their General Function

Experimentally observed calcium signals show a surprising variety of dynamics. In some cases, the meaning of the dynamics is known, but in many cases, it is still a point of speculation.

As long as activation of the receptor and release of IP_3 does not exceed a certain threshold, the calcium level in the cytosol remains basically at a low constant value of about 100-300 nM (Woods et al. 1986). This mechanism is called understimulation and ensures that only signals above a certain strength set the signalling machinery to work.

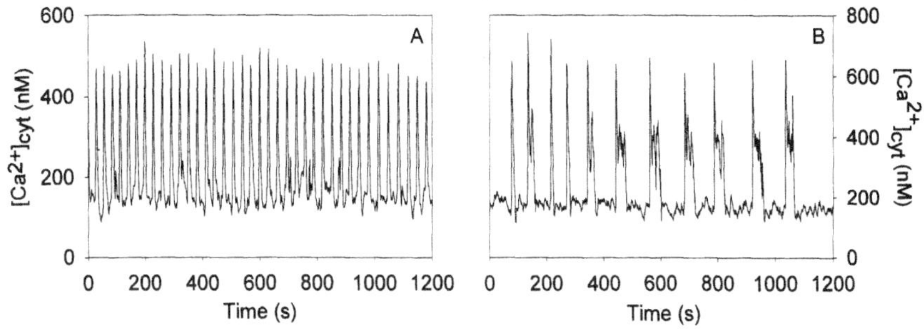

Fig. 2. Experimentally observed calcium oscillations in hepatocytes in response to 1.2 μM ATP (A) and 2 μM phenylephrine (B) (data kindly provided by Anne K. Green and C. Jane Dixon).

If the stimulation exceeds the threshold, usually oscillations are observed (again, we are restricting our discussion here to the hepatocyte). Depending on the kind of agonist, these oscillations are in general either simple periodic oscillations or so-called bursting oscillations, which can vary quite a lot in frequency and waveform. Agonists that evoke simple periodic oscillations are e.g. vasopressin and phenylephrine (Woods et al. 1986) and among the agonists evoking bursting one finds ATP and UTP (Dixon et al. 1990; Dixon et al. 2000). See Fig. 2 for an example. In both cases, however, the frequency of the oscillations reflects the amount of agonist used. Thus, within a certain range of oscillations, the higher the amount of agonist, the higher the frequency of the oscillations (Woods et al. 1986; Rooney et al. 1989). This finding is reflected in the term "frequency-coding". Combinations of agonist usually evoke a very complex calcium response, which seems to be a mixture of the individually observed calcium traces (Green et al. 1994). However, there is not too much data available on this topic. Even though much remains to be clarified concerning this, it seems to be obvious that the information about the amount and type of agonist is encoded in the frequency and shape of the calcium oscillations.

Raising the agonist concentration above a second threshold generally results in a non-oscillatory response meaning a constant elevated calcium concentration. The message of this phenomenon, called overstimulation seems to be apoptosis (Berridge et al. 1998). And not only hepatocytes, but also other cell types respond accordingly to this signal. However, this is only true, if the calcium concentration stays elevated for some time.

9.2.3. Computational Models – Different Types Answering Different Questions

Different kinds of computational models have evolved in the last 15 years and the question; why there is not a single valid one, might arise easily. The answer to that is that first of all; different cell types ask for different models since different details in calcium signal transduction will be present. However, apart from this obvious point, there is also the simple explanation that due to the above described complexity of the system, which involves a large number of unknown parameters, there is always the need to abstract some details and to focus on a subpart of the whole reaction network. Since a detailed description of models is given elsewhere in this book, we will only recapture some essential facts.

Many mathematical models focus on the explanation and study of oscillatory dynamics of calcium signals (for a review see Schuster et al. 2002). Due to some experimental evidence that oscillations could occur without an oscillating IP_3-concentration (Wakui et al. 1989) and due to the fact that the autocatalysis present with the CICR is sufficient to generate an oscillatory signal, most models have focused on the description of the events occurring at the calcium channel at the endoplasmic reticulum (ER) (see Chap.8). These models are summarized with the term "store-operated". They have been able to reproduce successfully many of the essential features of simple calcium spiking. In addition, some of the models are able to show complex oscillations, such as bursting. Store-operated models helped to clarify the mechanism of what is going on at the intracellular calcium store and are especially useful to study calcium waves, which propagate in the cell or across many cells and which most certainly are carried by the processes of activating and deactivating calcium channels in the ER.

However, leaving out the level of receptor activation in the plasma membrane results in models that cannot account for the different receptor-specific behaviour of calcium dynamics. Therefore, and due to the fact that recently it has been shown that IP_3 does indeed oscillate in the cell (Taylor et al. 2001), other models have evolved which include the receptor level as well. These fall into the category "receptor-operated", even though some of them include mechanisms like the CICR as well. They are able to explain some of the aspects associated with receptor-specific signalling. Both store-operated and receptor-operated models have also been enlarged to encompass calcium release and uptake by the mitochondria unravelling the role of these organelles for e.g. the almost constant amplitude of the calcium signal in the cell (Marhl et al. 1998; Grubelnik et al. 2001).

For the computational study of decoding calcium signals, it is not crucial which of the above models of calcium dynamics is chosen, as long as it is able to show the basic types of dynamics experimentally observed in calcium signal transduction, namely understimulation, spiking (with different frequencies), bursting, complex bursting (nonperiodic) and overstimulation. In the following, we will mainly focus on the usage of a receptor-oriented model, which we developed recently (Kummer et al. 2000; Larsen et al. 2002). This model was described in detail in the respective articles and we will therefore restrict our description to a general overview emphasizing a few important points.

It is important to stress that the model presented here is qualitative. It deals with levels of calcium in the different compartments, not concentrations. Therefore, we do not take into account the sizes of the different compartments. This could easily be done by multiplying the equations of the model with constant volume ratios (Marhl et al. 1998), but it does not change the qualitative behaviour of the model, and does not add to the information the model can provide. For the study of the en- and decoding of calcium oscillations we need a model, which faithfully reproduces the qualitative dynamics which are experimentally observed, but does not need to respond to every possible physiological event in detail.

The equations of the model are:

$$\frac{dG_\alpha}{dt} = k_1 + k_2 * G_\alpha - \frac{k_3 * G_\alpha * PLC}{G_\alpha + K_4} - \frac{k_5 * Ca_{cyt} * G_\alpha}{G_\alpha + K_6} \tag{9.1}$$

$$\frac{dPLC}{dt} = k_7 * G_\alpha - \frac{k_8 * PLC}{PLC + K_9} \tag{9.2}$$

$$\frac{dCa_{cyt}}{dt} = (Ca_{ER} - Ca_{cyt}) * \frac{k_{10} * Ca_{cyt} * PLC^4}{PLC^4 + K_{11}^4} + k_{12} * PLC + k_{13} * G_\alpha - \frac{k_{14} * Ca_{cy}}{Ca_{cyt} + K} - \frac{k_{16} * Ca_{cyt}}{Ca_{cyt} + K_{17}} - \frac{k_{18} * Ca_{cyt}^8}{Ca_{cyt}^8 + K_{19}^8} + (Ca_{mit} - Ca_{cyt}) * \frac{k_{20} * Ca_{cyt}}{Ca_{cyt} + K_{21}} \tag{9.3}$$

$$\frac{dCa_{ER}}{dt} = -(Ca_{ER} - Ca_{cyt}) * \frac{k_{10} * Ca_{cyt} * PLC^4}{PLC^4 + K_{11}^4} + \frac{k_{16} * Ca_{cyt}}{Ca_{cyt} + K_{17}} \tag{9.4}$$

$$\frac{dCa_{mit}}{dt} = \frac{k_{18} * Ca_{cyt}^8}{Ca_{cyt}^8 + K_{19}^8} - (Ca_{mit} - Ca_{cyt}) * \frac{k_{20} * Ca_{cyt}}{Ca_{cyt} + K_{21}} \tag{9.5}$$

The variables of the model are activated G_α subunit of the receptor complex (G_α), active phospholipase C (*PLC*), cytosolic calcium concentration (Ca_{cyt}), concentration of free calcium in the ER (Ca_{ER}) and calcium concentration in the mitochondria (Ca_{mit}). IP_3 is assumed in quasistationary state with *PLC*, and therefore IP_3 is left out as a variable in the model, and PLC is used instead wherever IP_3 is known to influence the dynamics of the system.

Different receptors have been shown to employ different dynamics for calcium signal transduction (Chatton et al. 1997; Green et al. 1999; Kummer et al. 2000; Nash et al. 2001). This is due to different properties of the different receptors as described in sect. 9.2.1. Therefore, we use receptor-specific parameters for G_α activation and receptor properties. These receptor-specific terms are receptor mediated calcium influx ($k_{13}*G_\alpha$) and a receptor-specific G_α activation term k_2*G_α. Both play a crucial role in determining the possible dynamic states of the model. If the terms $k_{13}*G_\alpha$ and k_2*G_α are included, the model is capable of showing a large variety of behaviours including understimulation, simple periodic spiking, periodic bursting, chaotic bursting and overstimulation (Rooney et al. 1991). This corresponds to receptors, which bind e.g. ATP or UTP.

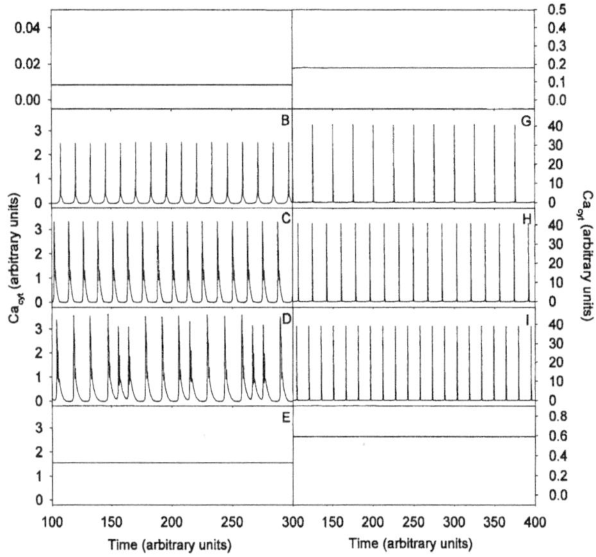

Fig. 3. Dynamics of the calcium model presented in the text. Timeseries A-E are simulations by the bursting model. The constants are; k_1=0.01, k_3=0.64, K_4=0.09, k_5=4.88, K_6=1.18, k_7=2.08, k_8=32.24, K_9=29.09, k_{10}=0.7, K_{11}=3, k_{12}=2.8, k_{13}=13.4, k_{14}=153, K_{15}=0.16, k_{16}=7, K_{17}=0.05, k_{18}=79, K_{19}=3.5, k_{20}=0.81 and K_{21}=4.5. k_2 is: A: k_2=1.03 (understimulation),. B: k_2=1.56 (periodic spiking), C: k_2=2.07 (periodic bursting), D: k_2=2.215 (complex bursting) and E: k_2=3 (overstimulation). Timeseries F-J show the dynamics of the spiking model, here the constants are; k_3=0.0001, K_4=0.783, k_5=1.24, K_6=0.7, k_7=5.82, k_8=32.24, K_9=29.09, k_{10}=0.93, K_{11}=2.667, k_{12}=0.768, k_{14}=149, K_{15}=0.16, k_{16}=20.9, K_{17}=0.05, k_{18}=79, K_{19}=2, k_{20}=1.5 and K_{21}=1.5. Here k1 models agonist influence; F: k_1=0.2 (understimulation), G: k_1=0.3, H: k_1=0.45 and I: k_1=0.55 (periodic spiking with increasing frequency), J: k_1=0.65 (overstimulation).

Omitting the terms and adjusting the other constants results in a system, which is able to show understimulation, periodic spiking and overstimulation, but no bursting behaviour. This corresponds to receptors binding e.g. vasopressin (Kummer et al. 2000). Therefore, in the following, whenever we refer to the "bursting model" we use the model including $k_{13}*G_\alpha$ and k_2*G_α ($k_{13} \neq 0$, $k_2 \neq 0$) and whenever we refer to the "spiking model", it is the model omitting the terms $k_{13}*G_\alpha$ and k_2*G_α (k_{13}=0, k_2=0). In Fig. 3 understimulation, spiking, bursting, complex bursting and overstimulation obtained from the model is presented.

The frequency encoding observed in hepatocytes covers a range of periods from 0.3 to 4 minutes (Gall et al. 2000). The frequency encoding ability of the spiking model is adequate for our purpose. An increase in agonist stimulation by a factor of 1.75 can result in an increase of the frequency by a factor of 1.7.

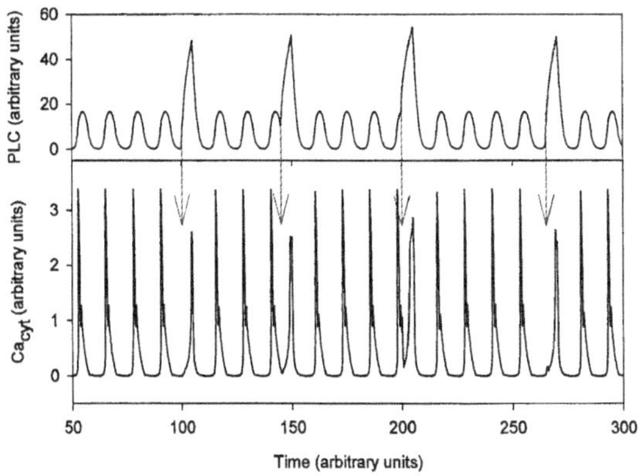

Fig. 4. Timeseries of *PLC* and Ca_{cyt} upon perturbations of PLC. At time =100, 145, 200 and 265 (arrows) square pulses of PLC were imposed on equation 9.2. The rest of the parameters are as in Figs. 9.3A-E. The effect of these pulses mimics the release of caged IP_3 presented in experiments and simulations conducted by Chatton et al. 1998.

In addition to reproducing the dynamics found in experimental literature the model is also very robust toward perturbations. This is a very important feature, which has also been shown experimentally (Chatton et al. 1998). The presented model (both bursting and spiking) is able to mimic the results presented by Chatton and co-workers in 1998 where large imposed pulses of IP_3 were shown not to disrupt calcium oscillation signals, but only reset their phase. In Fig. 4 we impose pulses of *PLC* activity on the calcium model, resulting in large transient increases in the *PLC* level (*PLC* is assumed in quasi-stationary state with IP_3). This pulse of PLC does not destroy or alter the dynamics of the calcium signal, but only resets the phase of the oscillation (Fig. 4).

9.3. Decoding of Signals

9.3.1. Experiments on Decoding Calcium Signals

Experiments on the decoding of oscillating calcium signals are scarce, the shortage of experimental data is mainly a result of the difficulties in measuring changes in activity of calcium regulated target proteins, both *in vitro* and *in vivo*. Most biochemical experiments on calcium dependent enzyme activity have been conducted under steady state conditions, using purified enzyme and constant concentrations of calcium. This type of experiment provides little information on how the target enzyme can translate the oscillating calcium signal, and even less information on how the enzyme is regulated *in vivo*. But within the last 5-10 years the methods have improved and there are now examples of experimental studies on oscillating calcium signal decoding both *in vitro* and *in vivo*.

The studies of decoding of calcium oscillations have exclusively focussed on how the cell decodes simple frequencies of spiking oscillations, despite the fact that the cell can display signals vastly more complex than periodic spiking (Robb-Gaspers et al. 1998; Dolmetsch et al.1998; Li et al 1998; Oancea and Meyer 1998; Tompa et al. 2001; Eshete and Fields 2001). This has several reasons, first of all the importance of information, held in the frequency of periodic spiking oscillations, has been accepted for years. Second, it is a lot more difficult to generate complex signals like bursting by e.g. caged IP_3-release *in vitro*, so that experimental studies are hampered. The above reasons have caused scientists to limit their experimental and theoretical studies to dealing with decoding of periodic spiking. In the following section we will present results from three systems in which decoding of calcium oscillation frequency has been experimentally explored.

Decoding by the Cytosolic CaMKII

An example of an experimentally investigated decoding system is the Ca^{2+}-calmodulin-dependent kinase II (CaMKII) which is activated by calcium through calmodulin and plays a central role in many cellular processes, ranging from controlling enzymes involved in metabolism and gene expression to memory formation in the brain. CaMKII is found in both excitable and non-excitable cells. The presence and use of the enzyme in both cell types imply that the resulting calcium oscillations can be assumed comparable and should be decoded employing the similar decoding mechanisms. CaMKII is activated in the following manner. As Ca^{2+} concentration is transiently raised in the cytosol during Ca^{2+} oscillations, calcium binds the two unoccupied calcium-binding sites of the Ca^{2+} binding protein calmodulin. Binding of Ca^{2+} to calmodulin is fast and cooperative, which leads to a large change in active calmodulin upon a small change in Ca^{2+} concentration. Binding of Ca^{2+} activates calmodulin, which in turn binds to one of the 8-10 subunits of CaMKII. The binding of calmodulin to the CaM kinase subunit increases the chance that the subunit phosphorylates its neighbouring subunit, presupposing that this has also a bound calmodulin. Phosphorylation activates the subunit and promotes binding of more calmodulin to the enzyme, which in turn promotes phosphorylation of the remaining subunits. When all the subunits of the enzyme have bound calmodulin and are phosphorylated, the enzyme is maximally active. Active CaM kinase then phosphorylates target enzymes in the cell.

How does this type of activation mechanism decode sustained calcium oscillations, and what is the advantage of this type of decoding? During successive calcium spikes, the amount of active calmodulin in the cell follows the Ca^{2+} concentration. When calmodulin activity is high it binds subunits of the CaM kinase where it may induce phosphorylation. Phosphorylation traps the calmodulin because of the increased affinity and this causes the calmodulin to stay bound even when the Ca^{2+} level and active calmodulin level decreases again (end of calcium spike). During the next pulse of Ca^{2+}, there is a larger chance that calmodulin binds next to an activated subunit and therefore gets phosphorylated and activated itself. This mechanism allows CaMKII to be fully activated in an additive process, although calcium and therefore calmodulin is present only in short pulses. The activation of CaMKII is dependent of the frequency of the calcium signal. If the frequency is below 10-20 seconds, then more calmodulin has time to dissociate from CaMKII and the probability that it dephosphorylates and is deactivated increases. This means that the autophosphorylation of CaMKII allows the decoding of calcium oscillation frequencies into discrete levels of activation, where higher frequency induces higher activity level. The above mechanism has been known for some time and CaMKII has been the target of investigations and discrete activity measurements *in vitro* (De Koninck and Schulman 1998). However, only recently it has been possible to conduct experiments on calcium decoding by CaMKII *in vivo*. The *in vivo* experiments in Ganglion neurons confirmed that CaM kinase activity is dependent on the frequency of the oscillations and also that oscillating signals *in vivo* evoked a much larger response than a comparable steady state calcium concentration *in vitro* (Eshete and Fields 2001). Although, these results were obtained for excitable cells, due to the similarity of the CaMKII in both cell types, they should also add to our understanding of decoding by CaMKII in non-excitable cells.

Ca^{2+} Dependent Regulation of Intramitochondrial Metabolism

An example of the role of Ca^{2+} signalling in metabolism is the control of the Ca^{2+} sensitive intramitochondrial dehydrogenases. In many tissues calcium signalling influences processes which increase the ATP demand of the cell. In order to maintain energy balance, the rate of ATP utilization must be balanced with ATP production. This means that regulation of ATP utilizing processes must be tightly coupled to ATP producing processes. The major ATP producing process is the oxidative phosphorylation in the mitochondria. Mitochondrial respiration is controlled in numerous ways both up- and downstream of mitochondrial NAD(P)H and it is not within the scope of this text to describe them all, but we focus on one example which is regulated by cytosolic calcium signals, namely the Ca^{2+} sensitive intramitochondrial dehydrogenases (CSMDHs). Intramitochondrial dehydrogenases include pyruvate dehydrogenase (PDH), oxoglutarate dehydrogenase (OGDH) and isocitrate dehydrogenase (ICDH). The effect of calcium on PDH is mediated through activating dephosphorylation by a Ca^{2+} sensitive phosphorylase, whereas Ca^{2+} directly binds OGDH and ICDH and lowers the K_M for their substrates. This activation by Ca^{2+} has the overall effect of speeding up the Krebs cycle leading to more NAD(P)H which can lead to the formation of more ATP through proton gradient driven oxidative phosphorylation (Hajnoczky et al. 1995).

Calcium oscillations are caused by agonist stimulation (as described earlier in the text). In addition to inducing a target response in the cytosol or nucleus, the transient rises in cytosolic calcium concentration cause the mitochondria to take up calcium and release it again when the cytosolic calcium level returns to resting level. This uptake and release from the mitochondria generates mitochondrial calcium oscillations, which follow the cytosolic pattern closely.

Robb-Gaspers et al. (Robb-Gaspers et al. 1998) conducted experiments similar to those of Hajnoczky et al (Hajnoczky et al. 1995), which showed that the mitochondrial dehydrogenases decode these mitochondrial calcium oscillations into a metabolic response. They measured $[Ca^{2+}]_{cyt}$, $[Ca^{2+}]_{mit}$ and [NAD(P)H] in single hepatocytes during exposure to different levels of the α_1-adrenergic hormone phenylephrine. As expected the binding of phenylephrine to membrane receptors induced calcium oscillations in the cytosol, which were closely followed by similar oscillations in the mitochondria. In Fig. 5 we present the $[Ca^{2+}]_{cyt}$ and mitochondrial NAD(P)H timeseries obtained in their experiment (Data kindly provided by Andrew P. Thomas).

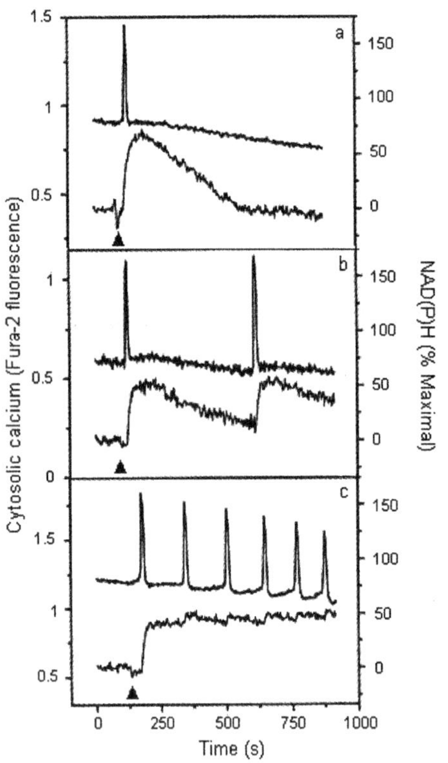

Fig. 5. Dependence of NAD(P)H level in the mitochondria on cytosolic calcium oscillation frequency. NAD(P)H and calcium responses to phenylephrine were monitored simultaneously measuring NAD(P)H and fura2/AM fluorescence (Robb-Gaspers et al. 1998). The top time series in each part of the Figure represents calcium whereas the bottom time series represents NAD(P)H. Due to the slow reoxidation phase of NAD(P)H, a rise in calcium signal frequency is decoded as an increasing average NAD(P)H level. At high frequencies, the oscillating calcium signal is decoded as a sustained rise in NAD(P)H. In this way the Ca^{2+} sensitive intramitochondrial dehydrogenases can translate the oscillating calcium signal into a time-averaged redox response. The data reproduced from Robb-Gaspers et al. 1998 was kindly provided by Andrew P. Thomas.

The frequency of the oscillations increases with increased amount of agonist, whereas the amplitude of the signals in both cytosol and mitochondria remains constant. The transient rises in mitochondrial calcium concentration causes transient rises in NAD(P)H levels, due to activation of the Ca^{2+} dependent dehydrogenases. The rise in NAD(P)H follows 2-3 seconds after the rise in $[Ca^{2+}]_{cyt}$, whereas it takes 5-10 minutes for NAD(P)H to reoxidise and return to baseline level, following a calcium peak (Hajnoczky et al. 1995; Robb-Gaspers et al. 1998). The amplitude of the NAD(P)H level remains constant independent of calcium spiking frequency (Fig. 5). Instead when the agonist level, and therefore the frequency of calcium oscillation, is raised, the slow reoxidation phase causes the NAD(P)H transients to fuse together. This results in a sustained rise in NAD(P)H when the Ca^{2+} spiking frequency exceeds 0.5 spikes/min, as opposed to oscillating levels of NAD(P)H at low calcium oscillation frequencies. The described mechanism lets the cell decode an oscillatory signal into a non-oscillatory target response and the decoding mechanism allows the cell to keep the target product concentration at any time below a certain level while raising the average concentration of the product.

Ca^{2+} Dependent Gene Expression

Ca^{2+} oscillations also play a major role in the control of gene expression in cells (Li et al. 1998; Hu et al. 1999; Dolmetsch et al. 1998). An example of an experimentally investigated system is the decoding of calcium signals by the proinflammatory transcription factors in T-lymphocytes. Activation of inflammatory responses in T-lymphocytes can be mediated by a cascade, in which Ca^{2+} is a second messenger. Antigen binds to T-cell antigen receptors (TCR) which causes generation of IP_3 through G-protein mediated PIP_2 cleavage and IP_3 releases calcium from the ER and onsets the oscillating calcium signals. The resulting Ca^{2+} transients activate NF-AT through Ca^{2+} dependent dephosphorylation and binding to a nuclear subunit, which has been induced by protein kinase C (PKC, which is also calcium regulated). Active NF-AT helps to regulate several immune-response genes including IL-2 (Interleukin-2 that induces differentiation and proliferation of T-cells), IL-4 (Interleukin-4 that activates B-cells) and TNF-α (tumour-necrosis factor-α that activates macrophages) (Janeway et al. 2001).

Dolmetsch et al. (Dolmetsch et al. 1998) investigated the dependence of NF-AT activation on calcium in T-cells. They compared signals employing sustained oscillations with constant calcium concentrations and found that oscillating calcium signals were able to activate NF-AT at a much lower average Ca^{2+} concentration than a constant rise in calcium. Thus, oscillations lower the effective Ca^{2+} threshold for activating NF-AT and thereby allow the cell to activate NF-AT without risking the potential dangers of a high calcium concentration in the cytosol (this was also shown by Li et al. 1998).

In addition, Dolmetsch et al. compared the activation of NF-AT by oscillatory signals with the activation of two other calcium dependent transcription factors namely NF-κB and Oct/OAP, which are also involved in the control of IL-2 transcription. The aim of the experiment was to show that the different transcription factors could decode the frequency of an oscillatory calcium signal, resulting in an activation of only a subset of targets. The experiments showed that when exposed to changing sustained elevations of calcium the activity of the three transcription factors could not be distinguished. When the transcription factors were exposed to calcium oscillations with changing frequency, the frequency was proportional to the resulting activity level, but whereas the activity of NF-AT and Oct/OAP disappears at frequencies less than 2.5 ms, NF-κB still shows activity at a frequency of 0.5ms. Thus calcium oscillation frequency can determine which transcription factors are active.

9.3.2. Modelling of Decoding

Similar to the experimental studies, most theoretical exploration of calcium signal decoding has been limited to frequency-encoded information, and only a few systems have been explored. In this section we present some of the results obtained by theoretical studies.

Glycogen Phosphorylase

Glycogen breakdown in liver cells is a physiological response, which can be mediated by Ca^{2+}. When a hormone such as vasopressin binds receptors on the surface of the hepatocyte and onsets the phosphoinositide signalling pathway, the resulting rise in cytosolic calcium causes activation of glycogen breakdown. The breakdown of glycogen to glucose is catalysed by Glycogen phosphorylase A. The regulation of Glycogen phosphorylase is as follows. Glycogen phosphorylase is converted from an inactive b-form to the active a-form by phosphorylation by phosphorylase kinase, and is inactivated (dephosphorylated) by phosphorylase phosphatase. The calcium dependence of the system is found in the calmodulin like δ subunit of phosphorylase kinase, which binds calcium cooperatively and activates the kinase. Glucose activates the phosphorylase phosphatase, and thereby inactivates glycogen breakdown.

Gall *et al.* (Gall *et al.* 2000) used the above system to investigate the effect of calcium oscillations of the regulation of cellular processes. They constructed a minimal model for the phosphorylation-dephosphorylation cycle, based on the model presented by Cardenas and Goldbeter (Cardenas and Goldbeter 1996), including Ca^{2+} activation of the phosphorylase kinase. Since they wanted to investigate the effect of calcium on the system, they chose to keep both the glucose level and the phosphorylation by cAMP activated kinases constant. The Ca^{2+} activation was modelled to both reduce the K_M and increase the V_{max} (maximum velocity of the reaction) of the phosphorylase kinase and Ca^{2+} activation was assumed cooperative. The resulting differential equation has only *Pha* (glycogen phosphorylase A) and Ca^{2+} as variables, whereas the rest of the effectors are held constant. Gall et al. then coupled the equation to at sinusoidal Ca^{2+} oscillator of the form $Ca^{2+} = A+B*\sin(2\pi vt)$, where A, B and v were chosen to match observed amplitudes and frequencies of calcium oscillations in hepatocytes. Although the sinusoidal oscillations do not resemble the oscillations observed in experiments, using this simple system allowed Gall et al. to easily control the dynamics of the oscillation clarifying the impact of frequency changes on the regulation of glycogen breakdown.

Gall et al. showed by these theoretical simulations one of the fundamental advantages observed of employing an oscillating signal. They found that the level of *Pha* was larger when an oscillating signal was employed compared to a steady state signal with the same average calcium concentration. They showed this for both the equation coupled to the simple sinusoidal oscillator, and coupled to a calcium minimal model showing physiological calcium spikes upon agonist binding (Goldbeter *et al.* 1990). The advantages of this dynamic property were discussed in sect. 9.2. Gal *et al.* also showed that due to the difference in the dynamics of phosphorylation and dephosphorylation of glycogen phosphorylase, a change in frequency of the calcium signal was decoded as a change of the average level of active glycogen phosphorylase (‹*Pha*›). That is, if the frequency of the calcium oscillations was increased, so was ‹*Pha*›.

Decoding of Calcium Oscillations by CaMKII

The decoding of calcium oscillations by CaMKII is probably the most thoroughly studied example of decoding, and several models have been presented which illuminate the additive mechanism of CaMKII activation (Bhalla and Iyengar 1999; Dosemeci and Albers 1996; Kubota and Bower 1999; Kubota and Bower 2001; Prank et al. 1998; Zhabotinsky 2000). Most CaMKII models decode calcium signals in excitable cells. But since CaMKII is also found in non-excitable cells, it is also interesting to investigate the decoding of agonist induced calcium signals in non-excitable cells.

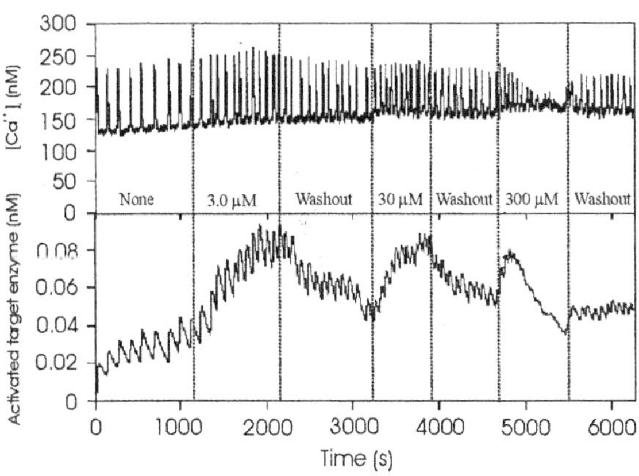

Fig. 6. Decoding of experimental calcium oscillations in a β-cell by a theoretical CaMKII model as an additive activation of target enzyme. The calcium spikes in the top panel were induced by bombesin and the frequency was increased by administrating increasing doses of tolbutamide (Prank et al. 1998). Decoding was simulated by the model presented in Prank et al. 1998. Similar to other experiments and models, the level of enzyme activation depends on the frequency, and activation is additive between pulses. Data was kindly provided by Klaus Prank and Leonard Läer.

We choose to exemplify the decoding by CaMKII with an example of theoretical decoding presented by Prank et al. (Prank et al. 1998). The CaMKII model presented by Prank et al. is qualitative and rather simple compared to other models (some of the terms do not represent realistic biochemical reaction kinetics). But we chose to use this as an example for modelling CaMKII for one particular reason, namely that their work illustrates a powerful use of mathematical modelling by coupling theoretical enzyme models to experimental calcium oscillation data. They do this in the recognition that CaM Kinase activity cannot be experimentally assessed continuously, and thus by combining experimental data such as calcium concentration and enzymatic constants, with knowledge about the mechanism of the enzyme, they obtain continuous data on CaMKII activity encoded in experimentally obtained calcium signals. The idea presented by Prank et al. can be used as a general method to obtain knowledge about enzymebehaviour when this is not accessible as experimental data *in vivo*, but discrete enzyme data and continuous effector data is. This idea may be very helpful, especially if good data and knowledge about the enzyme mechanism is present.

Figure 6 shows the Prank et al. 1998 model coupled to experimental calcium spike trains. Data was kindly provided by Klaus Prank and Leonard Läer. The model shows additive activation during calcium spike trains and a final level of activation, which is determined by the frequency of the signal. This corresponds well to both theoretical and experimental data in other studies of CaMKII, and shows the qualitative behaviour of the enzyme when decoding calcium oscillations.

9.3.3. Decoding by Simple Calcium Dependent Enzymes

In order to investigate information processing in calcium signal transduction, we modeled a calcium-activated enzyme and coupled this to the oscillating calcium model presented in sect. 9.2.3 (Larsen et al. 2002). Enzymes and their catalytic activity have been studied for decades, but most experimental enzyme studies employed *in vitro* methods with isolated enzymes and controlled conditions. But the conditions under which most experimental data is obtained poses a problem when using it in mathematical modelling. In principle it should be possible to insert experimental data (such as K_M's and V_{max}'s) into a know mechanism (such as the Michaelis-Menten equation), and reproduce experimental data quantitatively. But often the enzymatic data obtained for an isolated enzyme differ from the data obtained *in vivo*. Another problem when modelling enzyme kinetics is the need of all parameters of the system. The above problems make it very difficult to model target enzyme responses quantitatively, but this does not hinder a theoretical exploration of the enzymatic mechanisms of signal decoding. Instead of modelling a specific enzyme system, we therefore model an arbitrary enzyme, which employs a kinetic mechanism that is known to be used by several enzymes. This leads to results which we know to be true for the behaviour of some calcium binding proteins, without restricting ourselves to modelling a certain system like one of those described in the previous section.

We assume that our enzyme binds several calcium ions cooperatively at an allosteric activation site in the enzyme. This type of enzyme activation is common in eukaryotic cells although the number of calcium ions, which bind may vary (Carafoli et al. 2001). Calcium activation of our enzyme is modelled with a Hill term whereas a first order process only dependent on the amount of active enzyme models deactivation. We assume that the amount of calcium, which binds the enzyme is very small compared to the amount found in the cytosol, and therefore we do not subtract bound Ca^{2+} from Ca_{cyt}. Normally the amount of enzyme is small compared to the amount of substrate in the cell, and therefore an upper boundary of enzyme activity might seem appropriate. But simulation experiments showed that a limitation of enzyme was not necessary to explore the desired mechanisms, and did not add to the quality of the results. Therefore we assume that the total amount of enzyme (Enz_{total}) is large, and do not use the term (Enz_{total}-Enz) to describe the amount of enzyme available to Ca^{2+}, but assume this constant and implicitly included in k_{act}.

The resulting equation is:

$$\frac{dEnz}{dt} = \frac{k_{act} \cdot Ca_{cyt}^{p}}{K_{M}^{p} + Ca_{cyt}^{p}} - k_{inact} \cdot Enz \tag{9.9}$$

Where *Enz* is the amount of activated enzyme and K_M is the concentration of Ca^{2+}, which leads to half maximum activation of *Enz*. Finally k_{inact} is the velocity constant of the inactivation of enzyme. We have aimed to keep the equation simple, with as few parameters as possible. We have done this to show that the decoding of information held in calcium signals does not require elaborate mechanisms, but is a build in trade of allosteric enzymes, which bind calcium cooperatively. The above equation is far simpler than most enzyme models presented in decoding so far (Gall et al. 2000, Izu and Spangler 1995; Kubota and Bower 2001; Prank et al. 1998; Càrdenas and Goldbeter 1996), but still it is able to simulate all the aspects of decoding that have been shown with the more elaborate models.

We couple the enzyme activation equation (9.9) to equations 9.1-9.5 and study decoding at different values of k_{act}, K_M and k_{inact}. In order to show the versatility of this mechanism, we model two of the decoding patterns presented in sect. 9.3.

The CaMKII

The first example is that of CaM kinase activation. Several models of CaM activation have been presented over the years (Bhalla and Iyengar 1999; Dosemeci and Albers 1996; Kubota and Bower 1999; Kubota and Bower 2001; Prank et al. 1998; Zhabotinsky 2000). Ranging from simple qualitative models such as the one presented in sect. 9.3.2 (Prank et al. 1998) to very elaborate quantitative models such as the one presented by Kubota and Bower (Kubota and Bower 2001). These models have in common that they model the frequency dependency of CaM activation, and find that a higher frequency of calcium oscillations lead to a higher level of CaM activation, by an additive activation of CaMKII and slow deactivation by phosphorylases. With the very simple enzyme equation 9.9 we can show the same mechanism (Fig. 7, compare with Fig. 6). If we couple our enzyme equation to the spiking calcium model, and change the agonist influence (This case modelled by k_1), the model mimics the additive activation of CaM Kinase until reaching the level of activation, which corresponds to the chosen frequency. Each frequency results in a specific level of activation. And if the frequency is lowered, so is the CaM activity (Dosemeci and Albers 1996).

Figure 7 does not presume to show quantitative data on the activation of CaM kinase, but it shows that the simple enzyme equation contains the key dynamics, which is necessary to model key features of CaM kinase activation.

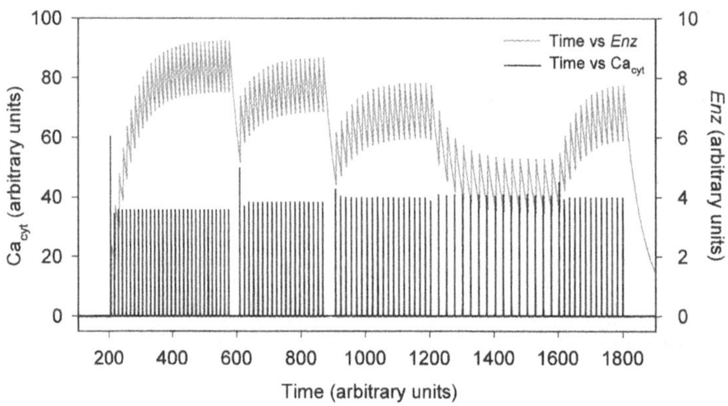

Fig. 7. Decoding of cytosolic calcium spike trains by *Enz* (simulating activation of CaMKII). Our enzyme equation is coupled to 1-5 (constants from Fig. 3.F-J). The parameter of Enz are; k_{act}= 2.85, K_M=3.7 , k_{inact}=0.017 and p=4. k_1 is changed over time assuming values between 0.5 and 0.9.

Ca^{2+} Dependent Dehydrogenases

The second example of decoding by a specific mechanism is that of the Ca^{2+} dependent dehydrogenases in the mitochondria. The experimental results on the decoding of calcium spikes by Ca^{2+} sensitive intramitochondrial dehydrogenases obtained by Robb-Gaspers et al (Robb-Gaspers et al 1998) were described in sect. 9.3.1. We wish to use the simple enzyme term of 9.9 to model this type of dynamics. We assume that the NAD(P)H changes follow the activity of the dehydrogenases, and therefore we compare theoretical periodic calcium spiking and the resulting CSMDHs activity (*Enz*) with measured cytosolic calcium and NAD(P)H level (Robb-Gaspers et al 1998). This assumption is a strong simplification of the actual regulation of NAD(P)H level, which is influenced not only by the Ca^{2+} activation of CSMDHss, but also by the reoxidation of

NAD(P)H, the ATP/ADP ratio and several other compounds. But we wish to show that the behaviour observed in the experiments can be simulated by a very simple mechanism, which only takes into account basic enzyme mechanisms. We couple 9.9 to equations 9.1-9.5 and changing k_1 (agonist influence) we obtain the results in Fig. 8.

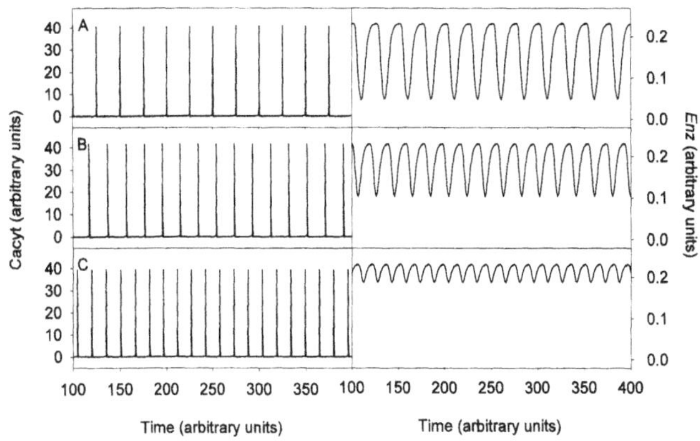

Fig. 8. Decoding of calcium spiking frequency by *Enz* (Modelling the calcium dependent dehydrogenases of the mitochondria). Agonist influence is modelled by k_1 with: A) k_1=0.3, B) k_1=0.4 and C) k_1=0.55. The parameters of the calcium model are as presented in Figure 3.F-J except for k_1, and constants of the *Enz* equation are: k_{act} = 0.1, K_M = 0.02, k_{inact} = 0.43 and p=4. The initial conditions for the active enzyme is *Enz*(0) = 0.

Figure 8 shows calcium spiking with increasing frequency following increases in k1. The increasing oscillation frequency leads to an increase in mean CSMDHs activity (Enz), with no change in the amplitude of the individual CSMDHs elevations. Thus, we show that simple cooperativity of calcium binding to an enzyme is sufficient to qualitatively reproduce the experimentally observed findings of the CSMDHs system.

Decoding of Periodic Bursting

In addition to modelling specific systems found in the literature, our model enzyme is also able to simulate and illuminate theoretical decoding that may be utilized by a wide variety of calcium regulated enzymes. In this section we deal with strictly theoretical results, we do not attempt to model a specific enzyme system, but instead try to reveal some of the possibilities bursting signals may encompass. We couple enzyme equation 9.9 with k_{act}=5, K_M=0.62 and k_{inact}=0.4 to the bursting calcium model ($k_2 \neq 0$ and $k_{13} \neq 0$) and vary the level of agonist influence (Modelled by k_2). Figure 9 shows timeseries of cytosolic calcium (Ca_{cyt}) and enzyme activity (*Enz*) at two different levels of agonist. In Fig. 9.A secondary oscillations in bursting of Ca_{cyt} is less pronounced than in Fig. 9.B, but both amplitude and frequency of the main oscillation remains all but constant. Although neither frequency nor amplitude of the calcium signal changes, the enzyme activity is almost tripled from 9.A to 9.B.

This allows the cell to change the activity of calcium dependent enzymes by varying the shape of the signal slightly leaving for example frequency controlled enzyme activities unaltered. This implies that a calcium-regulated enzyme, with the very simple kinetics modelled by 9.9 is able to decode information not only in frequency and amplitude, but also in a slight change of waveform.

An oscillatory signal can lower the calcium effective threshold for activation of a calcium dependent enzyme. This has been shown in previous experimental and theoretical studies employing periodic spiking only (Gall et al. 2000; Izu and Spagler 1995; Li *et al.* 1998; Dolmetsch and Lewis 1998). We show that this is also the case for periodic bursting in our model. We calculated the average Ca_{cyt} and *Enz* in Fig. 9.B and found that $\langle Ca_{cyt} \rangle \approx 0.9$ and $\langle Enz \rangle \approx 2.0$ during bursting with k_2=1.9, whereas a steady state level of calcium; $\langle Ca_{cyt} \rangle \approx 0.9$ only caused an enzyme activity $\langle Enz \rangle \approx 0.6$. This means that employing an oscillatory signal compared to a steady state concentration of calcium more than tripled the activity of *Enz*.

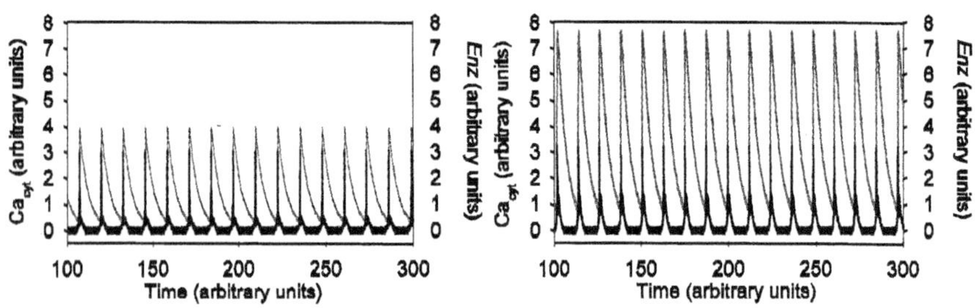

Fig. 9. Decoding of waveform. Two different levels of agonist (A: k_2=1.65 and B: k_2=1.9) induce two different calcium signals (Ca_{cyt}———), the amplitude and frequency of the two signals are practically identical, but are decoded by the enzyme (*Enz* ---------) very differently. A small change in waveform leads to a significant change in enzyme activity. The parameters of the calcium model are as in Figs. 3.A-E, whereas the parameters of equation 9.9 are: k_{act}=5, K_M=0.62, k_{inact}=0.4 and p=4.

The last theoretical example of decoding periodic bursting illustrates how an oscillating signal can contain information for more than one target process. We couple two enzymes (*Enz-1* and *Enz-2*), which both employ the kinetics of 9.9, to the bursting calcium model ($k_2 \neq 0$ and $k_{13} \neq 0$). *Enz-1* and *Enz-2* differ only in K_M and k_{act} of the activation term. In Fig. 10 the resulting timeseries from three k_2 values are plotted. At low agonist concentration (k_2=1.86) the resulting calcium signal shows periodic bursting with a very narrow secondary oscillation. This induces almost the same level of activity in both *Enz-1* and *Enz-2* as can be seen in Fig. 10.A. If the agonist influence is increased (k_2=2.21) the calcium signal changes slightly to higher amplitude, lower frequency and pronounced bursting oscillation. In Fig. 10.B this is decoded by *Enz-1* as a slight increase in activity whereas *Enz-2* activity is almost doubled. Finally nearly overstimulating the system with an agonist influence of k_2=2.239 causes the calcium oscillation to increase its frequency and lower its amplitude. This results in a high level of *Enz-1* activation, during which the enzyme activity does not have time to return to base level, and therefore *Enz-1* activity is almost sustained. *Enz-2* on the other hand is hardly activated by this calcium signal (Fig. 10.C).

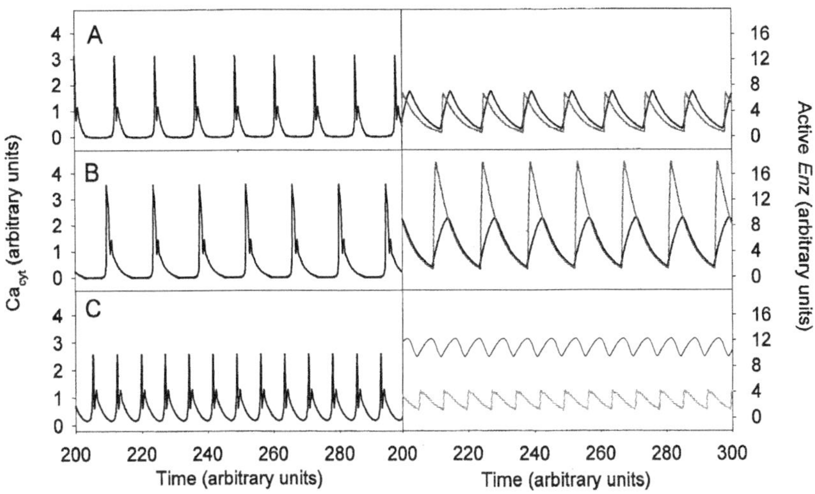

Fig. 10. Bursting oscillations simulated by the bursting model with the parameters of Fig. 3A and decoded by Enz-1(———) and Enz-2 (············) which both employ the dynamics of equation 9. Parameters of the enzymes are k_{inact}=0.2, p=4 and (*Enz*-1) $k_{act(1)}$= 3 , $K_{M(1)}$= 0.24, (*Enz*-2) $k_{act(2)}$= 70 , $K_{M(2)}$= 3.7. In panel A: k_2=1.86, panel B: k_2=2.21 and panel C: k_2=2.239. The different levels of agonist (k_2) induce different calcium signals, which in turn are decoded as very different enzyme activities depending on the kinetic parameters of the enzyme.

Decoding of signals more complex than periodic spiking, has so far been left relatively unexplored. However, plenty of experimental data show far more complex oscillation patterns than periodic spiking. Now we wish to investigate if the complexity of these signals can be an information carrier as well. By setting k_2=2.23 in the bursting model (Leaving the rest of the parameters unaltered), we obtain a calcium signal which is a periodic combination of high frequency/low amplitude and low frequency/high amplitude bursting oscillations.

These burst patterns are decoded by *Enz-1* and *Enz-2* from the previous section with very pronounced frequency locked oscillations (Fig. 11). Thus, oscillations in enzyme activity display large amplitude fluctuations superimposed on small amplitude oscillations. These large amplitude fluctuations lock with the calcium oscillations with a ratio of 6 - 10 calcium oscillations per large amplitude oscillation in enzyme activity. Both enzymes display frequency locking, but in very different ways. When the activity of *Enz-1* is high, the activity of *Enz-2* is low and *vice versa*. The large but slow fluctuations of enzyme activity are partly due to the difference in the small scale calcium oscillations, and partly due to additive activation of the enzymes.

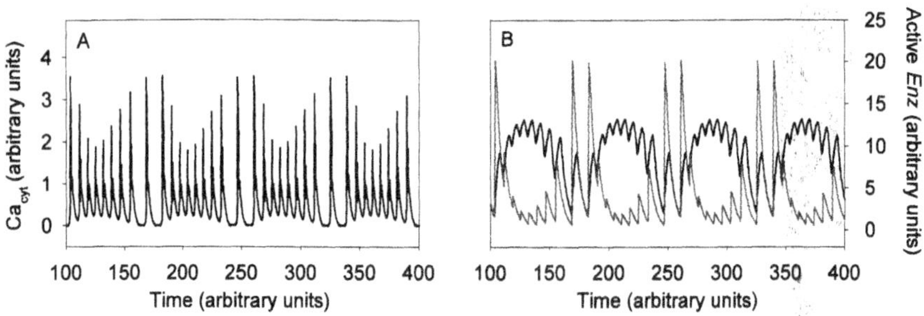

Fig. 11. Decoding of complex calcium signals. Panel A show complex bursting calcium signals simulated by equations 1-5 with the parameters of Figs. 3.A-E except for agonist influence which is k_2=2.23. These calcium signals are decoded by *Enz*-1(———) and *Enz*-2 (·········) in panel B. The enzyme activity shows phase locking with the calcium signals, and the two enzymes decode the calcium signal with intensities opposite of one another.

The phenomenon of additive behaviour of calcium regulated proteins has so far been observed only *in vitro* (De Koninck and Schulman 1998; Gunter et al. 1998; Sparagna et al. 1995) and frequency locking only in theoretical studies (Chay et al. 1995; Fan and Holden 1998). But the experimental systems explored so far have often shown complex irregular calcium signalling (Green et al. 1993; Kummer et al. 2000; Rooney et al. 1989; Rooney et al. 1996), so the question whether those complexities contain additional information is indeed relevant.

9.4. Discussion and Perspectives

Before the observation of calcium oscillations in hepatocytes, it was a real mystery, how approximately fifty different agonists which are responsible for different responses in the cell, are able to use one and the same second messenger, namely calcium to transfer their message to intracellular targets.

Even though there is still a lot to be solved, we presented both experimental and theoretical evidence, how different information can be encoded and decoded from different calcium oscillations. Thus, information can be stored in the frequency, amplitude, shape and complexity of the oscillations offering at least partial understanding of the information processing machinery in hepatocytes.

Amazingly enough, relatively simple enzymatic mechanisms (e.g. cooperativity) are sufficient to decode the respective information.

However, there are still a lot of open questions, e.g.: How is the information of combinations of different agonists encoded and decoded in the calcium oscillations? How is the information faithfully transferred through the cell and across tissues (encoding and decoding in spatial information)? How do other signalling pathways interfere with calcium signal transduction? These questions need to be addressed both experimentally and theoretically in the future to complete our picture of information processing in living cells.

Acknowledgements

We would like to thank Jane Dixon, Anne Green, Klaus Prank, Leonard Läer and Andrew P. Thomas for providing experimental data and Lars Folke Olsen for fruitful discussions. We also would like to acknowledge the Klaus Tschira Foundation and ESF for funding.

References

Berridge JM, Bootman MD and Lipp P (1998) Calcium - a life and death signal. Nature, 395:645-648

Bhalla US and Iyengar R (1999) Emergent properties of networks of biological signalling pathways. Science, 283:381-387

Bockaert, J, Claeysen, S, Becamel, C, Pinloche, S and Dumuis, A (2002) G protein-coupled receptors: dominant players in cell-cell communication. Int Rev Cytol, 212: 63-132

Bourne, HR and Stryer, L (1992) G proteins. The target sets the tempo. Nature, 358: 541-543

Carafoli E, Santella L, Brance D and Brini M (2001) Generation, control, and processing of cellular calcium signals. Crit Rev Biochem, Mol Biol 36:107-260

Cardenas ML and Goldbater A (1996) The glucose-induced switch between glycogen phosphorylase and glycogen synthase in the liver: Outlines of a theoretical approach. J theor Biol 182:421-426

Celio MR, Pauls T and Schwaller B (Eds)(1996) Guidebook to the calcium-binding proteins. Oxford University Press, Oxford

Chatton J, Cao Y and Stucki JW (1997) Agonist specific behaviour of the intracellular Ca^{2+} response in rat hepatocytes. Biochem J 328:573-579

Chatton J, Cao Y, Stucki JW (1998) Pertubation of myo-Inositol-1,4,5-Trisphosphate levels during Agonist-induced Ca^{2+} Oscillations. Biophys J 74: 523-531

Chay TR, Lee YS and Fan YS (1995) Appearance of Phase-locked Wenckebach-like rhythms, Devil's staicase and universality in intracellular calcium spikes in non-excitble cell models. J Theor Biol 174:21-44

Clapham DE (1995) Calcium signalling Cell, 80:259-268

De Koninck P and Schulman H (1998) Sensitivity of CaM Kinase II to the frequency of Ca^{2+} oscillations. Science, 279:227-230

Dixon CJ, Woods NM, Cuthbertson KSR and Cobbold PH (1990) Evidence for two Ca2+-mobilizing purinoreceptors on rat hepatocytes. Biochem J 269:499-502

Dixon CJ, Woods NM, Webb TE and Green AK (2000) Evidence that rat hepatocytes co-express functional P2Y1 and P2Y2 receptors. Br J Pharmacol 129:764-770

Dolmetsch RE, Xu E, Lewis RS (1998) Calcium oscillations increase the efficiency and specificity of gene expression. Nature 392:933-936

Dosemeci A and Albers RW (1996) a mechanism for synaptic frequency detection through autophosphorylation of CaM Kinase II. Biophys J 70:2493-2501

Dubyak GR and El-Motassim C (1993) Signal transduction via P2-purinergic receptors for extracellular ATP and other nucleotides. Am J Physiol 265: C577-606

Dupont G and Goldbeter A (1998) CaM kinase II as frequency decoder of Ca^{2+} oscillations. BioEssays 20: 607-610

Eshete F and Fields RD (2001) Spike frequency decoding and autonomous activation of Ca^{2+}-Calmodulin-dependent protein kinase II in dorsal root ganglion neurons. J Neurosci 21:6694-6705

Fain JN (1989) Regulation of phosphoinositide-specific phospholipase C. Biochim Biophys Acta 1053: 81-88

Fan YS and Holden AV (1998) An asymmetrical phase locking structure for a non-excitable cell model. Chaos, Solitons and Fractals 9:1637-1650

Gall D, Baus E and Dupont G (2000) Activation of the liver glycogen phosphorylase by Ca^{2+} oscillations: A theoretical study. J theor Biol 207:445-454

Goldbeter A, Dupont G and Berridge MJ (1990) Minimal model for signal induced Ca^{2+} oscillations and for their frequency encoding through protein phosphorylation. Proc Natl Acad Sci U.S.A. 87:1461-1465

Green AK, Dixon CJ, McLennan AG, Cobbold PH and Fisher MJ (1993) Adenine Dinucleotide-mediated Cytosolic Free Ca^{2+} Oscillations in Single Hepatocytes. FEBS Lett 322:197-200

Green AK, Cobbold PH and Dixon CJ (1994) Elevated intracellular cyclic AMP exerts different modulatory effects on cytosolic free Ca2+ oscillations induced by ADP and ATP in single rat hepatocytes. Biochem J 302:949-955

Green AK, Cobbold PH and Dixon CJ (1999) Thimerosal enhances agonist-specific differences between $[Ca^{2+}]_i$ oscillations induced by phenylephrine and ATP in single rat hepatocytes. Cell Calcium 25:173-178

Grubelnik V, Larsen AZ, Kummer U and Olsen LF (2001) Mitochondria regulate the amplitude of simple and complex calcium oscillations. Biophys Chem 94:59-74

Gunter TE, Butinas L, Sparagne GC and Gunter KK (1998) The Ca^{2+} transport mechanisms of mitochondria and Ca^{2+} uptake from physiological-type Ca^{2+} transients. Biochimica et Biophysica Acta 1366:5-15

Hajnoczky G, Robb-Gaspers LD, Seitz MB and Thomas AP (1995) Decoding of cytosolic calcium oscillations in the mitochondria. Cell, 82:415-424

Hansen CA, Yang L, Williamson JR (1991) Mechanisms of receptor-mediated Ca^{2+} signalling in rat hepatocytes. J Biol Chem 266:18573-18579

Hu Q, Deshpande S, Irani K and Ziegelstein RC (1999) $[Ca^{2+}]$i Oscillation frequency regulates agonist-stimulated NF-κB transcriptional activity. J Biol Chem 274:33995-33998

Izu LT and Spangler RA (1995) A class of parametrically excited calcium oscillation detectors. Biophys J 68:1621-1629

Janeway CA, Travers P, Walport M and Schlomchik M (2001) Immuno Biology, the immune system in health and disease. Garland Publishing

Kubota Y and Bower J (1999) Decoding time-varying calcium signals by the postsynaptic biochemical network: Computer simulations of molecular kinetics. Neurocomputing, 26-27:29-38

Kubota Y and Bower J (2001) Transient versus asymptotic dynamics of CaM kinase II: Possible roles of Phosphatase. J Comp Neuroscience 11:263-279
Kummer U, Olsen LF, Dixon CJ, Green AK, Bornberg-Bauer E and Baier G (2000) Switching From Simple to Complex Oscillations in Calcium Signalling. Biophysical J 79:1188-1195
Larsen AZ, Olsen LF and Kummer U (2002) On the encoding and decoding of calcium signals in hepatocytes. Biophys. Chem. (Submitted)
Li WH, Llopis J, Whitney M, Zlokarnik G and Tsien R (1998) Cell-permeant caged $InsP_3$-ester shows that Ca^{2+} spike frequency can optimize gene expression. Nature 392:933-936
Llopis J, Kass GEN, Gahm A and Orrenius S (1992) Evidence for two pathways of receptor-mediated Ca2+ entry in hepatocytes. Biochem J 284: 243-247
Marhl M, Schuster S and Brumen M (1998) Mitochondria as an important factor in the maintenance of constant amplitudes of cytosolic calcium oscillations. Biophys Chem 71:125-132
Nash MS, Young KW, Chaliss RAJ and Nahorski SR (2000) Receptor-specific messenger oscillations. Nature 413:381-382
Oancea E and Meyer T (1998) Protein Kinase C as a molecular machine for decoding calcium and diacylglycerol signals. Cell 95:307-318
Prank K, Läer L, von zur Mühlen A, Brabrant G and Schöfl C (1998) Decoding intracellular spike trains. Europhys Lett 42:143-147
Robb-Gaspers LD, Rutter GA, Burnett P, Hajnoczky G, Denton RM and Thomas AP (1998) Coupling between cytosolic and mitochondrial calcium oscillations: role in the regulation of hepatic metabolism. Biochem Biophys Acta 1966:17-32
Rooney TA, Sass EJ and Thomas AP (1989) Characterisation of Cytosolic Calcium Oscillations Induced by Phenylephrine and Vasopressin in Single Fure-2-loaded Hepatocytes. J Biol Chem 264:17313-17141
Rooney TA, Renard DC, Sass EJ and Thomas AP (1991) Oscillatory cytosolic calcium waves independent of stimulated 1,4,5-trisphosphate formation in the hepatocyte. J Biol Chem 19:12272-12282
Rooney TA, Suresh K, Joseph CQ and Thomas AP (1996) Cyclic GMP Induces Oscillatory Calcium Signals In Rat Hepatocytes. J Biol Chem 271: 19817-19825
Schuster S, Marhl M and Höfer T (2002) Modelling of simple and complex calcium oscillations. Eur J Biochem 269:1333-1355
Sparagna GC, Gunter K, Sheu S and Gunter TE (1995) Mitochondrial Calcium Uptake from Physiological-type Pulses of Calcium. J Biol Chem 270:27510-27515
Sternweis, PS (1994) The active role of $\beta\gamma$ in signal transduction. Curr Opin Cell Biol 6: 198-203
Striggow, F and Bohnensack R (1994) Inositol 1,4,5-trisphosphate activates receptor-mediated calcium entry by two different pathways in hepatocytes. Eur J Biochem 222:229-234
Taylor CW and Thorn P (2001) Calcium signalling: IP_3 rises again..... and again. Curr Biol 11:R352-R355
Tompa P, Toth-Boconadi R and Friedrich P (2001) Frequency decoding of fast calcium oscillations by calpain. Cell Calcium 29:161-170
Tsien, RW and Tsien RY (1990) Calcium channels, stores, and oscillations. Annu Rev Cell Biol 6: 715-760
Wakui M, Potter BVL and Petersen OH (1989) Pulsatile intracellular calcium release does not depend on fluctuations in inositol trisphosphate concentration. Nature 339:317-320

Woods NM, Cuthbertson, KSR and Cobbold PH (1986) Repetitive transient rises in cytoplasmic free calcium in hormone-stimulated hepatocytes. Nature 319: 600-602
Zerangue, N and Jan, LY (1998) G-protein signaling: Fine-tuning signaling kinetics. Curr Biol 8: R313-R316
Zhabotinsky AM (2000) Bistability in the Ca^{2+}/Calmodulin dependent protein kinase-phosphatase system. Biophys J 79:2211-2221

10. Modeling Calcium Waves

James Sneyd and Krasimira Tsaneva-Atanasova

Department of Mathematics, University of Auckland, Auckland, New Zealand

Abstract. We present an introduction to the mathematics of calcium waves, both intracellular and intercellular. After a discussion of the earlier models I shall discuss more recent work from a number of groups, and discuss in detail how the models have contributed to an understanding of calcium wave propagation in a variety of cell types. Bifurcations of travelling waves will be discussed briefly. I shall end with an overview of models of periodic intercellular waves in hepatocytes and pancreatic acinar cells, and discuss the major problems still to be solved by theoreticians in this area.

10.1. Introduction

Elsewhere in this volume we have seen how the local dynamics of calcium release and uptake can lead to oscillations in the free cytoplasmic calcium concentration. Such oscillations are not easy to understand by themselves, but the complexity of the problem is greatly increased when spatial aspects are brought into play. For a theoretician, one of the most interesting aspects of calcium dynamics is that local oscillations can be organised into complex spatio-temporal patterns such as oscillatory waves, spiral waves, and waves that travel from cell to cell. Not only are these structures physiologically important, they are also mathematically interesting, and challenging to understand. Reviews of the experimental literature on intracellular and intercellular calcium waves are given by Amundson and Clapham (1993), Sanderson *et al.* (1994), Thomas *et al.* (1996) and Røttingen and Iversen (2000).

The mathematical study of calcium waves has a lot in common with the study of wave propagation in other excitable models such as the Hodgkin-Huxley or FitzHugh-Nagumo models (Hodgkin and Huxley 1952; FitzHugh 1961; Britton 1986; Murray 1989; Grindrod 1991; Keener and Sneyd 1998). However, the field has matured to the extent that generic excitable models no longer give useful insight into the mechanisms of calcium wave propagation. One must instead construct models that are tailor-made to the particular type of cell, or type of wave, under consideration.

10.2. Generic Model Equations

The binding of an agonist to its receptor, the consequent activation of G-proteins, and the formation of inositol (1,4,5)-trisphosphate (IP_3) are the initial steps leading to calcium oscillations and waves (for reviews, see Putney and Bird 1993; Berridge 1993, 1997; Clapham, 1995; Parker *et al.* 1996). Most calcium wave models do not explicitly model these initial receptor-based reactions, although Kummer *et al.* (2000) include them in some detail. The rate of production of IP_3 can be modulated by the calcium concentration, as is assumed in some models (Meyer and Stryer 1988; Cuthbertson and Chay 1991; Keizer and DeYoung 1992). As yet it is not clear how important calcium modulation of IP_3 production is, although it is certainly of differing importance in different cell types. Because it is not our goal here to study all the ins and outs of various oscillation models we shall ignore this question, and just consider calcium waves that occur at a constant level of IP_3, as seems to be the case in, say, pancreatic acinar cells (Wakui *et al.* 1989).

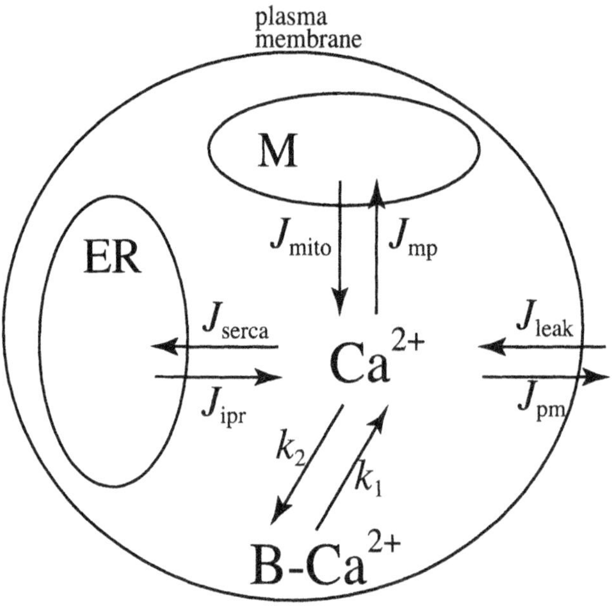

Fig. 1. Schematic diagram of the calcium fluxes in a general model. ER – endoplasmic reticulum. M – mitochondria. B – calcium buffer.

All calcium wave models are based upon underlying models for the dynamics of calcium release and reuptake, as shown in Fig. 1. At each point of the cell there are a number of calcium fluxes. For instance, calcium can flow into or out of the endoplasmic reticulum (ER) or mitochondria, or across the cell membrane in either direction. In addition, calcium can react with a variety of buffers. Other calcium fluxes can of course be included if desired. The fluxes of Fig. 1 lead to a system of ordinary differential equations of the general form

$$\frac{dc}{dt} = J_{ipr} - J_{serca} + J_{leak} - J_{pm} + J_{mito} - J_{mp} + k_1 b - k_2 c(b_t - b),$$

$$\frac{1}{\gamma_e}\frac{dc_e}{dt} = -J_{ipr} + J_{serca},$$

$$\frac{1}{\gamma_m}\frac{dc_m}{dt} = -J_{mito} + J_{mp},$$

$$\frac{db}{dt} = -k_1 b + k_2 c(b_t - b),$$

where c, c_e and c_m are the concentrations of calcium in the cytoplasm, ER and mitochondria respectively, b is the concentration of buffer with calcium bound, and b_t is the total concentration of buffer. Here we have modeled only a single buffer, but it is a simple matter to include multiple buffers just by adding in additional terms and equations. The constant γ_e is the ratio of the ER volume to the cytoplasmic volume, while γ_m is the ratio of the mitochondrial volume to the cytoplasmic volume. For simplicity we do not write down explicit equations for calcium buffering inside the ER or mitochondria. These terms will be similar to the cytoplasmic buffering terms, and treated the same way. In general these calcium equations will be coupled to other equations describing the dynamics of other parts of the system, such as the production and degradation of IP_3, the open probability of the IP_3 receptor (IP_3R) or the ryanodine receptor, the mitochondria, or the various calcium pumps.

10.2.1. Calcium Diffusion

To convert this model into a model of calcium waves requires a number of additional assumptions as to which species are able to diffuse, and how they diffuse. Most commonly one assumes that cytosolic calcium is free to diffuse, as are the mobile calcium buffers, but that calcium in the ER or the mitochondria is not. This leads to the equations

$$\frac{\partial c}{\partial t} = D_c \nabla^2 c + f(c) + k_1 b - k_2 c(b_t - b), \tag{1}$$

$$\frac{\partial b}{\partial t} = D_b \nabla^2 b - k_1 b + k_2 c(b_t - b), \tag{2}$$

where D_c and D_b are the diffusion coefficients of calcium and buffer respectively, and are assumed to be constant everywhere. The other differential equations remain unchanged. For simplicity we have summarised all the calcium fluxes by the single function $f(c)$: in general this function will be highly complicated, and coupled to the other equations in the model. It is important to note that such a formulation ignores the smaller spatial scales of puffs and sparks, the modeling of which is a large topic all by itself (Bugrim *et al.* 1997; Sneyd and Sherratt 1997; Keizer *et al.* 1998; Dawson *et al.* 1999; Swillens *et al.* 1999; Coombes 2001). However, at larger agonist concentrations the discrete nature of the calcium release sites becomes less important, with the cytoplasm acting more as a continuous medium. We are also not explicitly including the diffusion of other species such as inositol trisphosphate (IP_3). This is not to say that such diffusion does not exist and is not important. However, the principles of how to include other diffusing species are no different than for calcium. Indeed they are rather simpler, as buffering is usually not important for other species.

10.2.2. Boundary Conditions

The model equations are solved on some spatial domain, D, often in two spatial dimensions so that the computations do not take an excessive amount of time or memory, and usually with a simple geometry such as a square or a circle. Suppose the boundary of ∂D is denoted by ∂D. If D has three spatial dimensions then the calcium flux on ∂D is given by J_{leak}-J_{pm} , while, in the interior of D, the calcium fluxes include only those due to the ER, the mitochondria and the buffers. However, if ∂D has only two spatial dimensions the question becomes more complicated. For instance, if the cell is relatively flat, then every part of D is close enough to the plasma membrane to participate in trans-membrane calcium transport. In this case one just assumes that there is zero flux of calcium across ∂D while J_{leak} and J_{pm} are non-zero in the interior of D. The assumption of zero-flux across ∂D is just a mathematical fiddle, necessitated by the fact that we are trying to approximate a three-dimensional cell by a two-dimensional model. However, if the cell has a rounder shape, or is very large (like a *Xenopus* oocyte) this approximation will no longer be valid, and a full three dimensional model may be required.

10.2.3. Fast Buffering

If the buffer has fast kinetics, calcium is always in pseudo-steady-state with the buffer. In this case

$$k_1 b - k_2 c(b_t - b) = 0,$$

and thus

$$b = \frac{b_t c}{K + c},$$

where $K = k_1 / k_2$. From this it follows that

$$\frac{\partial c}{\partial t} + \frac{\partial b}{\partial t} = (1+\theta)\frac{\partial c}{\partial t},$$

where

$$\theta = \frac{b_t K}{(K+c)^2}.$$

If we now add equations (1) and (2) we get

$$\begin{aligned}\frac{\partial c}{\partial t} &= \frac{1}{(1+\theta)}\left(\nabla^2\left(D_c c + D_b b_t \frac{c}{K+c}\right) + f(c)\right)\\ &= \frac{D_c + D_b\theta}{1+\theta}\nabla^2 c - \frac{2D_b\theta}{(K+c)(1+\theta)}|\nabla c|^2 + \frac{f(c)}{1+\theta}.\end{aligned} \tag{3}$$

Here, we have assumed that b_t is independent of space and time. Equation (3) is the basic equation describing the dynamics of calcium when it can bind to a mobile buffer, and was first derived by Wagner and Keizer (1994).

We note that if the buffer is not mobile, i.e., $D_b = 0$, then (3) reduces to a much simpler form, in which the gradient of c plays no role. In this case, the dynamics of calcium are just scaled by the factor $1/(1+\theta)$ If K is much larger than c, this is essentially the same as scaling by the constant factor $K/(b_t+K)$, giving the equation

$$\frac{\partial c}{\partial t} = D_{\text{eff}}\nabla^2 c + f_{\text{eff}}(c), \tag{4}$$

where the effective diffusion coefficient and reaction rate, D_{eff} and f_{eff}, are just scaled versions of the original rates. In other words, a fast buffer with a large K does nothing but make the calcium dynamics and diffusion slower.

Not a great deal is known about how mobile buffers change the properties of calcium waves, although a few analytical results have been published (Sneyd and Kalachev 1994; Kupferman *et al.* 1997; Sneyd *et al.* 1998). There have also been detailed numerical studies of the effects of calcium buffers (Sala and Hernández-Cruz 1990; Nowycky and Pinter 1993; Jafri and Keizer 1995). The amount of experimental work on calcium buffering is far too extensive to summarize here. However, one particularly interesting recent paper is Naraghi *et al.* (1998). Most commonly, modelers just assume that the buffers are not mobile, and have K much larger than c. Of course, this is not always a good approximation, and more work is required to understand the significance of mobile buffers in any particular situation. We show below that the qualitative properties of the traveling waves in a particular model may remain relatively unaffected by the presence of mobile buffers, but this not not necessarily mean that similar conclusions can be drawn for other models.

10.3. Intracellular Calcium Waves

There have been a large number of models of calcium waves within single cells (Dupont and Goldbeter 1992, 1994; Atri *et al.* 1993; Othmer and Tang 1993; Bezprozvanny 1994; Jafri 1995; Falcke *et al.* 1999, 2000; Sneyd *et al.* 2000), most of them based on the assumptions of (4). Ultimately, the properties of the waves in the model depend heavily on the properties of the function $f(c)$, which controls the dynamics of calcium uptake and release at every point. In most cell types calcium release, whether through the inositol trisphosphate receptor (IP_3R) or through the ryanodine receptor (RyR), is excitable in nature. That is, the release of a sufficient quantity of calcium can stimulate the release of further calcium in an autocatalytic process that causes a large and rapid rise in the cytoplasmic concentration of free calcium. Because of this, the mathematical theory of wave propagation in calcium wave models is similar to that of wave propagation in other models of excitable media (Murray 1989; Grindrod 1991; Keener and Sneyd 1998), although some important differences exist (Sneyd *et al.* 1993; Sneyd and Atri 1993).

There are three particularly important types of waves in such models: isolated waves, periodic waves, and phase waves. An isolated wave occurs when the resting state of the model is stable, but a sufficiently large perturbation from equilibrium will cause a much larger release of calcium which can then travel across the cell as a single wave. A single nerve impulse is a wave of this type. Periodic waves occur when the resting state is unstable via growing oscillations, and a stable periodic solution exists in the local kinetics. Diffusion of calcium can then organise these local oscillations into trains of periodic waves (Kopell and Howard 1973a; Neu 1979; Maginu 1985; Sherratt 1994; Jafri and Keizer 1997). Many periodic calcium waves are of this type. Phase waves occur when a stable periodic solution exists in the local kinetics, but are not dependent on coupling via diffusion (Ortoleva and Ross 1973; Kopell and Howard 1973b). By starting each part of the domain at a different phase of the oscillations, waves can appear to be traveling across the cell. However, the waves are really just a result of a gradient in oscillation phase across the cell, and can occur in the absence of diffusion.

10.3.1. The Traveling Wave Variable

To study the first two kinds of wave it is easiest to begin by studying the behavior in one spatial dimension. One converts the partial differential equation to a related ordinary differential equation by defining the traveling wave variable $\xi = x + st$, where s is the wave speed. In the traveling wave variable we thus get

$$s\frac{dc}{d\xi} = D_{\text{eff}}\frac{d^2c}{d\xi^2} + f_{\text{eff}}(c), \qquad (5)$$

Isolated waves and periodic waves of (4) correspond, respectively, to homoclinic cycles (solutions that begin and end at the same steady state) and limit cycles of (5). Thus, a detailed study of the kinds of solutions that occur in (5) will give us a great deal of information about the traveling waves in the full model.

10.3.2. A Specific Model

To illustrate this approach, we consider a recent model of calcium wave propagation, the full equations of which are given in the Appendix. This model was originally developed to model calcium oscillations in the apical region of pancreatic acinar cells, and combines a modified version of the model of LeBeau *et al.* (1999) with the IP_3R model of Sneyd and Dufour (2002). In short, this model is based on the following assumptions:

1. Calcium flux is into and out of the ER only. Mitochondrial fluxes are ignored. In addition, there is no contribution to the ER calcium flux from ryanodine receptors. (Neither mitochondria nor ryanodine receptors are present in significant quantity in the apical region of pancreatic cells (Straub *et al.* 2000))
2. Buffering is assumed to be fast, with a large K, and thus can be incorporated by treating all fluxes as effective fluxes.
3. Only calcium in the cytoplasm can diffuse. Calcium in the ER is spatially restricted.
4. The principal calcium flux out of the ER is via a type-II IP_3R, which is modeled by the equations of Sneyd and Dufour (2002). These equations reproduce the correct dynamic behaviour of a hepatocyte type-II IP_3R. Although this is probably not such a good assumption for pancreatic acinar cells which have a predominance of type-III IP_3R (Wojcikiewicz 1995), there is presently no accurate dynamic model of the type-III IP_3R.
5. The principal calcium flux into the ER is via a SERCA ATP-ase calcium pump with saturating kinetics. This pump flux is assumed to be dependent on the ER calcium concentration (Favre *et al.* 1996).
6. Feedback of calcium on the rate of production of IP_3 is ignored. Thus, calcium oscillations can occur in the absence of IP_3 oscillations (Wakui *et al.* 1989).
7. There is a leak of calcium into the cytoplasm from the outside, and this leak increases as IP_3 increases (Shuttleworth 1999). Calcium is pumped out of the cell by plasma membrane ATP-ase pumps.

We begin by studying the behavior of the model in the absence of diffusion. Plotting the steady state of the model as a function of the concentration of IP_3, p, gives the curve labelled ss in Fig. 2. Stable portions of the curve are drawn as solid, while unstable portions are drawn dashed. As p increases the steady state value of c does also, due to the increase in calcium entry across the cell membrane. At two intermediate values of p Hopf bifurcations occur. The exact mathematical nature of these points is not crucial, we need only know that it is at such bifurcations that the steady state curve changes stability, and oscillations can arise. The curve labelled osc denotes the maximum value of c over an oscillation. Although it looks like the branch of periodic orbits begins at the lower HB point, this is not so. (For this particular model the lower HB point is irrelevant and we shall ignore it from now on.) In fact, as p decreases, the period of the oscillations gets larger, tending to infinity approximately at the point marked HC.

To look at the properties of traveling waves in this model we first convert to the traveling wave variable, ξ, as described above. We then fix the wave speed, s, at a

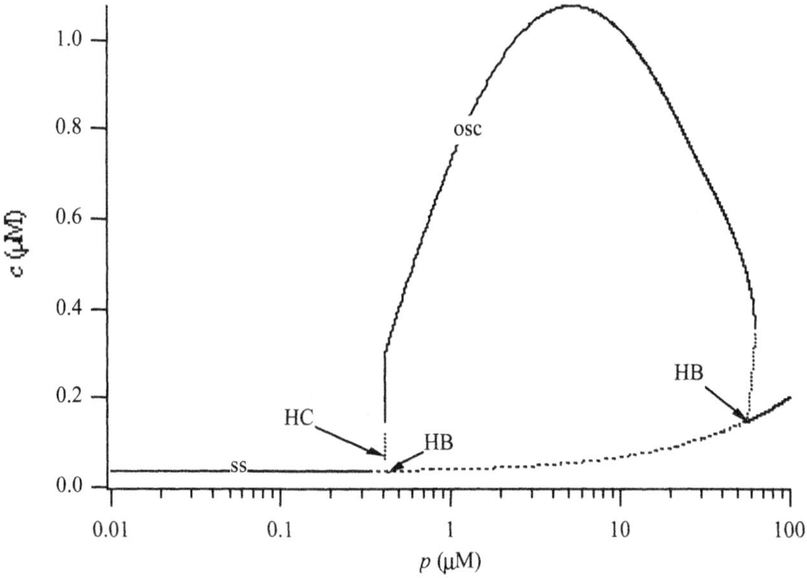

Fig. 2. Bifurcation diagram of the model with no diffusion; ss – steady state curve, osc – the maximum value of *c* over an oscillation, HB – Hopf bifurcation, HC – homoclinic bifurcation. Stable branches are shown as solid lines, unstable branches as dotted lines.

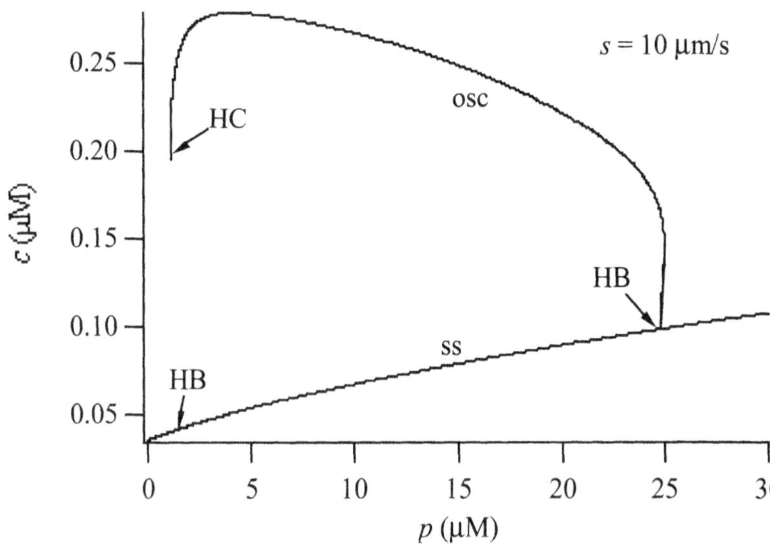

Fig. 3. Bifurcation diagram of the model for the constant wave speed s = 10 μμ s-1 ; ss – steady state curve, osc – the maximum value of *c* over an oscillation, HB – Hopf bifurcation, HC – homoclinic bifurcation.

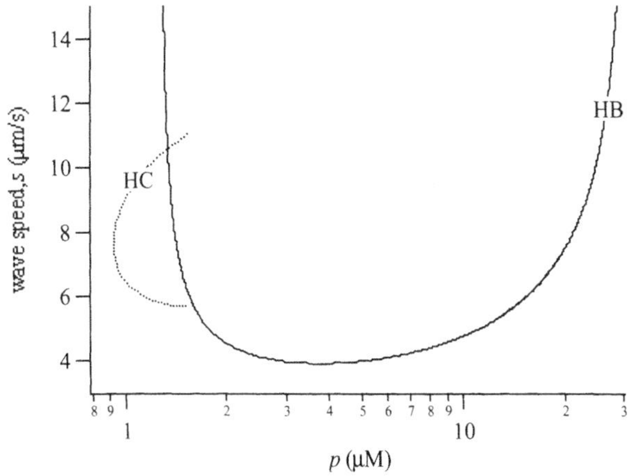

Fig. 4. Two-parameter bifurcation diagram of the model

particular value (in this case s =10 μm s^{-1}) and then plot the steady states and oscillatory solutions as functions of p (Fig. 3). The notation is the same as Fig. 2, except that now stability is not indicated. (The stability of the solutions to the traveling wave equations has no relevance to the stability of the solutions to the original partial differential equation.) As in Fig. 2, the branch of steady states has two Hopf bifurcations, and the branch of periodic orbits ends in a homoclinic bifurcation. Note that now each periodic orbit corresponds to a periodic wave in the original model, while the homoclinic orbit corresponds to an isolated traveling wave.

By themselves, the curves in Fig. 3 are of little interest. However, they do show that an isolated traveling wave of speed 10 μm s^{-1} occurs at approximately p=2. It is more interesting when we track the position of this isolated traveling wave as s varies, as shown in Fig. 4. Here, the curve labeled HC is the curve of homoclinic bifurcations, and thus corresponds to a branch of isolated traveling waves in the model. The curve labeled HB denotes the positions of the Hopf bifurcations as p and s vary. From this diagram we can see the following things. Firstly, isolated traveling waves occur in the model only at lower values of p (p approximately between 1 and 2), and secondly, as p increases, the speed of this isolated wave increases from about 8 to 11 μm s^{-1}. Usually, if an isolated wave exists at a value of p, then two exist, one faster than the other. Numerical computations show that, in general, the faster wave is the stable solution to the model partial differential equations, as observed in other models of excitable systems (Keener and Sneyd 1998). In fact, although there are often multiple isolated traveling wave solutions (some of which are discussed in Sneyd *et al.* 2000 for a different model), numerical computations show that it is always the one on branch HC that appears for realistic initial conditions. This wave has a single peak and is easily found numerically by holding p fixed at some constant level (for which branch HC exists) and raising c in one portion of the domain. If c is raised high enough, this stimulus will develop into a traveling wave and move across the domain. Behind the wave c returns to its resting state and stays there. Thus, a single stimulus gives a single traveling wave. Once p is large enough, we move into the region between the Hopf bifurcations, and thus into a region where limit cycles are expected to exist, although do not necessarily. For these larger values of p one then expects to see periodic waves. However, the precise period of these waves depends on, among other factors, the initial conditions; we shall not consider this difficult question any further here.

It is an interesting, but not well-understood, fact that practically all models of calcium waves behave in the same qualitative manner. There have been relatively few detailed bifurcation analyses of wave propagation in calcium wave models,

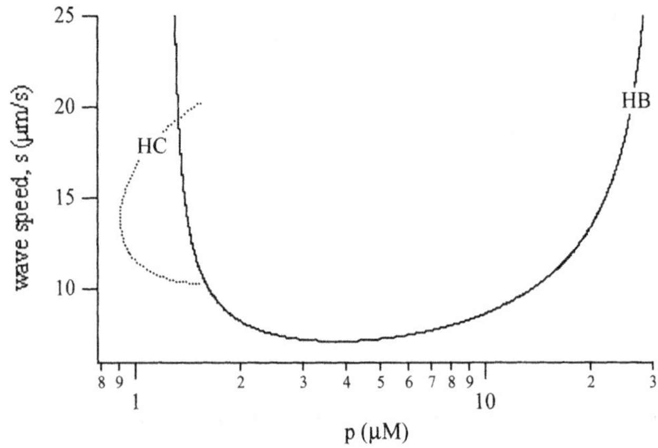

Fig. 5. Two-parameter bifurcation diagram of the model when fast calcium buffering is included explicity (i.e., using (3) as the calcium transport equation). Parameters are D_c=300 $\mu m^2\ s^{-1}$, D_b = 50 $\mu m^2\ s^{-1}$, b_t = 150 μM and K = 10 μM. = 300 $\mu m^2\ s^{-1}$, D_b = 50 $\mu m^2\ s^{-1}$, b_t = 150 μM and K = 10 μM.

but what results there are (Sneyd *et al.* 1993, 2000) show this clearly. It seems to be a generic property of calcium wave models that they have a U-shaped curve of Hopf bifurcations in the p-s plane, and a branch of isolated traveling waves whose speed increases as p does. Most likely this is a result of the fact that calcium release is an excitable system, and thus calcium wave models reflect many of the properties found in other excitable systems such as the FitzHugh-Nagumo or Hodgkin-Huxley models (Keener 1980; Rinzel and Terman 1982; Rinzel and Keener 1983).

It is interesting to investigate briefly the effect that diffusing calcium buffers will have on this wave structure. To study this we study the bifurcations of the full transport equation, (3), written in the traveling wave variable. There is one detail that we must be careful not to overlook, and that is that all fluxes must now be written as net fluxes, not as effective fluxes. Thus, each flux is multiplied by the term $1+b_t/K$, so that, when $c = 0$, the net flux is the same as in the previous model. Referring to (A1) in the appendix, we see that the model equations are then

$$s\frac{dc}{d\xi}=\left(\frac{D_c+D_b\theta}{1+\theta}\right)\frac{d^2c}{d\xi^2}-\frac{2D_b\theta}{(K+c)(1+\theta)}\left(\frac{dc}{d\xi}\right)^2 \tag{6}$$
$$+\left(\frac{1+b_t/K}{1+\theta}\right)\left[(k_fP_{ipr}+J_{er})(c_e-c)-J_{serca}(c,c_e)\right.$$
$$\left.+\delta(J_{in}(p)-J_{pm}(c))\right]$$
$$\frac{1}{\gamma}\frac{\partial c_e}{\partial t}=-(k_fP_{ipr}+J_{er})(c_e-c)+J_{serca}(c,c_e). \tag{7}$$

Note that we have cheated a little and assumed that calcium buffering inside the ER can be modeled by effective fluxes.

In the two-parameter bifurcation diagram, as shown in Fig. 5, we see that explicit inclusion of the diffusion of calcium buffers, and the changes in calcium dynamics caused by the presence of a calcium buffer, can have little qualitative effect on the wave structure of the model (at least at this low level of detail). The major difference is that the speed of the branch of isolated traveling waves, labeled HC, is faster than the previous model, due principally to an increased effective diffusion coefficient because of the diffusion of the buffer. A more detailed analysis of a different model (Slepchenko *et al.* 2000) shows that fast buffering can alter some of the qualitative wave properties, with two stable waves appearing for some parameter values. However, there is yet no comprehensive study of the bifurcation structure of the full transport equation.

10.3.4. Waves in Two Dimensions

There have been a number of modeling studies of the properties of calcium waves in two dimensions (Atri *et al.* 1993; Sneyd and Atri 1993; Dupont and Goldbeter 1994; Jafri and Keizer 1995; Wagner *et al.* 1998; Falcke *et al.* 1999; Fink *et al.* 2000; Wussling *et al.* 2001; for an early review see Sneyd *et al.* 1995b). One of the early goals of such modeling work was to understand how experimentally observed spiral calcium waves (Lechleiter *et al.* 1991) could be created and sustained. Spiral waves can be created in a number of ways, including the breaking of a wave front by a spatial heterogeneity (Atri *et al.* 1993), the bending of a calcium wave around a nucleus (Dupont *et al.* 1996) or by the diffusion of IP_3 in a homogenous medium (McKenzie and Sneyd 1998).

One particularly important calcium wave is the fertilization wave observed in eggs (Fontanilla and Nuccitelli 1998). A recent model of this wave (Wagner *et al.* 1998) indicates that this wave may be generated by a qualitatively different mechanism to that proposed above. In their model the calcium dynamics are bistable and thus the wave is a transition from the lower resting state to the upper resting state. Furthermore, the precise shape of the wave is closely dependent on the spatial distribution of IP_3R.

10.4. Intercellular Waves

As well as propagating through a single cell, calcium waves are also commonly observed traveling between cells, possibly spreading over large regions of tissue, or cultured cells (Sanderson *et al.* 1994; Sanderson 1996; Charles 1998; Thomas *et al.* 1996). The experimental literature on these waves is vast and it is not our intention to review it here. We shall concentrate instead on models of such waves. There are basically two hypotheses for how such waves spread; either by the diffusion of an intracellular messenger (usually either calcium or IP_3) from cell to cell through gap junctions, or by the diffusion of an extracellular messenger such as ATP. It is certain that different cell types and tissues propagate intercellular calcium waves by different mechanisms, and likely that both mechanisms coexist in most cell types, but are of differing relative importance.

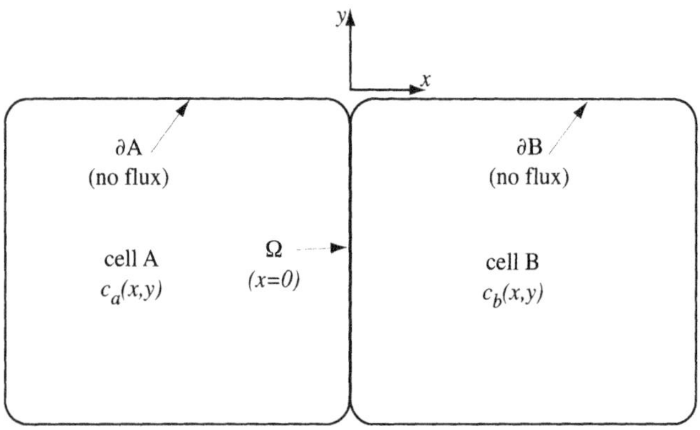

Fig. 6. Schematic diagram of two cells coupled by gap junctions along the curve Ω.

The first model of an intercellular calcium wave was constructed in an attempt to understand the propagation of a mechanically-stimulated wave over a field of cultured tracheal epithelial cells, and assumed that the wave was propagated by the passive diffusion of IP_3 through gap junctions (Sneyd *et al.* 1995a). The basic approach for models of this type is sketched in Fig. 6. Given two cells, A and B, the model assumes that the diffusing messenger (IP_3 in this case) can move from cell to cell at any point of the common boundary, and that the rate at which it diffuses is proportional to the difference in concentration between the cells. This is implicitly making the assumption that transport through gap junctions is a linear process, and that the gap junctions themselves are of infinitely small thickness. It is not clear how accurate these assumptions are but they are unlikely to introduce any errors more significant than those already present in the models.

The dynamics inside each of the cells are just given by the usual reaction-diffusion equations. For the purpose of the discussion we shall assume that the intercellular messenger is calcium, as this lets us write down specific equations more easily. However, it should be kept in mind that it is equally possible to assume that IP_3, or indeed any other substance, is diffusing between cells. Let c_a and

c_b denote the calcium concentrations in cells A and B respectively. We then get the equations

$$\frac{\partial c_a}{\partial t} = D_{\mathrm{a,eff}} \nabla^2 c_a + f_{\mathrm{a,eff}}(c_a),$$

$$\frac{\partial c_b}{\partial t} = D_{\mathrm{b,eff}} \nabla^2 c_b + f_{\mathrm{b,eff}}(c_b),$$

with the boundary conditions $\nabla c_a \bullet n_a = 0$ on the external boundary of cell A (labelled ∂A in Fig. 6), and $\nabla c_b \bullet n_b = 0$ on ∂B. Here, n_a and n_b are the normal vectors to ∂A and ∂B respectively. Note that we have allowed for the possibility that calcium could have a different effective diffusion coefficent, or different dynamics, in the two cells. Although we have written the problem in terms of diffusion of calcium through gap junctions, it is often the case that the intercellular diffusion of IP_3 is equally, or more, important. In this case the model equations are modified in the obvious way.

To finish the specification of the problem it remains only to specify the boundary condition on the intercellular border (labelled Ω in Fig. 6). For simplicity we assume this border is a straight line, and orient it along the $x = 0$ axis. Then, at each point of Ω, with coordinates $(0,y)$, the calcium flux from cell A to cell B is assumed to be $P(c_a(0,y)-c_b(0,y))$ for some constant P, called the permeability, with units of distance/time. Since the calcium flux at (x,y) in cell A is given by $-D_{\mathrm{a,eff}} \nabla c_a$, and similarly for the flux in cell B, we have that, on Ω,

$$-D_{\mathrm{a,eff}} \frac{\partial c_a(0,y)}{\partial x} = P(c_a(0,y) - c_b(0,y)) = -D_{\mathrm{b,eff}} \frac{\partial c_b(0,y)}{\partial x}. \qquad (8)$$

Thus, to solve the intercellular wave model involves the (numerical) solution of a set of partial differential equations inside each cell, coupled by the internal boundary condition (8). In general, the solutions will be discontinuous across the intercellular boundaries.

The model of Sneyd *et al.* (1995a) involved square cells laid out in a square grid, for simplicity of numerical computation. They concluded that these intercellular waves were generated by the release of large amount of IP_3 in the stimulated cell, and the passive diffusion of this IP_3 across the culture. Although they drew similar conclusions for mechanically-stimulated intercellular waves in glial cell cultures, more recent experimental data have shown that glial intercellular waves are propagated, at least in part, by an extracellular diffusing messenger (Charles 1998). Although the details of the model have been proven incorrect for other cell types, the original model has highlighted the importance of accurate measurements of the intercellular IP_3 permeability, as well as of the total amount of IP_3 made in the mechanically stimulated cell. There are still no data available on either of these questions, although the models have predicted the values that will be necessary to explain the waves. One of the most interesting open questions remaining is how intercellular diffusion can interact with extracellular diffusing messengers to propagate intercellular waves that partake of properties of both kinds of waves.

Another open question is whether or not intercellular waves in any given situation are propagated by a regenerative, active, mechanism. For the mechanically-stimulated waves in epithelial cell cultures there is a great deal of evidence, both experimental and theoretical, that the intercellular wave is the result of passive diffusion of IP_3, and is not actively propagated. Thus, the wave (which is not a true wave in the mathematical sense) propagates only a limited distance. However, it seems likely that, in other cell types, the intercellular wave is propagated, at least in part, by a regenerative mechanism. This is complicated by the fact that, in all cell types, the intercellular waves eventually stop. How a regenerative mechanism can result in a propagating wave that travels only a limited distance is not at all understood. Or rather, although one can propose mathematical explanations, none of them are physiologically plausible. Although phrases like "partial regeneration" have appeared in the literature, their precise meaning is entirely unclear, and their underlying mechanisms even more so.

More recent intercellular wave models have focused on the problem of how periodic intercellular waves in hepatocytes are coordinated. Experimental data have shown that intercellular waves of calcium sweep periodically across entire lobules of the liver (Robb-Gaspers and Thomas 1995), as well as being closely coordinated across hepatocyte doublets and triplets (Tordjmann *et al.* 1997, 1998). It is still not entirely clear how such periodic waves are organised. Höfer (1999; Höfer *et al.* 2001) has proposed a model in which the diffusion of calcium through gap junctions is sufficient to synchronise the intracellular oscillations into coordinated intercellular waves. Although this is theoretically plausible, the available experimental data seem to lend stronger support to the alternative hypothesis that it is the intercellular diffusion of IP_3 that synchronises the waves (Dupont *et al.* 1999; 2000a,b; Clair *et al.* 2001). Nevertheless, many questions still remain as to how the latter mechanism could lead to such well-organised waves over such long distances. The controversy is by no means resolved.

Appendix

The model equations in one spatial dimension are

$$\frac{\partial c}{\partial t} = D_{\text{eff}}\frac{\partial^2 c}{\partial x^2} + (k_f P_{ipr} + J_{er})(c_e - c) - J_{serca}(c, c_e) + \delta(J_{in}(p) - J_{pm}(c)), \tag{A1}$$

$$\frac{1}{\gamma}\frac{\partial c_e}{\partial t} = -(k_f P_{ipr} + J_{er})(c_e - c) + J_{serca}(c, c_e). \tag{A2}$$

Here, c denotes the cytoplasmic calcium concentration, c_e the concentration of calcium in the ER, and p denotes the IP_3 concentration. Although not strictly necessary, the constant δ is useful to control the overall rate of plasma membrane transport compared to the rate of transport across the ER. $J_{\text{pm}}(c)$ is the rate at which calcium is pumped out of the cell and is given by

$$J_{pm}(c) = \frac{V_{pm}c^2}{(K_{pm})^2 + c^2}. \tag{A3}$$

Similarly, $J_{serca}(c,c_e)$ is the rate at which calcium is pumped from the cytoplasm into the ER. It is assumed to be dependent on c_e also, using a simplified version of the model of Favre *et al.* (1996). The SERCA calcium pumps are assumed to have a higher sensitivity than the plasma membrane pumps and different kinetics. Thus,

$$J_{serca}(c,c_e) = \left(\frac{V_{serca}c}{K_{serca} + c}\right)\left(\frac{1}{c_e}\right). \tag{A4}$$

The question of exactly how to model calcium pumping is not an easy one. Just a few numerical experiments with different types of pumping kinetics shows quickly that the properties of oscillations and waves are highly sensitive to the way in which calcium pumping is modeled. This is not surprising, of course, as the efflux mechanisms are just as important as the influx mechanisms when it comes to generating oscillatory behavior. However, although a great deal of effort has been expended on modeling the dynamics of calcium flux through IP_3 or ryanodine receptors, there has been much less work done on exactly how to model the calcium ATPase pumps in any given cell type. To add to the complications, it appears that, in some cell types, the activity of the calcium ATPases is actively regulated by various agonists (Zhang *et al.* 1992). The above functions for the calcium ATPases were chosen to agree as closely as possible with experimental data, while still giving oscillatory behaviour.

The rate of influx of calcium into the cell is not modeled from first principles. It is known to be an increasing function of p, but the exact mechanism for this dependence is still controversial (Shuttleworth 1999). We thus just use the simplest increasing function that gives reasonable agreement with experimental data;

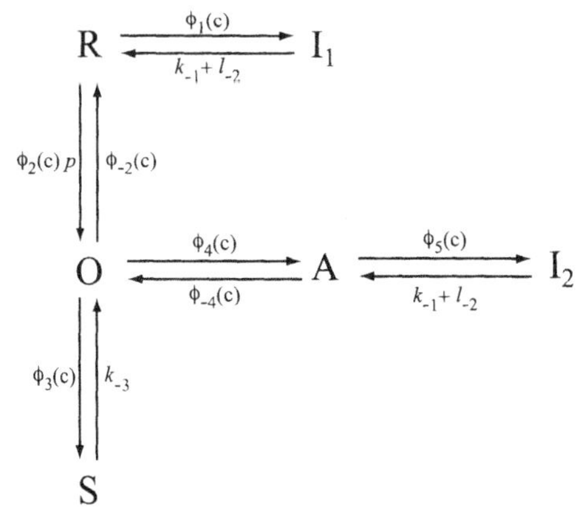

Fig. A1. The state diagram of the IP_3R, after Sneyd and Dufour (2002).

$$J_{in}(p) = 0.2 + 0.05p. \tag{A5}$$

Finally, P_{ipr} is the open probability of the IP_3R, and is governed by its own model, as derived by Sneyd and Dufour (2002). This model is based on experimental data from a hepatocyte type-II IP_3R, and reproduces the experimentally observed dynamic behavior of the IP_3R. In brief summary, the IP_3R is assumed to exist in one of five different states; O - open, S – shut; A – activated; I_1 – first inactivated state; I_2 – second inactivated state. These states are shown in Fig. A1. Movement between the states is governed by mass action kinetics in a more complicated model, which can then be simplified to give the scheme shown in Fig. A1. This simplification results in transitions between states that, although calcium-dependent, nevertheless have a saturating velocity as calcium increases.

We assume that

$$P_{ipr} = (0.1O + 0.9A)^4, \tag{A6}$$

where

$$\frac{dR}{dt} = \phi_{-2}O - \phi_2 pR + (k_{-1} + l_{-2})I_1 - \phi_1 R, \tag{A7}$$

$$\frac{dO}{dt} = \phi_2 pR - (\phi_{-2} + \phi_4 + \phi_3)O - \phi_{-4}A + k_{-3}S, \tag{A8}$$

$$\frac{dA}{dt} = \phi_4 O - \phi_{-4}A - \phi_5 A + (k_{-1} + l_{-2})I_2, \tag{A9}$$

$$\frac{dI_1}{dt} = \phi_1 R - (k_{-1} + l_{-2})I_1, \tag{A10}$$

$$\frac{dI_2}{dt} = \phi_5 A - (k_{-1} + l_{-2})I_2. \tag{A11}$$

The functions ϕ are derived in detail in Sneyd and Dufour (2002), but we give them here for completeness.

$$\phi_1(c) = \frac{(k_1 L_1 + l_2)c}{L_1 + c(1 + L_1/L_3)}, \tag{A12}$$

$$\phi_2(c) = \frac{k_2 L_3 + l_4 c}{L_3 + c(1 + L_3/L_1)}, \tag{A13}$$

$$\phi_{-2}(c) = \frac{k_{-2} + l_{-4}c}{1 + c/L_5}, \tag{A14}$$

$$\phi_3(c) = \frac{k_3 L_5}{L_5 + c}, \tag{A15}$$

$$\phi_4(c) = \frac{(k_4 L_5 + l_6)c}{L_5 + c}, \tag{A16}$$

$$\phi_{-4}(c) = \frac{L_1(k_{-4} + l_{-6})}{L_1 + c}, \tag{A17}$$

$$\phi_5(c) = \frac{(k_1 L_1 + l_2)c}{L_1 + c}. \tag{A18}$$

All the *k*s, *l*s and *L*s are positive constants, and their values are given, together with those of the rest of the model parameters, in Tables 1 and 2.

Table 1. IP_3R parameters

k_1	0.64 μM^{-1} s^{-1}	k_{-1}	0.4 s^{-1}
k_2	37.4 μM^{-1} s^{-1}	k_{-2}	1.4 s^{-1}
k_3	0.11 μM^{-1} s^{-1}	k_{-3}	29.8 s^{-1}
k_4	4 μM^{-1} s^{-1}	k_{-4}	0.54 s^{-1}
L_1	0.12 μM	l_2	1.7 s^{-1}
L_3	0.025 μM	l_4	1.7 μM^{-1} s^{-1}
L_5	54.7 μM	l_6	4707 s^{-1}

Table 2. Cell parameters

J_{er}	0.002 s^{-1}	γ	5.405
V_{serca}	120 μM^2 s^{-1}	K_{serca}	0.18 μM
V_{pm}	28 μM s^{-1}	K_{pm}	0.425 μM
k_f	0.4 s^{-1}	δ	0.1

References

Amundson J, Clapham D (1993) Calcium waves. Curr Opin Neurobiol 3: 375 382

Atri A, Amundson J, Clapham D, Sneyd J (1993) A single-pool model for intracellular calcium oscillations and waves in the *Xenopus laevis* oocyte. Biophys J 65: 1727--1739

Berridge MJ (1993) Inositol trisphosphate and calcium signalling. Nature (Lond) 361: 315-325

Berridge MJ (1997) Elementary and global aspects of calcium signalling. J Physiol 499: 291-306

Bezprozvanny I (1994) Theoretical analysis of calcium wave propagation based on inositol (1,4,5)-trisphosphate ($InsP_3$) receptor functional properties. Cell Calcium 16: 151-66

Bugrim AE, Zhabotinsky AM, Epstein IR (1997) Calcium waves in a model with a random spatially discrete distribution of Ca^{2+} release sites. Biophys J 73: 2897-906

Charles A (1998) Intercellular calcium waves in glia. Glia 24: 39-49

Clair C, Chalumeau C, Tordjmann T, Poggioli J, Erneux C, Dupont G, Combettes L (2001) Investigation of the roles of Ca^{2+} and $InsP_3$ diffusion in the coordination of Ca^{2+} signals between connected hepatocytes. J Cell Sci 114: 1999-2007

Clapham D (1995) Calcium signaling. Cell 80: 259-268

Coombes S (2001) The effect of ion pumps on the speed of travelling waves in the fire-diffuse-fire model of Ca^{2+} release. Bull Math Biol 63: 1-20

Cuthbertson KSR, Chay TR (1991) Modelling receptor-controlled intracellular calcium oscillators. Cell Calcium 12: 97-109

Dawson SP, Keizer J, Pearson JE (1999) Fire-diffuse-fire model of dynamics of intracellular calcium waves. Proc Natl Acad Sci USA 96: 6060-6063

Dupont G, Goldbeter A (1992) Oscillations and waves of cytosolic calcium: insights from theoretical models. BioEssays 14: 485-493

Dupont G, Goldbeter A (1994) Properties of intracellular Ca^{2+} waves generated by a model based on Ca^{2+}-induced Ca^{2+} release. Biophys J 67: 2191-2204

Dupont G, Pontes J, Goldbeter A (1996) Modeling spiral Ca^{2+} waves in single cardiac cells: role of the spatial heterogeneity created by the nucleus. Am J Physiol (Cell Physiol) 271: C1390-C1399

Dupont G, Tordjmann T, Clair C, Swillens S, Claret M, Combettes L (2000a) Mechanism of receptor-oriented intercellular calcium wave propagation in hepatocytes. FASEB J 14: 279-289

Dupont G, Swillens S, Clair C, Tordjmann T, Combettes L (2000b) Hierarchical organization of calcium signals in hepatocytes: from experiments to models. Biochim Biophys Acta 1498: 134-52

Falcke M, Hudson JL, Camacho P, Lechleiter JD (1999) Impact of mitochondrial Ca^{2+} cycling on pattern formation and stability. Biophys J 77: 37-44

Falcke M, Or-Guil M, Bar M (2000) Dispersion gap and localized spiral waves in a model for intracellular Ca^{2+} dynamics. Phys Rev Lett 84: 4753-6

Favre CJ, Schrenzel J, Jacquet J, Lew DP, Krause KH (1996) Highly supralinear feedback inhibition of Ca^{2+} uptake by the Ca^{2+} load of intracellular stores. J Biol Chem 271: 14925-30

Fink CC, Slepchenko B, Moraru II, Watras J, Schaff JC, Loew LM (2000) An image-based model of calcium waves in differentiated neuroblastoma cells. Biophys J 79: 163-183

FitzHugh R (1961) Impulses and physiological states in theoretical models of nerve membrane. Biophys J 1: 445-466

Fontanilla RA, Nuccitelli R (1998) Characterization of the sperm-induced calcium wave in Xenopus eggs using confocal microscopy. Biophys J 75: 2079-87

Hodgkin AL, Huxley AF (1952) A quantitative description of membrane current and its application to conduction and excitation in nerve. J Physiol, Lond 117: 500-544

Hofer T (1999) Model of intercellular calcium oscillations in hepatocytes: synchronization of heterogeneous cells. Biophys J 77: 1244-1256

Hofer T, Politi A, Heinrich R (2001) Intercellular Ca^{2+} wave propagation through gap-junctional Ca^{2+} diffusion: a theoretical study. Biophys J 80: 75-87

Jafri MS, Keizer J (1995) On the roles of Ca^{2+} diffusion, Ca^{2+} buffers and the endoplasmic reticulum in IP_3-induced Ca^{2+}waves. Biophys J 69: 2139-2153

Jafri MS (1995) A theoretical study of cytosolic calcium waves in *Xenopus* oocytes. J Theor Biol 172: 209-216

Jafri MS, Keizer J (1997) Agonist-induced calcium waves in oscillatory cells: a biological example of Burger's equation. Bull Math Biol 59: 1125--1144

Keener JP (1980) Waves in excitable media. SIAM J Appl Math 39: 528-548

Keizer J, Young GWD (1992) Two roles of Ca^{2+} in agonist stimulated Ca^{2+} oscillations. Biophys J 61: 649-60

Keizer J, Smith GD, Ponce-Dawson S, Pearson JE (1998) Saltatory propagation of Ca^{2+} waves by Ca^{2+} sparks. Biophys J 75: 595-600

Kopell N, Howard LN (1973a) Plane wave solutions to reaction-diffusion equations. Stud Appl Math 52: 291-328

Kopell N, Howard LN (1973b) Horizontal bands in the Belousov reaction. Science 180: 1171--1173

Kummer U, Olsen LF, Dixon CJ, Green AK, Bornberg-Bauer E, Baier G (2000) Switching from simple to complex oscillations in calcium signaling. Biophys J 79: 1188-95

Kupferman R, Mitra PP, Hohenberg PC, Wang SS (1997) Analytical calculation of intracellular calcium wave characteristics. Biophys J 72: 2430-44

LeBeau AP, Yule DI, Groblewski GE, Sneyd J (1999) Agonist-dependent phosphorylation of the inositol 1,4,5-trisphosphate receptor: a possible mechanism for agonist-specific calcium oscillations in pancreatic acinar cells. J Gen Physiol 113: 851-871

Lechleiter J, Girard S, Peralta E, Clapham D (1991) Spiral calcium wave propagation and annihilation in *Xenopus laevis* oocytes. Science 252: 123-126

Maginu K (1985) Geometrical characteristics associated with stability and bifurcations of periodic travelling waves in reaction-diffusion equations. SIAM J Appl Math 45: 750-774

McKenzie A, Sneyd J (1998) On the formation and breakup of spiral waves of calcium. Int J Bif Chaos 8: 2003-2012

Meyer T, Stryer L (1988) Molecular model for receptor-stimulated calcium spiking. Proc Natl Acad Sci USA 85: 5051-5055

Naraghi M, Muller TH, Neher E (1998) Two-dimensional determination of the cellular Ca^{2+} binding in bovine chromaffin cells. Biophys J 75: 1635-47

Neu JC (1979) Chemical waves and the diffusive coupling of limit cycle oscillators. SIAM J. Appl. Math. 36: 509--515

Nowycky MC, Pinter MJ (1993) Time courses of calcium and calcium-bound buffers following calcium influx in a model cell. Biophys J 64: 77-91

Ortoleva P, Ross J (1973) Phase waves in oscillatory chemical reactions. J. Chem. Phys. 58: 5673-5680

Parker I, Choi J, Yao Y (1996) Elementary events of $InsP_3$-induced Ca^{2+} liberation in Xenopus oocytes: hot spots, puffs and blips. Cell Calcium 20: 105-21

Putney JW, Bird GSJ (1993) The inositol phosphate-calcium signaling system in nonexcitable cells. Endocrine Rev 14: 610-631

Rinzel J, Terman D (1982) Propagation phenomena in a bistable reaction-diffusion system. SIAM J Appl Math 42: 1111-1137

Rinzel J, Keener JP (1983) Hopf bifurcation to repetitive activity in nerve. SIAM J. Appl. Math. 43: 907-922

Robb-Gaspers LD, Thomas AP (1995) Coordination of Ca^{2+} signaling by intercellular propagation of Ca^{2+} waves in the intact liver. J Biol Chem 270: 8102-8107

Røttingen J-A, Iversen J-G (2000) Ruled by waves? Intracellular and intercellular calcium signalling. Acta Physiol Scand 169: 203-219

Sala F, Hernández-Cruz A (1990) Calcium diffusion modeling in a spherical neuron: relevance of buffering properties. Biophys J 57: 313-324

Sanderson MJ, Charles AC, Boitano S, Dirksen ER (1994) Mechanisms and function of intercellular calcium signaling. Mol Cell Endocrin 98: 173-187

Sanderson MJ (1996) Intercellular waves of communication. News Physiol Sci 11: 262-269

Sherratt JA (1994) On the evolution of periodic plane waves in reaction-diffusion systems of λ–ω type. SIAM J Appl Math 54: 1374-1385

Shuttleworth TJ (1999) What drives calcium entry during $[Ca^{2+}]_i$ oscillations? - challenging the capacitative model. Cell Calcium 25: 237-246

Slepchenko BM, Schaff JC, Choi YS (2000) Numerical approach to fast reactions in reaction-diffusion systems: application to buffered calcium waves in bistable models. J Comput Phys 162: 186-218

Sneyd J, Girard S, Clapham D (1993) Calcium wave propagation by calcium-induced calcium release: an unusual excitable system. Bull Math Biol 55: 315-344

Sneyd J, Atri A (1993) Curvature dependence of a model for calcium wave propagation. Physica D 65: 365-372

Sneyd J, Kalachev L (1994) A profile analysis of propagating calcium waves. Cell Calcium 15: 289-296

Sneyd J, Wetton B, Charles AC, Sanderson MJ (1995a) Intercellular calcium waves mediated by diffusion of inositol trisphosphate: a two-dimensional model. American Journal of Physiology (Cell Physiology) 268: C1537-C1545

Sneyd J, Keizer J, Sanderson MJ (1995b) Mechanisms of calcium oscillations and waves: a quantitative analysis. FASEB J 9: 1463-1472

Sneyd J, Sherratt J (1997) On the propagation of calcium waves in an inhomogeneous medium. SIAM J Appl Math 57: 73-94

Sneyd J, Dale P, Duffy A (1998) Traveling waves in buffered systems: applications to calcium waves. SIAM J Appl Math 58: 1178-1192

Sneyd J, LeBeau A, Yule D (2000) Traveling waves of calcium in pancreatic acinar cells: model construction and bifurcation analysis. Physica D 145: 158-179

Sneyd J, Dufour JF (2002) A dynamic model of the type-2 inositol trisphosphate receptor. Proc Natl Acad Sci USA 99: 2398-403

Straub SV, Giovannucci DR, Yule DI (2000) Calcium wave propagation in pancreatic acinar cells: functional interaction of inositol 1,4,5-trisphosphate receptors, ryanodine receptors, and mitochondria. J Gen Physiol 116: 547-560

Swillens S, Dupont G, Combettes L, Champeil P (1999) From calcium blips to calcium puffs: theoretical analysis of the requirements for interchannel communication. Proc Natl Acad Sci USA 96: 13750-5

Thomas AP, Bird GSJ, Hajnóczky G, Robb-Gaspers LD, Putney JWJ (1996) Spatial and temporal aspects of cellular calcium signaling. FASEB J 10: 1505--1517

Tordjmann T, Berthon B, Claret M, Combettes L (1997) Coordinated intercellular calcium waves induced by noradrenaline in rat hepatocytes: dual control by gap junction permeability and agonist. Embo J 16: 5398-407

Tordjmann T, Berthon B, Jacquemin E, Clair C, Stelly N, Guillon G, Claret M, Combettes L (1998) Receptor-oriented intercellular calcium waves evoked by vasopressin in rat hepatocytes. Embo J 17: 4695-703

Wagner J, Keizer J (1994) Effects of rapid buffers on Ca^{2+} diffusion and Ca^{2+} oscillations. Biophys J 67: 447-456

Wagner J, Li Y-X, Pearson J, Keizer J (1998) Simulation of the fertilization Ca^{2+} wave in *Xenopus laevis* eggs. Biophys J 75: 2088-97

Wakui J, Potter BVL, Petersen OH (1989) Pulsatile intracellular calcium release does not depend on fluctuations in inositol trisphosphate concentration. Nature 339: 317-320

Wojcikiewicz RJH (1995) Type I, II, and III inositol 1,4,5-trisphosphate receptors are unequally susceptible to down-regulation and are expressed in markedly different proportions in different cell types. J Biol Chem 270: 11678-11683

Wussling MH, Krannich K, Drygalla V, Podhaisky H (2001) Calcium waves in agarose gel with cell organelles: implications of the velocity curvature relationship. Biophys J 80: 2658-66

Zhang BX, Zhao H, Loessberg P, Muallem S (1992) Activation of the plasma membrane Ca^{2+} pump during agonist stimulation of pancreatic acini. J Biol Chem 267: 15419-25

11. Mathematical Modelling and High Resolution Fluorescence Imaging for Studying Calcium Regulation

Dietmar Uttenweiler

Medical Biophysics Unit, Institute of Physiology and Pathophysiology, Ruprecht-Karls-Universität Heidelberg, Im Neuenheimer Feld 326, D-69120 Heidelberg, dietmar.uttenweiler@urz.uni-heidelberg.de, http://www.compbiophysics.uni-hd.de

Abstract. The aim of this chapter is to describe and discuss a fundamental concept for studying cellular signalling. It is the combination of state of the art experimental techniques, in particular high resolution fluorescence imaging, with detailed spatio-temporal mathematical models of intracellular calcium regulation. This approach provides a powerful tool to elucidate the very complex mechanisms involved in cellular Ca^{2+}-signalling and allows to correct for inherent experimental limitations and biases. It has proved to be particularly successful, since experimental findings can be directly incorporated into mathematical models. In principle, this allows to use the most accurate assumptions and parameters for the model calculations and vice versa, model predictions can be directly tested in experiments.

11.1. Introduction

The advances in modern fluorescence microscopy have led to a much deeper understanding of cellular function on the basis of the underlying molecular architecture. Especially, the possibility of measuring intracellular ion concentrations with high spatial and temporal resolution has significantly contributed to our knowledge about the mechanism of cellular signalling and the very central role of Ca^{2+}-ions as the ubiquitous 'second messenger' in cells.

The advances in experimental techniques have helped to reveal many new and exciting concepts of cellular Ca^{2+}-signalling, as described in the various chapters of this book. In the past decade for example, the importance of spatial gradients in Ca^{2+}-concentration and localized sub-cellular Ca^{2+}-release events, in addition to the many important aspect of temporal changes in global Ca^{2+}-concentration, have been shown to be of fundamental importance in cellular signalling.

The aim of this chapter is to describe and discuss a fundamental concept for studying cellular signalling. It is the combination of state of the art experimental techniques, in particular high resolution fluorescence imaging, with detailed spatio-temporal mathematical models of intracellular calcium regulation. This approach provides a powerful tool to elucidate the very complex mechanisms involved in cellular Ca^{2+}-signalling and has proved to be particularly successful, since experimental findings can be directly incorporated into mathematical models. In principle, this allows to use the most accurate assumptions and parameters for the model calculations. Vice versa, model predictions can be directly tested in experiments and thereby this approach can in principle be used for closed loop systems to optimise the studies on cellular signalling and Ca^{2+}-regulation.

In the following this concept will be presented by describing some fundamental properties of cellular systems with respect to Ca^{2+}-ion regulation as well as by shortly discussing modern high resolution fluorescence imaging techniques to measure Ca^{2+}-ion concentrations. Subsequently, different modelling approaches for the interpretation of the experimental data will be presented. This spans from simple compartment models for Ca^{2+}-ion handling, which are particularly useful for the effective simulation of the temporal behaviour of cellular systems, to spatially resolved mathematical modelling approaches for the most detailed analysis of cellular regulation processes. Additionally, an example will be given for the treatment of geometrically complex cellular structures in mathematical models. It is a model which was developed to study the depletion of Ca^{2+}-ions in the spatially restricted diffusion space of the transverse tubular system in skeletal muscle fibres. Then, spatio-temporal models for elementary Ca^{2+}-release events will be presented. Particular emphasis will be laid on the combination of experiments using high resolution confocal fluorescence microscopy with the model approaches and on the limitations of ion concentration determinations with fluorescence indicators in general.

11.2. Basic Principles of Ca^{2+}-Ion Regulation

Ca^{2+}-ions inside a cell are present in various states as shown in Fig. 1. In contrast to most other ions, free Ca^{2+}-ions are present in very low cytosolic concentrations, ranging between 10-100 nM. Additionally, Ca^{2+}-ions are bound to intrinsic and extrinsic intracellular buffers with various amounts of Ca^{2+}-binding sites. In experiments using fluorescence techniques, the fluorescence indicator very often acts as the main extrinsic Ca^{2+}-ion buffer itself. Also, membrane bound organelles, as e.g. the endoplasmatic reticulum (ER) or mitochondria can sequester Ca^{2+}-ions to a very large extend. These different states of Ca^{2+}-ions interact by complex processes. In addition to the transport of Ca^{2+}-ions by diffusion, Ca^{2+}-ions are released from intracellular stores and actively transported by cellular proteins.

Ca^{2+}-signalling mostly involves transient increases in the intracellular free Ca^{2+}-ion concentration (Ca^{2+}-transients). Ca^{2+}-ions are derived from two major sources. Either they enter the cell from the extracellular space through ion channels in the external cell membrane driven by the large concentration gradient. Or they are released from the internal Ca^{2+}-ion stores, the ER and the sarcoplasmic reticulum (SR) in muscle cells, where the free Ca^{2+}-ion concentrations are in the mM-range, similar to the extracellular space. Ryanodine receptors (RyRs) and inositol 1,4,5-triphosphate receptors (IP_3Rs) have been identified as the two major classes of Ca^{2+}-release channels in intracellular ion stores known to date (Berridge, 1997). In this chapter we will focus on the RyRs, which exist as homotetrameric protein complexes composed of four very large subunits of ≈ 560 kDa and four small 12 kDa FK506 binding proteins (Meissner, 1994). Three RyR isoforms have been found in mammals, RyR1 the skeletal, RyR2 the cardiac and RyR3 the brain type isoform. Despite this nomenclature the isoforms are widely expressed in excitable cells (Giannini et al., 1995) with RyR1 being the most abundant in skeletal and RyR2 the most abundant in cardiac muscle. Three-dimensional reconstructions have shown that about 80 % of the mass forms a cytoplasmic assembly (Sharma et al, 2000) which in the case of skeletal muscle can also be identified as the foot-structure in cryo-electron micrographs (Franzini-Armstrong et al., 1999). RyRs are activated by micromolar concentrations of Ca^{2+}-ions via an activating (high-affinity) binding site, and are inhibited by millimolar concentrations of Ca^{2+}-ions via an inhibiting (low-affinity) binding site. Mg^{2+} -ions competitively bind to both sites. Binding to the high-affinity site does not result in a channel opening in contrast to binding to the low-affinity site, which results in a channel inactivation. Thus, the free intracellular Mg^{2+}-concentration of about 0.5-1 mM has a strong inhibitory effect preventing excessive spontaneous opening of RyRs. In addition to Mg^{2+}-ions, ATP and caffeine have also been shown to significantly affect RyRs (Meissner, 1994).

During Ca^{2+}-ion release from internal stores and thereafter numerous processes are involved in the interaction of the ion store and the cytosol. Since release and re-uptake of Ca^{2+}-ions is equivalent with the movement of electrical charges, a simultaneous counter ion movement is necessary to preserve electro-neutrality (Fink and Veigel, 1996). Active transport proteins (Ca^{2+}-ATPases) are responsible for the re-filling of the stores and under resting conditions also for the constant maintenance of the concentration gradient across the SR membrane. The required energy is taken from ATP hydrolysis. Mostly, Ca^{2+}-ions are also highly buffered inside the stores, ensuring a high release capacity even in the small volumes of these stores.

For a detailed understanding of Ca^{2+}-regulation one has to consider all the above mentioned processes and interactions. In addition to the movement of free Ca^{2+}-ions, the soluble Ca^{2+}-buffers, either in their free or Ca^{2+}-bound form, undergo diffusional transposition. It should also be noted that binding of Ca^{2+}-ions to buffers, including the fluorescence indicator, is kinetically limited by the kinetic on- and off-rate constants of the buffer. Therefore, for very fast changes in free Ca^{2+}-ion concentration, buffers may not be in equilibrium with free Ca^{2+}-ions. This is especially important to consider and to compensate for when using fluorescence indicators to determine intracellular Ca^{2+}-ion concentrations.

It should also be noted that fluorescence indicators themselves affect the free Ca^{2+}-ion concentration due to their own binding affinity. Also, the distribution of the dye inside the cell is very often not known in detail. Binding to the surface or clustering inside membrane bound organelles can significantly alter the fluorescence signal obtained in the various experimental situations.

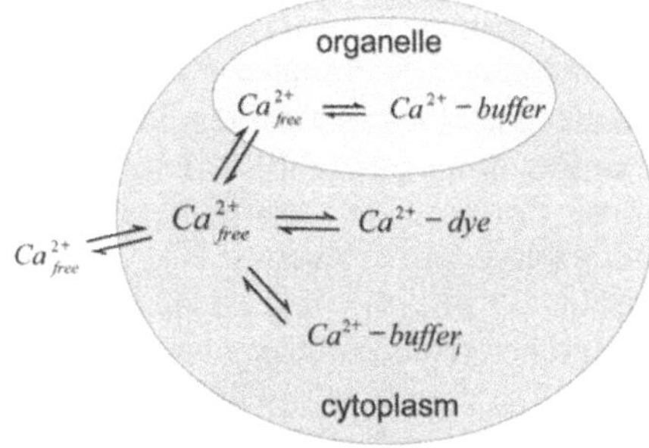

Fig. 1. The complex interplay of calcium ions in a cellular environment. Ca^{2+}-ions are present as either free ions or bound to various buffer systems, including the fluorescence indicators used in fluorescence microscopic experiments. Additionally, Ca^{2+}-ions are sequestered and bound in organelles.

11.3. High Resolution Fluorescence Imaging Techniques

In the past two decades fluorescence techniques have become the most popular tools for the determination of intracellular ion concentrations. In addition to the high spatial and temporal resolution of modern fluorescence imaging techniques, sensitivity and selectivity for specific ions are among the benefits. Generally, fluorescence indicators for measuring Ca^{2+}-ion concentrations need to have high affinities due to the abovementioned low cytoplasmic free Ca^{2+}-ion concentrations. Additionally, they need a high selectivity against Mg^{2+}-ions as these ions competitively bind to the dye. This effect is reinforced due to the higher cytoplasmic Mg^{2+}-concentrations in the mM-range in comparison to 100 nM for Ca^{2+}-ions.

11.3.1. Linear Fluorescence Microscopy

Figure 2 shows the three most popular fluorescence imaging techniques for high spatial and temporal resolution ion concentration measurements. Wide field epi-fluorescence microscopy (Fig. 2A) has particularly benefited from the development of ratiometric fluorescence indicators (Grynkiewicz et al., 1985), which was mainly done by the lab of R.Y. Tsien. These dyes exhibit a strong shift in their excitation or emission spectrum upon binding of Ca^{2+}-ions. This allows dual-excitation or dual-emission ratio measurements. The most popular ratiometric Ca^{2+}-indicator, Fura-2, is designed to shift its maximum excitation wavelength towards shorter wavelengths upon binding of Ca^{2+}-ions. Sampling the fluorescence intensities at two excitation wavelengths (mostly at 340 nm and at

380 nm) the ratio of the two emission intensities is independent of such critical quantities like dye concentration, illumination intensities etc. This allows the quantitative measurement of global cellular Ca^{2+}-ion concentrations with high accuracy.

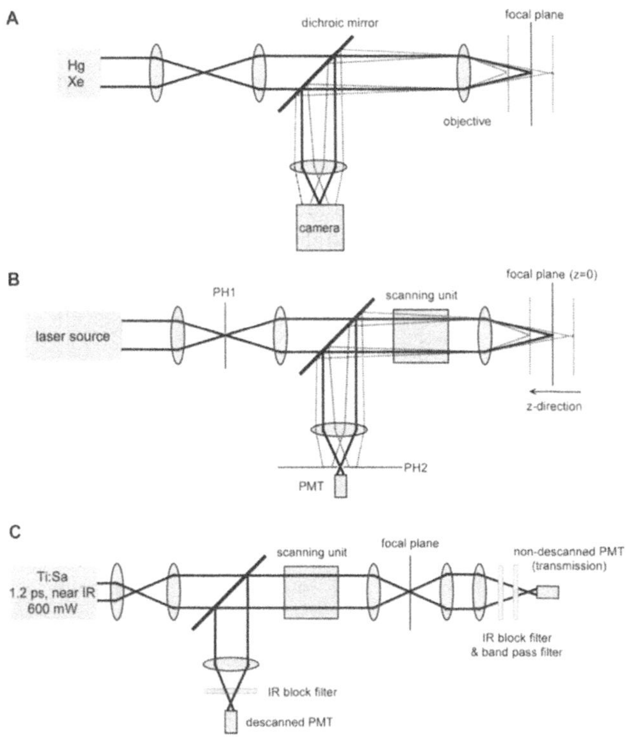

Fig. 2. Principle of epi-fluorescence, confocal laser scanning microscopy and non-linear microscopy. **A** Epi fluorescence techniques are widely used for the determination of global cellular changes in ion concentrations. The Stokes-shifted fluorescence emission can be separated from the excitation light by a dichroic mirror and is reflected onto the detector, typically a CCD-camera. **B** Confocal microscopy uses a focused laser beam for exciting the fluorescence indicator. Fluorescence emission from the focal plane is focused onto the detection pinhole (PH2) and detected via a photomultiplier (PMT). Fluorescence light emitted from out-of-focus planes is effectively blocked by the detection pinhole. **C** In non-linear fluorescence microscopy ultra-short laser pulses are focused into the preparation. In the case of multiphoton fluorescence microscopy, the fluorescent indicator is excited by two or more photons which are absorbed simultaneously. Only a small focal volume is excited and axial resolutions comparable to confocal microscopy can be obtained without a detection pinhole. The fluorescence emission can either be detected in the transmitted beam (non-descanned PMT) or in the backwards direction (descanned PMT). In addition other non-linear techniques can be used, as e.g. 'Second Harmonic Generation (SHG)'-microscopy.

Generally, the lateral resolution in epi-fluorescence microscopy is in the range of 250-350 nm, depending on the numerical aperture (N.A.) of the microscope objective used and the wavelength of the excitation light.

If spatial cellular concentration gradients and sub-cellular processes are of interest, fluorescence imaging techniques with high axial resolution have to be used. Using epi-fluorescence techniques the acquired image inherently contains information from the entire depth (in the axial microscopic z-direction) of the preparation. Therefore sub-cellular events can only be recorded with appropriate high axial optical resolution.

Confocal laser scanning microscopy and multiphoton fluorescence microscopy are the two most widely used methods when axial resolution is required. Figure 2B shows the principle of confocal laser scanning microscopy. A laser beam is focused on the specimen and the emitted fluorescence is detected with a photomultiplier. The axial resolution is achieved by effectively blocking out-of-focus fluorescence by the detection pinhole. The intensity distribution I in the axial z-direction of the optical system is theoretically given by (Wilson, 1990):

$$I(u) = \left(\frac{\sin(u/4)}{u/4}\right)^4 \text{ with } \quad u = \frac{8\pi}{\lambda}\sin^2(\alpha/2)\cdot z, \qquad (1)$$

with u being the normalized axial optical coordinate in z-direction, λ the wavelength and $\sin(\alpha/2)$ the numerical aperture (N.A.) of the objective. In comparison to a conventional epi-fluorescence microscope the lateral and axial resolution as derived from the full width at half maximum of the intensity distributions of a confocal microscope is improved by a factor $\sqrt{2}$. In order to fully understand the optical sectioning capability one has to look at the integrated intensity distribution. It is given by (Wilson, 1990):

$$I_{\text{int}}(u) = \int_0^\infty I(u,\mathrm{v})\,\mathrm{v}\,d\mathrm{v}, \qquad (2)$$

with v being the normalized optical unit in lateral direction. Figure 3 shows the integrated intensity as a function of the normalized axial coordinate u. For a conventional epi-fluorescence microscope the integrated intensity remains constant, in contrast to a decline according to an inverse square law in the confocal case. This fact is the basis for the optical sectioning capability of confocal microscopes in contrast to conventional epi-fluorescence microscopes. Thus, axial resolutions of about 0.8-1.2 μm can be achieved, again depending on the N.A. of the objective and the wavelength of the excitation light. The lateral resolution is only slightly improved in comparison to epi-fluorescence microscopy.

As confocal microscopy is a point scanning method an additional scanning unit has to be used in order to allow image formation. In most cases this is realized by using galvanometric mirrors for the deflection of the excitation light in lateral (x- and y-) direction, respectively. For the acquisition of 3D-data either the objective is moved or the specimen is mounted on a piezo-driven microscope stage.

The temporal resolution of both linear fluorescence microscopy methods is dependent on the image acquisition method. In the case of epi-fluorescence illumination the temporal resolution is only limited by the read-out speed of the camera system. If dual excitation ratiometric measurements are used the temporal resolution is also limited by the interval of changes in excitation light (Uttenweiler et al., 1995). Using modern CCD technology, frame rates of up to 100 Hz can be achieved with this approach.

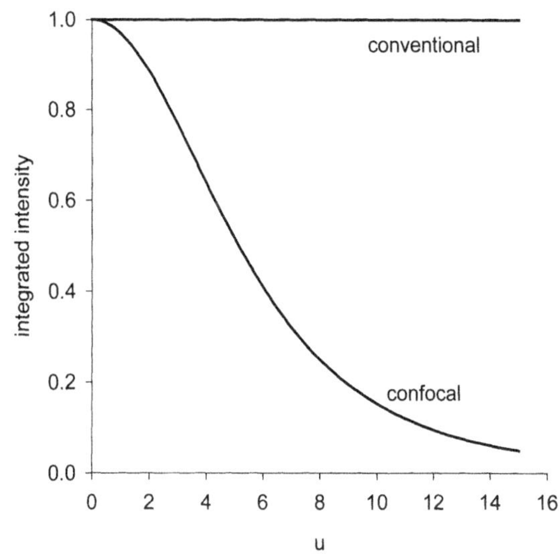

Fig. 3. Principle shape of the integrated intensity of a confocal and a normal epi-fluorescence microscope. u is the normalised axial z-coordinate. The decline of the integrated intensity for larger distances relative to the focal plane is the basis for the optical sectioning capability of confocal microscopy. According to Wilson (1990).

For confocal microscopy the temporal resolution is mainly limited by the galvanometric scanning units most commonly used. Typical frame rates of up to 10 Hz can be achieved. For the measurement of high dynamic physiological processes in the ms-range one has to restrict the measurement to a single spatial dimension, resulting in so called line-scan images. A typical temporal resolution of up to 0.5 ms per line is achieved by modern instruments.

Two general aspects of fluorescence imaging should also be considered. Photobleaching, the irreversible destruction of fluorophores, and photodamage, the unwanted interaction of the excitation photons with the specimen, are both side effects which should be avoided. Generally both effects are dependent on the excitation energy and therefore one should reduce the illumination intensity to the minimal level at which a reasonable signal to noise ratio in the detected fluorescence is obtained.

11.3.2. Non-Linear Microscopy

Multiphoton fluorescence microscopy (Denk et al., 1990) shown in Fig. 2C has become a popular technique especially for measuring ion concentrations in thick preparations. In principle a multiphoton set-up is very similar to a confocal microscope with two main modifications: i.) the detection pinhole can be removed or fully opened ii.) instead of a visible c_w-laser for the excitation of the fluorophore an ultra-short pulsed laser source has to be used. In multiphoton fluorescence the fluorophore is excited by the simultaneous absorption of two or more photons instead of one. Therefore the excitation wavelengths for two-photon excitation of typical fluorophores are in the near infrared. For two photon excitation, the non-linear optical absorption process is limited to the focal region of the objective due to an I^2-dependence of the absorption probability. Hence, this method has an intrinsic axial resolution by avoiding the excitation of background fluorescence and consequently there is no need of rejecting excited fluorescence photons from out-of-focus planes by a detection pinhole. In order to achieve a reasonable two-photon absorption probability, high pulse intensities are required, which are typically accomplished with mode-locked lasers producing ultrashort pulses of 100 fs-10 ps in the near infrared at repetition rates around 100 MHz. The main benefits of multiphoton microscopy are the larger penetration depth of infrared light into thick specimen or tissue and reduced out-of-focus photobleaching and photodamage. For future applications it is important to note that with minor modifications multiphoton setups can also be used for other non-linear optical measurements techniques like for example 'Second Harmonic Generation' (SHG) imaging (Campagnola et al., 2001, 1999), which promises a high potential especially for physiological measurements in living cells.

11.3.3. Combination of Experiments and Modelling

For fluorescence techniques in general it is important to note that all fluorescent indicators have an intrinsic kinetic delay to changes in their environment. Therefore, one has to explicitly consider and correct for the kinetic on- and off-rate constants in a subsequent analysis of high temporally resolved fluorescence data. The kinetic on-rate constant (k_{on}) and the off-rate constant (k_{off}) of a chemical reaction is related to the dissociation constant by

$$K_D = k_{off} / k_{on} \,. \tag{3}$$

The complex internal structure of cells and the elaborate mechanism of Ca^{2+}-regulation as well as the inherent limitations of experimental techniques emphasize the necessity to use mathematical models in order to fully understand how Ca^{2+}-ions precisely regulate so many different cellular and molecular processes. The full benefit of this combination is especially given, when the tight interplay of experimental measurements and mathematical models allows to use the advantage of a feedback based approach.

To summarize the necessity of using mathematical models in conjunction with high resolution fluorescence imaging techniques, the following issues have to be carefully considered:

- Ca^{2+}-ions are present in various states inside cellular systems. In addition to free ions they are bound to mobile and immobile intrinsic and extrinsic buffers and are sequestered in organelles.
- fluorescence indicators affect the free Ca^{2+}-ion concentrations by their own buffering capacities.
- fluorescence indicators have an inherent kinetic delay, which is especially important to consider when monitoring dynamic physiological processes.
- due to the complex structure of many cells the cytoplasm and organelles do not represent solutions with isotropic unhindered diffusion properties. Often, depletion, sequestration and accumulation are physiologically important.
- Many quantities are not directly accessible by experimental techniques. Release fluxes from intracellular ion stores, ion concentrations inside membrane-bound organelles are sometimes very difficult or even not accessible.
- The contribution of single processes to global Ca^{2+}-signals are very hard to determine from measurements alone.

Therefore, mathematical models help to obtain the unbiased Ca^{2+}-ion concentrations in much more detail. Vice versa, mathematical models need the direct experimental input and feedback in order to fully exploit their power. Model predictions should be tested under controlled experimental conditions, input parameters should be as much validated by experiments as possible. Last but not least, the models should be generally dealing with current biophysical and physiological questions, and therefore make significant contributions to the field of biophysics and physiology.

11.4. Compartment Models

The most straightforward approach to a mathematical model of Ca^{2+}-ion regulation, especially for a relative simple situation as e.g. described in Fig. 1, is to use compartment models. These types of models are particularly useful when dealing with situations where intracellular spatial concentration gradients can be neglected or if gradients are not present and when the main interest is focused on the temporal behaviour of Ca^{2+}-ion exchange between the different cellular components. Also, for very complex regulation processes, where spatially resolved models still generate an excessive computational load, this is a useful and valid approach.

Let us for example consider the release of Ca^{2+}-ions from an internal store into the cytoplasm without complex interactions. In a first approximation this situation is given for caffeine induced Ca^{2+}-transients in permeabilised skeletal muscle fibres. Single isolated skeletal muscle fibres are permeabilised by a chemical detergent in order to allow a direct diffusional access to the myoplasm. Addition of caffeine to the surrounding bath solution leads to the release of Ca^{2+}-ions from the sarcoplasmic reticulum and the subsequent activation of the contractile protein system, resulting in an isometric force development (Endo and Iino, 1988). Since the sarcolemma is permeabilised, the Ca^{2+}-ions additionally diffuse into the surrounding bathing solution.

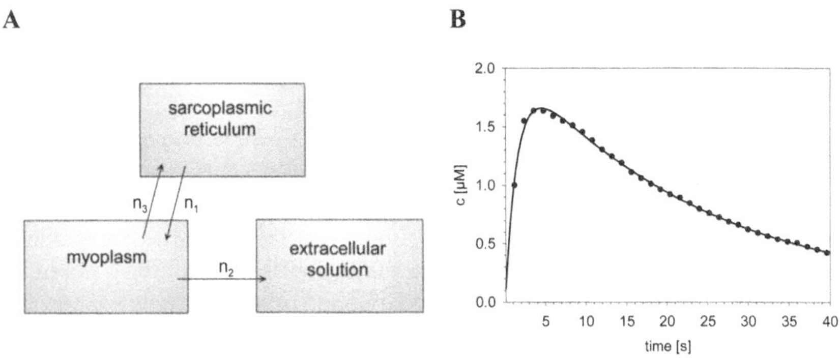

Fig. 4. A Compartment model for a caffeine induced Ca^{2+}-release in permeabilised muscle fibres. The Ca^{2+}-ion exchange between the compartments is described by rate equations with rate constants n_1, n_2 and n_3. If the sarcoplasmic reticulum ATPase is blocked n_3 can be set equal zero. For the case of the special experimental situation given in these experiments the extracellular compartment can be assumed to be clamped at zero Ca^{2+}-concentration. For details see Makabe et al. (1996). **B** A caffeine induced Ca^{2+}-transient (dotted points). The simulated time course according to Eqs. 7-10 is given by the solid line. The experimentally observed time course of Ca^{2+}-transients can be very well reproduced. (Uttenweiler, unpublished data).

These experiments are typically performed to assess the influence of substances on the Ca^{2+}-release channels of the sarcoplasmic reticulum (SR), the ryanodine receptors (RyRs), and the SR Ca^{2+}-ATPases under strictly controlled myoplasmic conditions. The transients are recorded as isometric force transients and converted to Ca^{2+}-transients by the individually determined pCa-force relations of each muscle fibre.

To extend the quantitative interpretation of caffeine induced Ca^{2+}-transients compartment models have been used (Makabe et al., 1996). This approach allows a more detailed insight into the bulk properties of Ca^{2+}-compartments and their Ca^{2+}-exchange under various experimental conditions. In a first and simplified approach this situation can be mathematically described by a three compartment model as seen in Fig. 4A (for details see Makabe et al., 1996). The first compartment represents the sarcoplasmic reticulum, the second compartment the myoplasm and the third compartment the surrounding bath solution. Since the volume of the bath solution is several orders of magnitude larger than the volume of the myoplasm, the free Ca^{2+}-ion concentration of the third compartment can be set equal to zero.

The exchange between the compartments is described by the rate constants n_1 for Ca^{2+}-release from the SR into the myoplasm, n_2 for the loss of Ca^{2+}-ions from the myoplasm into the bath solution and n_3 for the active re-transport of Ca^{2+}-ions from the myoplasm into the sarcoplasmic reticulum by the SR Ca^{2+}-ATPases. In the case of a blocked SR Ca^{2+}-ATPase, e.g. by adding cyclopiazonic acid (CPA) to the solution, n_3 equals zero and the differential equations describing the Ca^{2+}-exchange between the compartments are given by (quantities in brackets denote the free ion concentrations):

$$\frac{d[Ca^{2+}]_{SR}}{dt} = -n_1\left([Ca^{2+}]_{SR} - [Ca^{2+}]_{myo}\right) \tag{4}$$

for the sarcoplasmic reticulum and

$$\frac{d[Ca^{2+}]_{myo}}{dt} = n_1 v_{SR}\left([Ca^{2+}]_{SR} - [Ca^{2+}]_{myo}\right) - n_2[Ca^{2+}]_{myo} \tag{5}$$

for the myoplasm. v_{SR}=1/10 is a volume factor which compensates for the fact that the SR occupies approximately 10 % of the cell volume.

In this simplified situation these coupled differential equations can be solved analytically and the solution is a double exponential function given by

$$[Ca^{2+}]_{SR}(t) = a\left(\frac{n_1}{n_1+\lambda_1}e^{\lambda_1 t} - \frac{n_1}{n_1+\lambda_2}e^{\lambda_2 t}\right) \tag{6}$$

and

$$[Ca^{2+}]_{myo}(t) = a\left(e^{\lambda_1 t} - e^{\lambda_2 t}\right) \tag{7}$$

with

$$a = \frac{C_{SR}^{t=0}}{\dfrac{n_1}{n_1+\lambda_1} - \dfrac{n_1}{n_1+\lambda_2}}, \tag{8}$$

where $C_{SR}^{t=0}$ denotes the initial free Ca^{2+}-concentration inside the SR,

$$\lambda_1 = -\frac{1}{2}\left(n_1(1+v_{SR}) + n_2 - \sqrt{(1+v_{SR})^2 n_1^2 + n_2^2 + 2(v_{SR}-1)n_1 n_2}\right), \tag{9}$$

and

$$\lambda_2 = -\frac{1}{2}\left(n_1(1+v_{SR}) + n_2 + \sqrt{(1+v_{SR})^2 n_1^2 + n_2^2 + 2(v_{SR}-1)n_1 n_2}\right). \tag{10}$$

Figure 4B shows an experimentally recorded caffeine induced Ca^{2+}-transient (dotted points), superimposed with the simulated time course (solid line) as obtained with the compartment model according to Eqs. 7-10. It can be seen that the time course of the experimentally obtained Ca^{2+}-transient can be very well reconstructed. For the case that the rate constant n_3 for the active re-uptake of Ca^{2+}-ions from the myoplasm is not equal to zero the differential equations have to be solved numerically. As shown in Makabe et al. (1996) the dose dependent effects of CPA on the SR Ca^{2+}-ATPase could be simulated with the compartment model. A graded increase in the maximum Ca^{2+}-concentration during the Ca^{2+}-transients with increasing concentrations of CPA was observed both in the experiments and the model calculations.

However, a detailed understanding of the complex Ca^{2+}-ion fluxes and transpositions on the basis of the underlying molecular interactions and physiological parameters during a Ca^{2+}-transient can not be obtained with such a simple model. This has to be done with a detailed spatio-temporal mathematical model of Ca^{2+}-regulation as described in Uttenweiler et al. (1998).

11.5. Mathematical Models for Complex Cellular Geometries

Generally, cells have very complex structures and in most cases the structure directly reflects the functional purpose of the cell. Therefore, the simple sketch of a cellular geometry as shown in Fig. 1 is not valid for the majority of cells. The geometry of skeletal muscle fibres as shown in Fig. 5 is an example for the complexity of differentiated cells. Skeletal muscle cells have approximately cylindrical shape and the smallest contractile units, the so called myofibrils, contain the contractile proteins actin and myosin arranged in parallel filament systems, which can slide past each other during contraction. The force necessary for the contraction is the result of the cyclic conformational changes in the myosin heads forming cross bridges with the actin filaments. The contractile force development is regulated via changes in the intracellular free Ca^{2+}-concentration.

In contrast to most other excitable cells the transient increase in Ca^{2+}-concentration is not due to a Ca^{2+}-influx from the extracellular space, but the Ca^{2+}-ions are almost exclusively released from the membraneous network of the sarcoplasmic reticulum (SR), the intracellular Ca^{2+}-ion store. Following release, the Ca^{2+}-ions subsequently diffuse into the myoplasm and bind to Ca^{2+}-sensitive proteins. The Ca^{2+}-binding to troponin-C implies a rearrangement within the entire troponin molecule, which in conjunction with tropomyosin forms the regulatory protein system associated with the actin filament. This conformational change in the troponin structure allows the tropomyosin molecule to move on the actin filament from an inhibitory position at the edge of the filament to a position in the centre of the groove (Rüegg, 1996). That enables the binding of myosin heads to the actin filament and finally initiates the cross bridge cycle, leading to the development of contractile force development.

The initial activation of skeletal muscle fibres is achieved by a travelling action potential across the sarcolemma, the outer cell membrane. In order to ensure a homogenous contraction of a muscle fibre the sarcolemma forms transverse invaginations into the cylindrical interior of the cell, the so called transverse tubular system (TTS). This structural geometry assures a simultaneous activation of the cell over its entire cross section. The simultaneous activation is crucial for the synchronized contraction over the entire cell cross section in order to avoid damage due to shear forces.

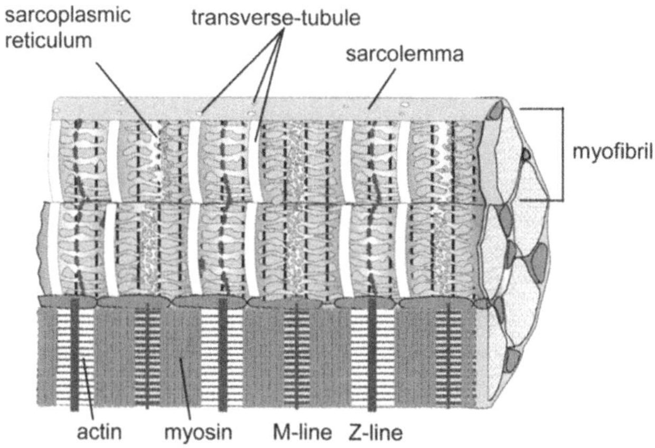

Fig. 5. Structure of a skeletal muscle fibre. The smallest functional units of the contractile protein system, the myofibrils, which consist of the actin and myosin filaments are surrounded by the membranous network of the sarcoplasmic reticulum (SR). This assures small diffusion distances for the Ca^{2+}-ions released from the SR to their target binding proteins, the regulatory protein system associated with the actin filaments. The cell membrane (sarcolemma) has transverse invaginations deep into the cylindrical cell interior, which allows a homogeneous spread of an action potential over the entire cellular cross section. These transverse tubules (marked white) form the transverse tubular system (TTS) and are filled with extracellular solution. The diameter of the tubules is approximately 100 nm.

At the triadic regions, the TTS and the terminal cisternea of the SR are in very close contact, only separated by a 10 nm gap, and the electrical signal is converted to the release of Ca^{2+}-ions from the SR via a direct protein-protein-interaction (Peachy, 1965; Ríos and Brum, 1987). A detailed description will be given in the next section.

In the following the effect of complex cellular geometries and morphologies on physiological cell functions and its treatment with mathematical models will be exemplarily discussed for the TTS in skeletal muscle. The TTS consists of a mesh of small tubules with diameters around 100 nm and is distributed over the entire muscle fibre. Due to the small diameter, it is a restricted diffusion space and therefore depletion or accumulation effects of ions during membrane current flows are important to consider. Such depletion effects influence the driving forces according to the Nernst-Planck equation and therefore have a major influence on the time course of membrane currents. Due to the non-accessibility of the TTS by classical electrophysiological techniques, mathematical models are a very important tool for studying these effects.

Using the vasiline gap technique, Almers et al. (1981) showed that Ca^{2+}-depletion in the TTS of frog skeletal muscle is an important factor influencing the decay of slow L-type Ca^{2+}-currents (I_{Ca}). However, the issue was controversially discussed in the following, since other studies proposed that the decay of I_{Ca} was mainly due to a voltage dependent inactivation process rather than due to a Ca^{2+}-depletion in the TTS (Cota et al., 1984). A recent study (Friedrich et al., 2001) using a combined experimental and mathematical modelling approach finally showed that Ca^{2+}-ion depletion in the TTS is a major contributing factor to the decline of I_{Ca} during maintained depolarisation.

The approach, which was taken in this study is summarized in Fig. 6. The geometry which was used for modelling the TTS, with its limited and restricted diffusion space was a modification of a model originally proposed by Barry and Adrian (1973). In this cylindrical symmetric model the TTS is reduced to a homogenous disk with an equivalent radial conductance (G_L) and an equivalent wall conductance (G_W). The hindered diffusion accessibility of the individual tubules due to the small diameter is included by the access resistance r_S at the mouth of the tubules at the outer cell surface. The spatial discretisation of the TTS disk is accomplished by dividing the disk into m concentric shells. Ca^{2+}-ion concentration changes are then calculated numerically for discrete time intervals δt. The small lumen of the TTS also leads to hindered diffusion of Ca^{2+}-ions and therefore effective diffusion constants D_{Ca}^{eff} have to be used.

For a detailed description of the mathematical model and the experimental protocol see Friedrich at al. (2001). Very shortly, two main assumptions were made for the model:

i) For the situation of a Ca^{2+}-buffered solution no ion concentration change is present in the TTS.

ii) Ca^{2+}-channels are exclusively located in the TTS, therefore the Ca^{2+}-currents (I_{Ca}) solely flow through the TTS membrane (see Fig. 6B) and not through the surface membrane (Almers et al., 1981).

The Ca^{2+}-concentration change in the TTS is assumed to follow the equation of continuity, which is given by

$$\frac{\partial c(r,t)}{\partial t} = D_{Ca}^{eff} \frac{1}{r}\frac{\partial}{\partial r}\left(r \frac{\partial c(r,t)}{\partial r} \right) - \frac{4\pi \cdot a \cdot I_{Ca}}{\rho \cdot zF}, \tag{11}$$

with the calcium current flow across the wall of ring n in the tubule given by

$$I_{Ca} = -\frac{G_W}{4\pi \cdot a}\left(V_0 - V(r,t) - \frac{RT}{zF}\ln\frac{c(r,t)}{c_0} \right), \tag{12}$$

with the radial coordinate r, the fibre radius a, time t, the Ca^{2+}-concentration c, the extracellular free Ca^{2+}-concentration c_0. R denotes the gas constant, T the absolute temperature, F is the Faraday constant, z=2 the Ca^{2+}-ion valency, V_0 the potential of the myoplasm with respect to the external solution, V(r,t) the potential at radial distance r in the TTS lumen with respect to the extracellular space, and ρ denotes the fraction of TTS volume to the total muscle fibre volume. Using appropriate expressions for the wall conductance, given by Barry and Adrian (1973), for the effective diffusion constants and the voltage dependence of the Ca^{2+}-conductance, as suggested by Dietze et al. (1998), the differential equations can be integrated using a fourth-order Runge-Kutta method.

The general approach of this study was to use the experimentally recorded Ca^{2+}-currents in solution, which are strongly Ca^{2+}-buffered as the input quantity for the numerical simulation of Ca^{2+}-currents for non Ca^{2+}-buffered solutions. Thus, the effect of Ca^{2+}-ion depletion in the TTS could be predicted for every individual muscle fibre used in the experiment. In the next step, the model predictions were compared to the experimentally recorded Ca^{2+}-currents in non-Ca^{2+}-buffered solution, which are influenced by Ca^{2+}-depletion. When using the experimentally obtained Ca^{2+}-current in buffered solution $I_{Ca}^{buff-\exp}(t)$ (circles in Fig. 6E), the model predictions for the Ca^{2+}-current in unbuffered solution $I_{Ca}^{unbuff-sim}(t)$ (solid trace) corresponds very well with the experimentally obtained Ca^{2+}-current in unbuffered solution $I_{Ca}^{unbuff-\exp}(t)$ (squares). These findings strongly suggest that Ca^{2+}-ion depletion in the transverse tubular system is of significant importance for the decline of the slow L-type Ca^{2+}-currents during maintained depolarisation.

Generally, this approach very nicely demonstrates how the tight interplay of experiments and mathematical models allow the powerful use of mathematical modelling approaches. By using the experimentally obtained data as the model input, and in a second step by comparing the model predictions with experimental findings for the altered conditions, the most powerful feedback of experiment and mathematical model is achieved.

11.6. Spatially and Temporally Resolved Models for Ca^{2+}-Ion Regulation

The mathematical models described so far are based on the assumption that cellular and molecular constituents (e.g. organelles) and processes (e.g. release and diffusion) can be lumped into bulk quantities, as e.g. generalized compartments and generalized rate equations. This approach is very efficient in the sense that it reduces the complexity of the biological system to a minimal extend. Therefore, these approaches help to understand which minimal set of different functional units and which generalized rate equations are at least necessary to reproduce the experimentally observed behaviour.

However, if a detailed knowledge about the cellular and molecular processes is required one has to use spatio-temporal mathematical models in order to fully understand the complex interactions. Spatial gradients have been shown to be of significant importance for a variety of cellular regulation processes (Berridge et al., 1998). Therefore, spatio-temporal models are the most comprehensive approach for a detailed description of Ca^{2+}-regulation. In the following this type of mathematical modelling will be exemplarily discussed for elementary Ca^{2+}-release events (ECRE) in permeabilised mammalian skeletal muscle fibres. ECRE is a general expression for Ca^{2+}-release events through single or clusters of RyRs in mammalian skeletal muscle. It comprises Ca^{2+}-sparks and also ember-type Ca^{2+}-release as discussed in Kirsch et al. (2001) and González et al. (2000).

ECRE studies in muscle fibre preparations help to understand the central mechanism of excitation-contraction (EC) coupling. In skeletal muscle fibres, the transformation of an action potential into the release of Ca^{2+}-ions from the SR is achieved by a direct mechanical protein-protein interaction. The voltage sensing dihydropyridine receptors (DHPR), which are located in the TTS (Ríos and Brum 1987) are able to directly open the RyRs, which serve as the Ca^{2+}-release channels of the SR (Dulhunty 1992). This direct mechanical interaction is based on the structure of the triadic junction, where the TTS and the terminal cisternae of the SR are only separated by a tiny gap of approx. 10 nm. Every second RyR in the SR membrane is facing a tetrade of DHPR (skip pattern, Block et al. 1988) in the TTS. The other half of RyRs can be activated by Ca^{2+}-ions via calcium induced calcium release (CICR), which is the main trigger for calcium release in heart muscle as described in Chap.6 of this book. Although this basic principle is known for some time, many centrally important questions still remain unclear. One of the two most important questions is how mammalian ECRE differ from ECRE in amphibian skeletal muscle fibres. Ca^{2+}-sparks were known in heart (Cheng et al. 1993), smooth (Nelson et al. 1995), and amphibian skeletal (Tsugorka et al. 1995; Klein et al. 1996) muscle fibre preparations. Only controversial results were reported for mammalian skeletal muscle fibre preparations ranging from non-existence of sparks in adult rat fibres under voltage clamp conditions (Shirokova et al. 1998) to a scarce appearance of sparks in intact adult mouse muscle fibres (Conklin et al. 1999). In a recent study (Kirsch et al., 2001) the existence of ECRE in adult mammalian skeletal muscle fibre preparations, which were chemically or mechanically skinned, was demonstrated. In addition to the existence of ECRE in these preparations, the study also revealed that mammalian ECRE consist of both spark-like and ember-like events, which could additionally occur in combination at the same location. Another centrally important question, the mechanism of spark termination and RyR coupling in heart muscle, will be discussed in Chap.6.

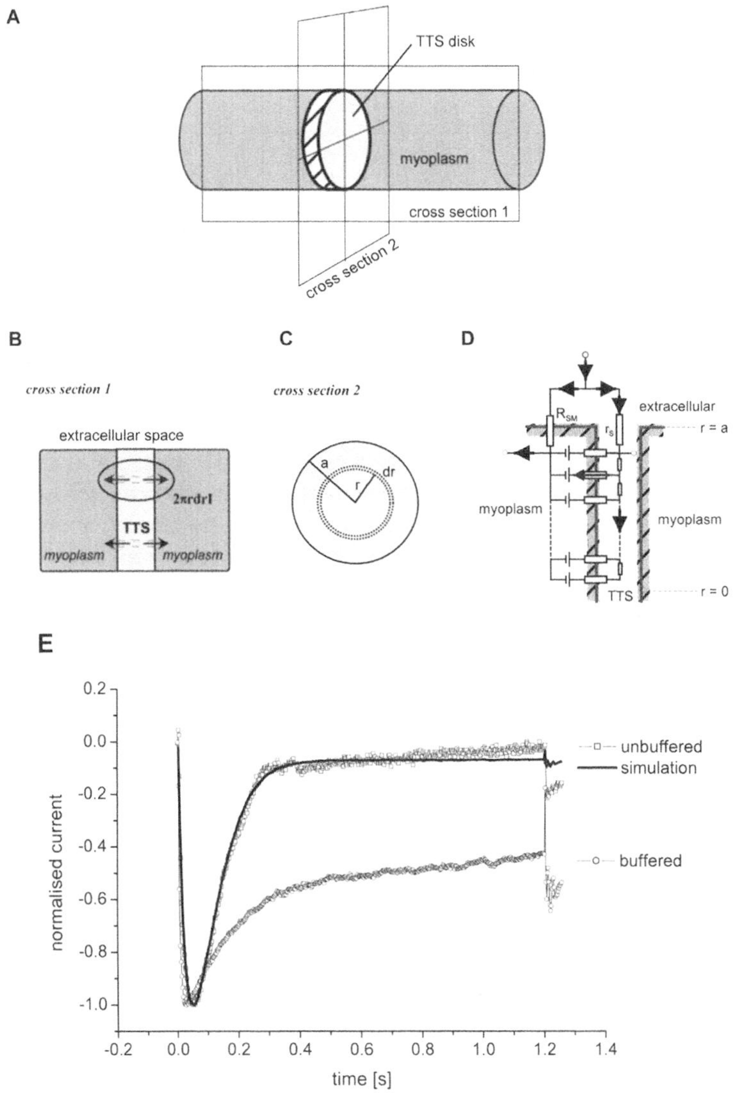

Fig. 6. Mathematical model for a complex cellular geometry. **A** The transverse tubular system (TTS) in skeletal muscle is reduced to a disk in a cylindrical symmetric model of the muscle fibre. **B-C** The two cross sections show the geometrical discretisation used in the model. **D** The equivalent electrical circuit that is used in the simulations. **E** For a fibre with a diameter of d=70 μm the Ca^{2+}-current I_{Ca} was recorded for a 1.2 s depolarising voltage clamp pulse (E_R=-42 mV) in unbuffered solution and after a solution change in buffered solution. The time course of the experimentally determined I_{Ca} in buffered solution (circles) is used as the input for the model calculations for I_{Ca} in unbuffered solution. The model prediction (solid trace) for the depletion governed I_{Ca} is in good accordance with the experimentally observed I_{Ca} decline (square symbols). According to Friedrich et al. (2001).

11.6.1. ECRE Measurements with High Resolution Dynamic Confocal Microscopy

For recording ECRE high resolution confocal fluorescence laser scanning microscopy (see Fig. 2B) is used. Since ECRE are sub-cellular events with a spatial extension of about 1-2 μm, the high axial resolution of confocal or multi-photon microscopy is necessary in order to detect these events against the constant background of the myoplasmic Ca^{2+}-concentration. As these events have a typical time scale of tens of milliseconds one has to restrict the scanning to one spatial dimension in order to follow the dynamics, resulting in line scan images as shown in Fig. 7 (see colorplate). The time resolution is 2.05 ms per line and the spatial resolution is 0.35±0.02 μm in lateral and 1.06±0.05 μm in axial direction, respectively. The spatial resolution was measured by recording the microscopic point spread function (PSF) with sub-resolution fluorescent beads.

Permeabilised skeletal muscle fibres were mounted in an experimental chamber which was placed on an inverted microscope. A 40x water immersion objective (UAPO40xW/340/1.15, Olympus) was used to image the fluorescence emission of the Ca^{2+}-sensitive fluorescence indicator fluo-4, which was excited with the 488 nm line of a 20 mW Kr/Ar-laser. The fluorescence emission was measured at wavelengths larger than 510 nm using a barrier filter and a pinhole size of 150 μm. Typical ECRE in permeabilised mammalian skeletal muscle fibre preparation are shown in Fig. 7 (see colorplate).

The fluorescence in the line scan images is expressed in normalized units F/F_0. It is obtained from the ratio of the raw line scan data of fluorescence F(x,t) and the average baseline fluorescence $F_0(x)$ prior to the event for every value of the temporal coordinate t, respectively. The most important morphological parameters of ECRE are maximum amplitude, full width at half maximum (FWHM), rise time and full duration at half maximum (FDHM). ECRE in permeabilised mammalian skeletal muscle fibres consist of spark-like events (panel A-C), ember-like events (panel D-G) and combinations thereof (Kirsch et al., 2001). This is a major difference to amphibian skinned muscle fibre preparations, where only Ca^{2+}-sparks could be detected. Typical values of the morphological parameters of mammalian ECRE are in the range of 1.5-4.5 F/F_0 for the amplitude and 1.2-3.0 μm for the FWHM. For spark-like events the typical temporal morphological parameters are 8-18 ms for the rise time and 10-49 ms for the FDHM, whereas for ember-like events rise times are in the range of 15-250 ms and FDHM in the range of 50-350 ms. From the greater variability of mammalian ECRE time courses and spatial properties, it can be assumed that additional control factors of Ca^{2+}-release are involved in mammalian compared to amphibian skeletal muscle preparations.

In the following a spatio-temporal mathematical model of ECRE will be presented. The necessity of analysing these fast dynamic sub-cellular events will be evident when analysing the results of the model. Shortly, the exclusive use of fluorescence data would lead to a dramatic underestimation of the free Ca^{2+}-ion concentrations during the event peak and also the time course of the Ca^{2+}-ion concentration inside the SR is different from what would be expected from fluorescence measurements of ECRE alone.

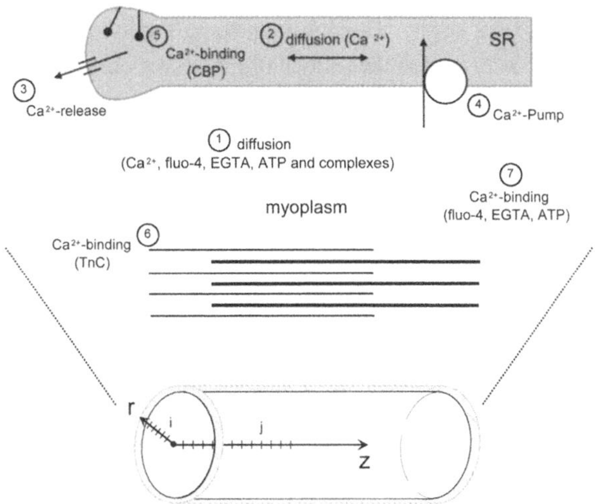

Fig. 8. Spatio-temporal model of Ca^{2+}-ion regulation in muscle fibres. First, a simplified geometry is defined. In the case of a muscle fibre a cylindrical coordinate system is used. In this case, symmetry with respect to the angular coordinate is assumed. Thus, Ca^{2+}-ion concentrations are calculated as a function of the radial coordinate r and the longitudinal coordinate z. The processes which are explicitly included are: 1.) diffusion of free Ca^{2+}-ions, diffusion of free buffers as well as diffusion of the Ca^{2+}-bound buffers in the myoplasm. 2.) diffusion of free Ca^{2+}-ions inside the SR. 3.) Ca^{2+}-release from the SR. 4.) active re-uptake of Ca^{2+}-ions into the SR by the SR Ca^{2+}-ATPase. 5.) Ca^{2+}-binding inside the SR to Ca^{2+}-binding proteins. 6.) Ca^{2+}-binding to troponin-C. 7.) Ca^{2+}-binding to the mobile buffers. Subsequently the differential equations of these processes are solved on the spatio-temporal grid with an explicite finite difference method. From Uttenweiler et al. (2002).

11.6.2. Mathematical Modelling of ECRE with a Spatio-Temporal Model of Ca^{2+}-Ion Distribution

The general approach using spatio-temporal mathematical models is to define a simplified geometry of the cellular system explicitly considering existing symmetries. This helps to reduce computational load. In a second step, the main cellular processes have to be identified and described by their respective differential equations. Finally, these differential equations have to be solved with effective numerical methods on a spatio-temporal grid, which is adopted to the simplified geometry defined in step one. When working with a reduced spatial coordinate set due to symmetry simplifications, it is sometimes necessary to reconstruct the full three dimensional Ca^{2+}-ion distribution in time. In a final step, the microscopic recording conditions (e.g. the PSF) have to be included into the model in order to assure a maximum correlation of the measured fluorescence signal with the theoretically predicted fluorescence signal. Once this has been done, the model subsequently allows to determine many quantities, which are not directly accessible with experimental techniques or which are inherently biased by the experiment itself.

The model used for the simulation of ECRE (Uttenweiler et al., 2002) is based on a mathematical model, which was originally developed for global Ca^{2+}-transients in skinned muscle fibres (Uttenweiler et al., 1998). As already mentioned before, for the global cellular caffeine induced Ca^{2+}-transients spatio-temporal mathematical models are also necessary to fully understand the Ca^{2+}-ion re-distribution during this type of release.

For ECRE simulations the existing model has been expanded to a 2-D cylindrical geometry with radial symmetry as shown in Fig. 9. In this geometry, ion concentrations are calculated as a function of the radius r, the longitudinal coordinate z and time t. The following processes are explicitly included into the model (see Fig. 8):

1. diffusion of free Ca^{2+}-ions, the fluorescent indicator fluo-4, EGTA, ATP and the respective Ca^{2+}-complexes
2. diffusion of free Ca^{2+}-ions in longitudinal (z-) direction inside the SR
3. Ca^{2+}-release by the SR through a source of variable extension
4. Ca^{2+}-reuptake by the SR Ca^{2+}-ATPases
5. Binding of Ca^{2+}-ions inside the SR to calcium binding proteins (CBP), pooled as a calsequestrin-buffer
6. Binding of Ca^{2+}-ions to troponin-C
7. Binding of Ca^{2+}-ions to the mobile buffers fluo-4 , EGTA and ATP

The model geometry has additionally been adopted such that the SR forms a compartment surrounding a myofibril which is used as the simulation volume. The SR is treated in a fully functional and geometrical context. It is assumed to interact with the myofibrillar space by Ca^{2+}-release from the Ca^{2+}-source and an active Ca^{2+}-reuptake by the SR Ca^{2+}-ATPases. Longitudinal diffusion of free Ca^{2+}-ions inside the SR and binding to SR Ca^{2+}-binding proteins with the kinetic properties of calsequestrin is also included in the model. Again, as in the other model types certain properties of the cellular structure are lumped together. The complex structure of the triadic region is expressed by a symmetric Ca^{2+}-source centred at the cylinder axis. Additionally, the size of the source can be adjusted in both the radial and longitudinal spatial dimensions. A more detailed treatment of the triadic region can be found in a model for Ca^{2+}-sparks in heart muscle (Sobie et al., 2002). The dimensions of the model volume are 1 μm in radial and 2 μm in longitudinal direction, respectively. The symmetry additionally leads to a mirroring of the model volume at z=0 μm.

The main process of Ca^{2+}-ion movement in the cytosol is diffusion. The diffusion equation with the various source and sink terms in the 2-D cylindrical geometry can be written as (Crank 1975):

$$\frac{\partial c(r,z,t)}{\partial t} = \frac{1}{r}\frac{\partial}{\partial r}\left(rD\frac{\partial c(r,z,t)}{\partial r} \right) + \frac{\partial}{\partial z}\left(D\frac{\partial c(r,z,t)}{\partial z} \right) + h(r,z,t) , \tag{13}$$

with the concentration c, the diffusion coefficient D and h(r,z,t) being the sum of all source and sink terms of the various processes involved.

Note: $[Ca^{2+}]$ denotes the free Ca^{2+}-concentration in the following equations.

11.6.2.1. Ca^{2+}-Release from the SR:

The release of Ca^{2+}-ions from the SR is assumed to be proportional to the concentration gradient across the SR membrane:

$$\frac{d[Ca^{2+}]}{dt} = a_1\left([Ca^{2+}]_{SR} - [Ca^{2+}]_{myoplasm} \right) \text{ with } a_1 = a_1(r,z,t) , \tag{14}$$

where $a_1(r, z, t)$ is a factor with dimension s^{-1}, which can be used for adjusting the extend of SR Ca^{2+}-ion release per unit time.

11.6.2.2. Ca^{2+}-re-Uptake into the SR by the Ca^{2+}-ATPase:

The active removal of Ca^{2+}-ions from the cytosol by the SR Ca^{2+}-pump is modelled with a Hill-type relation, assuming a Ca^{2+}-dependent second order saturable pump. The uptake of Ca^{2+}-ions into the SR can then be written as:

$$\frac{d[Ca^{2+}]}{dt} = \frac{v_{max}[Ca^{2+}]^n}{[Ca^{2+}]^n + K_m^{\ n}} , \tag{15}$$

where v_{max} is the maximum uptake velocity, K_m is the half maximal uptake rate and n=2 is the Hill-coefficient.

11.6.2.3. Ca^{2+}-Binding to Buffers:

Calcium is assumed to bind to all buffers in a 1:1 stochiometry and without cooperativity, so that the following equation holds

$$\frac{d[Ca^{2+}]}{dt} = \sum_{\ell} k_{on}^{(\ell)} [Ca^{2+}]_{free} \cdot [buffer^{(\ell)}]_{free} - k_{off}^{(\ell)} \cdot [Ca^{2+}buffer^{(\ell)}] , \tag{16}$$

where ℓ denotes the various buffers, $k^{(\ell)}_{on}$ is the kinetic on-rate constant, $k^{(\ell)}_{off}$ is the kinetic off-rate constant of the $buffer^{(l)}$-Ca^{2+} binding. For the buffering of troponin-C, only the regulatory sites are considered. As the troponin concentration in skeletal muscle is roughly 70 μM (Gillis 1985) and each molecule has two regulatory binding sites, as a first approximation with no cooperativity, the effective concentration can be taken as 140 μM. The high affinity binding sites and the effect of Mg^{2+}-competition are neglected in the model.

11.6.2.4. Discretization

The differential equations described above are solved on the cylindrical grid shown in Fig. 8 using the method of explicit finite-differences. Let i and j denote the radial and the longitudinal grid position, respectively, k the discrete time index, δr and δz the step-sizes in radial and longitudinal directions, respectively and δt the temporal step-size. Then, for the case of the diffusion equation, the finite difference approximation is given by

$$c_{i,j,k+1} = D\frac{\delta t}{2i(\delta r)^2}\left\{(2i+1)c_{i+1,j,k} + 2ic_{i,j+1,k} - 8ic_{i,j,k} + (2i-1)c_{i-1,j,k} + 2ic_{i,j-1,k}\right\} + c_{i,j,k} \quad \text{for} \quad i \neq 0,$$

$$c_{0,j,k+1} = D\frac{\delta t}{(\delta r)^2}\left\{4c_{1,j,k} - 6c_{0,j,k} + c_{0,j+1,k} + c_{0,j-1,k}\right\} + c_{0,j,k} \quad \text{for} \quad i = 0 \tag{17}$$

if additionally δr=δz, i.e. equal step-size in radial and longitudinal direction. The model volume is divided into 50 intervals in radial and 100 intervals in longitudinal direction, leading to an effective spatial step-size of 20 nm for both spatial dimensions. The boundary of the system is treated with Neumann boundary conditions, i.e. zero perpendicular flux. For that case the explicit finite difference formula is given by

$$c_{i,j,k+1} = D\frac{\delta t}{2i(\delta r)^2}\left\{2ic_{i,j+1,k} - (6i-1)c_{i,j,k} + (2i-1)c_{i-1,j,k} + 2ic_{i,j-1,k}\right\} + c_{i,j,k} \quad \text{for} \quad i = i_{max}. \tag{18}$$

The discretization for all other differential equations are implemented according to Uttenweiler et al. (1998).

11.6.2.5. Computing

The finite-difference algorithm is programmed in ANSI-C language and the calculations were performed using a R10000 workstation (Silicon Graphics, IRIX 6.4). The step-size for the time iteration is 4.000.000 time steps per second. The error in calculating the solution of the differential equations with the finite difference algorithm was controlled by monitoring the amount of total Ca^{2+} in the model. Errors were kept below 1 % in all calculations.

11.6.2.6. Simulation of the Effect of the Microscopic PSF

As mentioned before, the model also has to account for the microscopic recording conditions. Therefore, the simulated distribution of Ca^{2+}-ions bound to the fluorescence indicator is additionally corrected for the microscopic PSF by convolution of the modelled Ca^{2+}-dye distribution, explicitly calculated from the 2-D data of the model output, with the experimentally determined PSF:

$$F(x,y,z) = \frac{(2\pi)^{-3/2}}{\sigma_x \sigma_y \sigma_z} \iiint [\mathrm{Ca^{2+}fluo4}](x',y',z') e^{-\left(\frac{(x-x')^2}{2\sigma_x^2} + \frac{(y-y')^2}{2\sigma_y^2} + \frac{(z-z')^2}{2\sigma_z^2}\right)} dx'\,dy'\,dz' \cdot \quad (19)$$

Note: x, y, z are now Cartesian coordinates with z being the axial direction of the microscope. The PSF can be chosen as a 3-D Gaussian kernel with the spatial FWHMs corresponding to the measured FWHMs of the PSF of the microscopic set-up (i.e. x-direction: 350 nm, y-direction: 350 nm, axial-direction: 1060 nm). The convolution in the spatial domain can be simplified by additionally exploiting the separability of the 3-D Gaussian kernel into three independent 1-D Gaussian kernels in order to reduce the computational load (Jähne 1999):

$$F(x,y,z) = \frac{(2\pi)^{-3/2}}{\sigma_x \sigma_y \sigma_z} \int e^{-\frac{(x-x')^2}{2\sigma_x^2}} \int e^{-\frac{(y-y')^2}{2\sigma_y^2}} \int e^{-\frac{(z-z')^2}{2\sigma_z^2}} [\mathrm{Ca^{2+}fluo4}](x',y',z')\, dz'dy'dx' \cdot \quad (20)$$

Finally, the theoretical fluorescence F(x,y,z) is normalized with the baseline fluorescence at time t=0s, resulting in the normalized theoretical fluorescence F/F_0.

11.6.2.7. Modelled ECRE

The results of the mathematical model are shown in Figs. 9 and Fig. 10. In Fig. 9 modelled line scan images are summarized, which are obtained with various assumptions for the size of the release source and the time course of Ca^{2+}-ion release from the SR. The simulations in panels S1-S3 correspond to spark-like events. The source size in the model calculations for S1 and S2 is 40 nm in both radial and longitudinal direction and the period of Ca^{2+}-release is 10 ms in S1 and 25 ms in S2. In the simulation shown in panel S3, the source size is 40 nm in ra dial and 1 µm in longitudinal direction and the source is open for 10 ms. With these settings spark-like events with different time courses and spatial properties can be effectively simulated.

The simulation of ember-like events, corresponding to the experimentally observed events in panels D-G in Fig. 7 (see colorplate) is shown in panels S4-S6. The source size is 40 nm in both radial and longitudinal direction for all three simulations and Ca^{2+}-release is modelled either as a two stage release (S4, S5) or a one stage release (S6) with a source open time of 62.5 ms. The two stage release is obtained by an initial opening of the source for 10 ms (S4) and 20 ms (S5) followed by the second opening of 90 ms (S4) and 180 ms (S5) with a reduced factor a_1 in (14). In simulation S4, a_1 was reduced by a factor of 10 and in simulation S5 it was reduced by a factor of 14. The situation of a two stage source basically corresponds to either changes in single channel conductances or changes in channel recruitment during ECRE as discussed by González et al. (2000b) for experiments with drug-modulated amphibian Ca^{2+}-sparks.

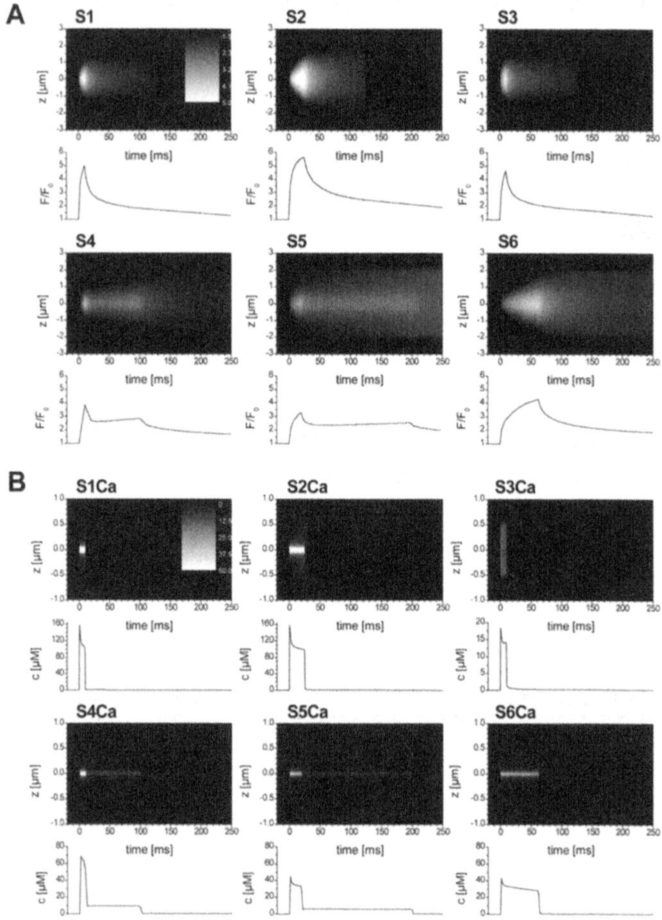

Fig. 9. Output of the mathematical model for ECRE. **A** Simulated line scan images. Simulations S1-S3 correspond to spark-like ECRE and S4-S6 to ember-like ECRE. **B** The distribution of the free Ca^{2+}-ion concentrations are given in panels S1Ca-S6Ca. Note that there are profound differences in the spatial extension and the temporal characteristics between the free Ca^{2+}-concentration and the simulated line scans, which arise from the non-equilibrium of the fluorescence indicator with the underlying free Ca^{2+}-concentration. From Uttenweiler et al. (2002).

From the model output it can be seen that spark-type ECRE can be very well reconstructed with the model based on the underlying assumptions and initial conditions. The variety of experimentally observed spark-like ECRE can be modelled by varying the open times of the Ca^{2+}-source in a range of 5-30 ms. Interestingly, the duration of Ca^{2+}-release shows a larger effect on ECRE spatial dimensions (FWHM) than the actual source size in longitudinal (z)-direction. This can be seen in Fig. 10 A where the FWHM of the ECRE with a 40 nm point source and 25 ms release time is 2.12 µm (dotted line) in comparison to a FWHM of 1.56 µm for the ECRE with a 1 µm source opened for 10 ms (dashed line). The simulations of the ember-like ECRE were accomplished by either a one-stage or a two-stage source as explained before. With these assumptions ember-like ECRE could be in principle reproduced as seen in panels S4-S6 of Fig. 9. However, it was not possible to simulate ember-like events with steady fluorescence levels at low amplitudes, as e.g. shown in Fig. 7G (see colorplate). Also, the simulated ember-like events were all preceded by an overshoot in the rising phase. From the time courses in Fig. 9, it is difficult to distinguish if experimentally observed ember-like events are due to a one-stage, a two-stage or even a different release mode.

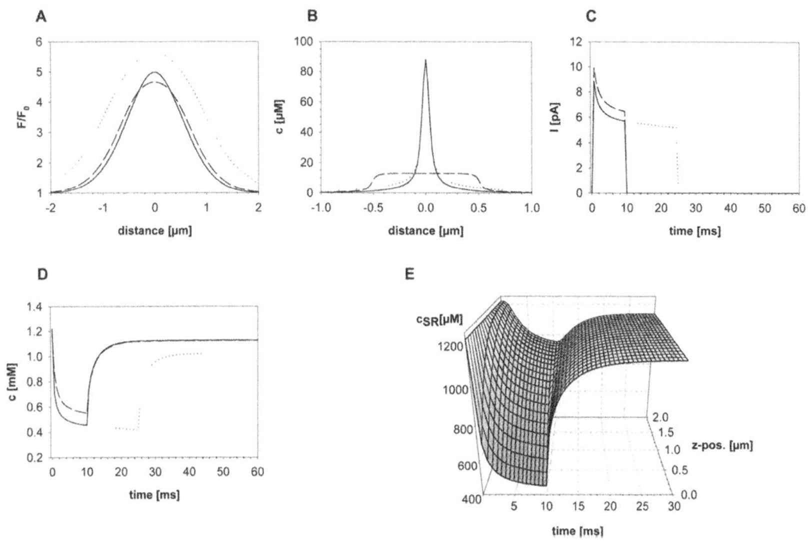

Fig. 10. Spatial and temporal properties of ECRE. **A** Spatial profiles of spark-like ECRE measured at the temporal coordinate of maximum amplitude. The solid lines correspond to simulation **S1** in Fig. 9, the dotted lines to **S2** and the dashed lines to **S3**, respectively. **B** Spatial profile of the free Ca^{2+}-ion concentration measured at the temporal coordinate of maximum amplitude of the simulated line scan images. Note that there is a distinct difference of the time in peak amplitude for the line scan images and the free Ca^{2+}-ion concentration, which can be seen in the respective time courses of Fig. 9. **C** Ca^{2+}-release current from the SR. **D** Average free Ca^{2+}-ion concentration inside the SR. The average was calculated from the SR concentrations between z=0 µm and z= 0.6 µm. **E** The detailed spatio-temporal free Ca^{2+}-ion concentration inside the SR during the spark-like ECRE of simulation **S1**. From Uttenweiler et al. (2002).

11.6.3. The Necessity of Using Spatio-Temporal Models

The importance of using spatio-temporally resolved mathematical models for an accurate quantitative analysis of high resolution fluorescence microscopic data will be evident from the following analysis of the ECRE model output. As already mentioned before, mathematical models not only provide more quantitative information about the directly accessible quantities in experiments, but they also provide exciting new insight into the biological system by giving access to quantities, which are hard or even not at all measurable.

A first very important issue can be seen when examining the time course of the free Ca^{2+}-concentrations in panels S1Ca-S6Ca. The free Ca^{2+}-ion distribution turns out to be much more confined in space and it also exhibits more rapid decay characteristics than the fluorescence signal. Moreover, the peaks in free Ca^{2+}-ion concentration are in the 100 µM range during ECRE. Estimations of peak free Ca^{2+}-concentrations from experimental data alone, which were obtained by various classical calibration methods, suggested peak concentrations in the 500 nM range. These significant differences are due to the lack of equilibrium between the fluorescent indictor and the free Ca^{2+}-ions and due to the large spatial concentration gradients, as discussed in Smith et al. (1998). The fluorescent indicator can not follow the fast dynamic rising phase in Ca^{2+}-release due to its limited kinetic on- rate constant.

A second very important issue is that the model provides information about quantities, which are hard or even not at all experimentally accessible. Figure 10 shows two examples for quantities, which could be derived from the model directly. Fig. 10 C shows the time course of the Ca^{2+}-release current from the SR. Interestingly, the release current is not constant in time. This seems to be in controversy with most estimations so far, which assumed a constant release current. Examining the free Ca^{2+}-ion concentration inside the SR yields the explanation for the declining release current. The averaged free Ca^{2+}-ion concentration inside the SR, as shown in Fig. 10 D is not constant in time, but there is a massive depletion of free Ca^{2+}-ions inside the SR even during short spark-type ECRE. The spatio-temporal free Ca^{2+}-concentration profile inside the SR is shown in Fig. 10 E. Even at a distance of 2 µm from the release source there is a depletion of free Ca^{2+}-ions inside the SR.

These findings explicitly demonstrate the necessity of spatio-temporal modelling approaches. So far, Ca^{2+}-depletion in the SR was neglected. Only an explicit treatment of the SR with its small volume in the terminal cisternae and the kinetically limited uptake capacity in addition with the limited dissociation characteristics of the intra-SR Ca^{2+}-buffers explain the decline in luminal free SR Ca^{2+}-ion concentration even during a short spark-like event.

11.7. Summary and Conclusions

The aim of this chapter was to discuss the concept of analysing the dynamic process of Ca^{2+}-regulation with a combination of state of the art experimental techniques and detailed mathematical models of cellular systems and their underlying molecular architecture. The main points raised can be summarized as follows.

The strength of mathematical models comprises two major points: i.) mathematical models help to quantitatively analyse experimentally obtained data and ii.) the models are able to yield insight into processes which are not experimentally accessible. Two very important examples were discussed in this chapter. First, for fast dynamic Ca^{2+}-regulation processes only spatio-temporal models of Ca^{2+}-ion regulation provide the unbiased physiologically important free Ca^{2+}-ion concentration. Second, these models also allow studying intracellular organelles and release currents, which are normally very hard or even not at all accessible by experimental techniques.

Generally, there are numerous levels of simplification, which can be used in modelling approaches. The use of the different approaches mainly depends on the type of information that has to be extracted and also on the specific properties of the biological system that is studied. Compartment models offer high numerical efficiency especially for very complex signalling processes. Bulk models allow the reduction of complex cellular geometries to simple systems that can be mathematically treated. Spatio-temporal models offer the highest level of detail, only limited by computational power and the complexity of generating the model.

Finally, there are numerous biomedical applications where this combined approach is essential for a detailed understanding of pathological diseases or for the characterization of the effects of clinically relevant substances on cellular properties. One example, which underlines the necessity of understanding the mechanism of clinically relevant substances, are studies about the effects of anesthetics on cytosolic Ca^{2+}-regulation, centrally important in neurons of the central nervous system, intra-cardiac ganglia neurons and in skeletal muscle cells (Kunst et al., 1999). The knowledge of the cellular mechanisms of anesthetics is most important for the understanding of the state of narcosis. Further, it is known that a severe, life-threatening genetic disorder, Maligne Hyperthermia, is caused by a rise of the myoplasmic Ca^{2+}-concentration, which is induced by volatile anesthetics during narcosis (Kunst et al., 1999). It is clear that the very complex interactions of anesthetics with the various molecular processes responsible for the cellular regulation of the cytosolic Ca^{2+}-concentration can only be adequately and comprehensibly described by mathematical modelling approaches (Uttenweiler et al., 2003).

This leads to another central challenge for the future. Theoretical models, which are able to describe cellular processes in terms of the underlying molecular interactions, have to be developed at different levels of spatio-temporal resolution. For example, very detailed molecular models are necessary to describe the function of individual ion channels (Chung and Kuyucak, 2002). More macroscopic models calculate the spatio-temporal changes in Ca^{2+}-concentration inside cells on the basis of the partial differential equations describing the bulk interaction of the different processes involved, e.g. diffusion, Ca^{2+}-release and reuptake and buffering (Uttenweiler et al., 1998, 2002). Therefore, it is necessary to find effective concepts which consider the scale transition between these levels of abstraction. This also requires new and effective concepts for sophisticated spatio-temporal mathematical models, including the possibility of solving differential equations on the exact cellular and sub-cellular 3D-topology and methods of scale transitions from molecular processes in the Å- and fs-range to the cellular responses in the µm- and ms-range. Many more important questions are accompanied with this transition. Is it necessary to describe individual channel behavior in order to understand the dynamics of Ca^{2+}-concentration in the cell? How many channels are needed to allow a more concentration based view instead of a consideration of each channel? How can the different scales of processes be treated within a single mathematical model for cellular systems?

In conclusion, the combined approach of experiments and mathematical models is only at the very beginning of its development. In the future we will certainly see many important studies in the entire field of life science using this methodology.

Acknowledgements

The author would like to thank T. Ehmer, R.H.A. Fink, O. Friedrich, W.G. Kirsch and U. Knobloch for their kind permission to reproduce figures.

References

Almers W, Fink RHA, Palade PT (1981) Calcium depletion in frog muscle tubules: the decline of calcium current under maintained depolarisation. J Physiol 312:177-207

Barry P H, Adrian RH (1973) Slow conductance changes due to potassium depletion in the transverse tubules of frog muscle fibers during hyperpolarizing pulses. J Membr Biol 14:243–292.

Berridge MJ, Bootman MD, Lipp P (1998) Calcium-a life and death signal. Nature 395:645-648

Berridge MJ (1997) Elementary and global aspects of calcium signalling. J Physiol 499.2:291-306

Block BA, Imagawa T, Campbell KP, Franzini-Armstrong C (1988) Structural evidence for direct interaction between the molecular components of the transverse tubule/sarcoplasmic reticulum junction in skeletal muscle. J Cell Biol 107:2587-600

Campagnola PJ, Clark HA, Mohler WA, Lewis A, Loew LM (2001) Second-harmonic imaging microscopy of living cells. J Biomed Opt. 6(3):277-286

Campagnola PJ, Wei MD, Lewis A, Loew LM (1999) High resolution non-linear optical microscopy of living cells by second harmonic generation. Biophys J 77:3341-3349

Cheng H, Lederer WJ, Cannell MB (1993) Calcium sparks: elementary events underlying excitation–contraction coupling in heart muscle. Science 262:740-744

Chung S-H, Kuyucak S (2002) Ion channels: recent progress and prospects. Eur Biophys J DOI 10.1007/s00249-002-0216-4

Conklin MW, Barone V, Sorrentino V, Coronado R (1999) Contribution of ryanodine receptor type 3 to Ca(2+) sparks in embryonic mouse skeletal muscle. Biophys J **77**:1394-403

Cota G, Nicola-Siri L, Stefani E (1984) Calcium channel inactivation in frog (Rana pipiens and Rana montezuma) skeletal muscle fibres. J Physiol 354:99 –108

Crank J (1975) The mathematics of diffusion. Oxford University Press

Denk W, Strickler J, Webb W (1990) Two-photon laser scanning fluorescence microscopy. Science 248:73-76

Dietze B, Bertocchini F, Barone V, Struck A, Sorrentino V, Meltzer W (1998) Voltage-controlled Ca21 release in normal and ryanodine receptor type 3 (RyR3)-deficient mouse myotubes. J Physiol 513.1:3–9

Dulhunty AF (1992) The voltage-activation of contraction in skeletal muscle. Progress in Biophysics & Molecular Biology 57:181-223

Endo M, Iino M (1988) Measurement of Ca^{2+} release in skinned fibres from skeletal muscle. Methods in Enzymology 157:12-26.

Fink RHA, Veigel C (1996) Calcium uptake and release modulated by counter-ion conductances in the sarcoplasmic reticulum of skeletal muscle. Acta Physiol Scand 156: 387-396

Franzini-Armstrong C, Protasi F, Ramesh V (1999) Shape, size and distribution of Ca^{2+} release units and couplons in skeletal and cardiac muscles. Biophys J 77:1528-1539

Friedrich O, Ehmer T, Uttenweiler D, Barry P, Fink RHA (2001) Numerical analysis of Ca^{2+}-depletion in the transverse tubular system (TTS) of mammalian muscle. Biophys J 80(5):2046-2055

Giannini G, Conti A, Mammarella S, Scrobogna M, Sorrentino V (1995) The ryanodine receptor/calcium channel genes are widely and differentially expressed in murine brain and peripheral tissues. J Cell Biol 128:893-904

Gillis JM (1985) Relaxation of vertebrate skeletal muscle. A synthesis of the biochemical and physiological approaches. Biochimica et Biophysica Acta 811:97-145

González A, Kirsch WG, Shirokova N, Pizarro G, Stern MD, Ríos E (2000) The spark and its ember: separately gated local components of Ca(2+) release in skeletal muscle. J Gen Physiol 115:139-58

González A, Kirsch WG, Shirokova N, Pizarro G, Brum G, Pessah IN, Stern MD, Cheng H, Ríos E (2000b) Involvement of multiple intracellular release channels in calcium sparks of skeletal muscle. PNAS 97:4380-4385

Grynkiewicz G, Poenie M, Tsien RY (1985) A new generation of Ca^{2+} indicators with greatly improved fluorescence properties. J. Biol. Chem. 260:3440-3450

Jähne B (1999) Spatial and Fourier Domain. In: Handbook of Computer Vision and Applications. Jähne B, Haussecker H, Geissler P (eds) Academic Press, San Diego. Vol. 2. pp 35-66

Kirsch WG, Uttenweiler D, Fink RHA (2001) Spark- and ember-like elementary Ca^{2+} release events in skinned fibres of adult mammalian muscle. J Physiol 537.2:379-389

Klein MG, Cheng H, Santana LF, Jiang YH, Lederer WJ, Schneider MF (1996). Two mechanisms of quantized calcium release in skeletal muscle. Nature 379:455-8

Kunst G, Graf BM, Schreiner R, Martin E, Fink RHA (1999). Differential Effects of Sevoflurane, Isoflurane and Halothane on Ca 2+-Release from the Sarcoplasmic Reticulum of Skeletal Muscle. Anesthesiology 91:179-86

Makabe M, Fink RHA (1996) The contribution of the sarcoplasmic reticulum Ca^{2+}-transport ATPase to caffeine induced Ca^{2+}-transients of murine skinned skeletal muscle fibres. Pflügers Arch 432:717-726

Meissner G (1994) Ryanodine receptor/Ca^{2+}- release channels and their regulation by endogenous effectors. Annu Rev Physiol 56:485-508

Nelson MT, Cheng H, Rubart M, Santana LF, Bonev AD, Knot HJ, Lederer WJ (1995) Relaxation of arterial smooth muscle by calcium sparks. Science 270:633-637

Peachy LD (1965) The sarcoplasmic reticulum and transverse tubules of the frog's satorius. J Cell Biol 25:209-231

Ríos E, Brum G (1987) Involvement of dihydropyridine receptors in excitation-contraction coupling in skeletal muscle. Nature 325:717-720

Rüegg JC (1996) Muscle contraction: Molecular and cellular physiology. *In* Comprehensive Human Physiology, Vol.1, Springer Verlag, 935-957

Sharma MR, Jeyakumar LH, FleischerS, Wagenknecht T (2000) Three-dimensional Structure of Ryanodine Receptor Isoform Three in Two Conformational States as Visualized by Cryo-electron Microscopy. J Biol Chem 275:9485-9491

Shirokova N, Garcia J, Ríos E (1998). Local calcium release in mammalian skeletal muscle. J Physiol 512:377-84

Smith GD, Keizer JE, Stern MD, Lederer WJ, Cheng H (1998) A simple numerical model of calcium spark formation and detection in cardiac myocytes. Biophys. J. 75:15-32

Sobie EA, Dilly KW, dos Santos Cruz J, Lederer WJ, Jafri MS (2002) Termination of Cardiac Ca2+ Sparks: An Investigative Mathematical Model of Calcium-Induced Calcium Release. Biophys J 83:59-78

Tsugorka A, Ríos E, Blatter L (1995) Imaging elementary events of calcium release in skeletal muscle cells. Science 269:1723–1726

Uttenweiler D, Wojciechowski R, Makabe M, Veigel C, Fink RHA (1995) Combined analysis of intracellular calcium with dual excitation fluorescence photometry and imaging. Opt Eng 34 (10):2864-2871

Uttenweiler D, Weber C, Fink RHA (1998) Mathematical modelling and fluorescence imaging as a new approach to study the Ca^{2+}-turnover in skinned muscle fibres. Biophys J 74(4):1640-1653

Uttenweiler D, Kirsch WG, Schulzke E, Both M, Fink RHA (2002) Model based analysis of elementary Ca^{2+}-release events in skinned mammalian skeletal muscle fibres. Eur Biophys J 31:331-340

Uttenweiler D, Both M, Zink W, Sinner B, Martin E, Graf BM, Fink RHA (2003) High resolution fluorescence microscopy in combination with mathematical modelling: first demonstration of the subcellular effects of anesthetics on Ca^{2+}-sparks in situ. Anaesthesist 52: 162-168.

Wilson T (1990) Confocal Microscopy. Academic Press London

12. Langevin Modeling of Intracellular Calcium Dynamics

J.W Shuai and P. Jung

Department of Physics and Astronomy and Institute for Quantitative Biology, Ohio University, Athens, OH 45701, USA

Abstract. Recent progress in fluorescent imaging technology has revealed that Ca^{2+} is released from the endoplasmic reticulum (ER) through clusters that are about 0.1μm in extent and comprise no more than 20-50 release channels. The calcium flux released by such small channel clusters exhibits significant stochasticity due to thermal fluctuations. These fluctuations have been simulated successfully by using Markov models for the binding and unbinding processes at the subunits based on the DeYoung-Keizer model (DeYoung and Keizer 1992).

For whole cell and intercellular modeling of Ca^{2+} signaling the DeYoung-Keizer model with its numerous variables is computationally too demanding. Hence, computationally less expensive but yet accurate algorithms are desirable. In this chapter we report on recent work of our group towards a Langevin approach for calcium dynamics. The starting point is the stochastic DeYoung-Keizer model. In a first step, we simplify towards a stochastic Li-Rinzel approach where the slow inactivation process is treated stochastically by a Markov process, but the other faster processes are approximated by their mean values. In a second step, the Markov process is described by a master equation that is then approximated by a Fokker-Planck equation and subsequently by a Langevin equation. The Langevin equation has the same form as the deterministic 2-variable Li-Rinzel model except that a noise-term is added where the *strength* is related to the *size of the release cluster*. We compare characteristic quantities of the calcium release such as mean values and variances, puff amplitude distributions, puff-width distributions and inter-puff interval distributions generated by an isolated cluster of release channels with all three methods and discuss conditions under which the stochastic Li-Rinzel model and the Langevin approach produce results consistent with the DeYoung-Keizer model. The Langevin approach produces results consistent with the other methods although the necessary compute time is reduced by more than a factor of 10 even for cluster sizes of 20 channels. Furthermore, the compute time required by the Langevin approach to mimic a single cluster does not increase with the size of the cluster.

12.1. Introduction

Recently, high-resolution recordings have shed new light on the elementary intracellular Ca^{2+} release events. It has been observed that the Ca^{2+} release channels are spatially organized in clusters with only 20-50 release channels and a size of about 100nm. The calcium release through such small clusters is subject to random fluctuations due to thermal open-closed transitions of individual release channels. After Ca^{2+} is released, it rapidly diffuses within the cluster (in a few μs) and into the cytosol. There, Ca^{2+} is being absorbed by buffers and pumped back into the ER and into the extracellular space resulting in a spatially and temporally limited event that is called calcium puff or spark (Cheng et al. 1993; Callamaras et al., 1998; Melamed-Book et al. 1999; Gonzalez et al. 2000; Mak et al. 2001). Ca^{2+} blips arising from the opening of single release channels have been observed as well (Bootman et al. 1997; Lipp and Niggli 1998; Sun et al. 1998). Binding of Inositol Trisphosphate (IP_3) activates the calcium release channels and additional binding of Ca^{2+} opens the channels via Ca^{2+} induced Ca^{2+} release (Bezprozvanny et al. 1991). Puffs remain spatially restricted at low concentration of IP_3 stimulus, whereas at high levels of IP_3, neighboring clusters become functionally coupled by Ca^{2+} diffusion and Ca^{2+}-induced Ca^{2+} release so as to support intracellular Ca^{2+} waves that propagate in a saltatory manner throughout the cell. Therefore, Ca^{2+} puffs serve as elementary building blocks of intracellular Ca^{2+} waves. Moreover, puffs can arise spontaneously before a wave is initiated and can act as the triggers to initiate waves (Bootman et al., 1997). Calcium puffs provide a unique window on the dynamics of local calcium release.

In this chapter we report on recent work on stochastic modeling of the Ca^{2+} release through small clusters of release channels (IP_3Rs). In Sect. 2 we review various stochastic models that have been applied recently to account for the fluctuations in small release clusters. These include the stochastic DeYoung-Keizer model, the stochastic Li-Rinzel model and a recently applied Langevin method. In Sect. 3 we discuss results obtained for the Ca^{2+} release of isolated clusters and compare predictions of all three methods. Predictions include, puff amplitude distributions, puff lifetime distributions, inter-puff interval distributions and the approach to the deterministic limit of large channel numbers.

12.2. Stochastic Models for Ca^{2+} Release from ER

12.2.1 The Stochastic DeYoung-Keizer Model

The first theoretical model for agonist-induced Ca^{2+} oscillations based on the microscopic kinetics of IP_3 and Ca^{2+} gating of the IP_3R was proposed by DeYoung and Keizer (1992). The model assumes that three equivalent and independent subunits are involved in conduction of an IP_3R. Each subunit has one binding site for IP_3 (site 1) and two binding sites for Ca^{2+}; one Ca^{2+}-binding site (site 2) for activation, the other (site 3) for inhibition. The subunit is conducting only when site 1 and site 2 are bound and site 3 is unbound. Thus, each subunit may exist in eight states with transitions governed by second-order and first-order rate constants. The reaction scheme shown in Fig. 1 (DeYoung and Keizer 1992) illustrates the processes at a single subunit:

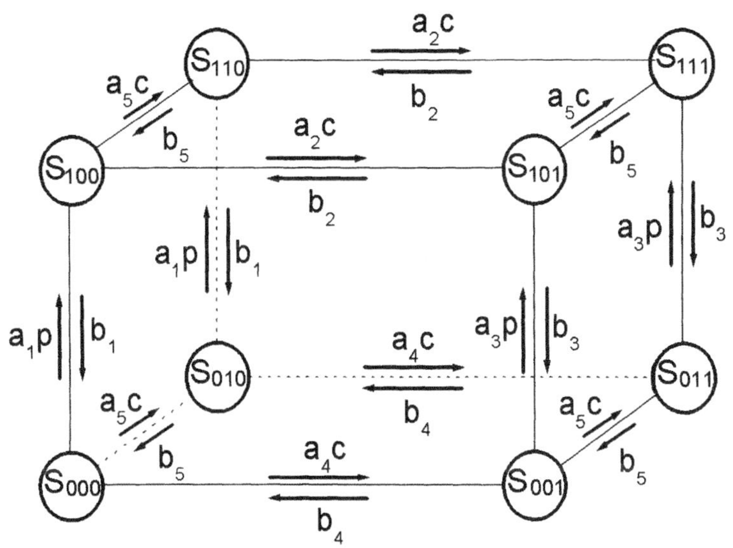

Fig. 1. The binding scheme of a single unit of the IP_3 receptor. S_{nml} denotes the probability for the subunit to be in one of the states [nml] where n,m,l can assume the values 0 and 1. The first index indicates the state of the IP_3 binding site, the second the state of the activating Ca^{2+} binding site and the third index the state of the inhibiting Ca^{2+} binding site. If an index is 0 the respective binding site is unbound; if it is 1 the respective binding site is bound. The values for the rate constants a_s and b_s are the original values used in (DeYoung and Keizer, 1992). The concentration of IP_3 is denoted by p and the intracellular concentration of Ca^{2+} is denoted by c.

For the rate constants the same rates as in (DeYoung and Keizer, 1992) are used, see the table below.

Table 1. Parameter values for the DeYoung-Keizer model of the IP_3 receptor. For the meaning of the rate constants, see **Fig. 1.**

$a_1 = 400/(\mu M\ s)$	$b_1 = 52/s$
$a_2 = 0.2/(\mu M\ s)$	$b_2 = 0.21/s$
$a_3 = 400/(\mu M\ s)$	$b_3 = 377.2/s$
$a_4 = 0.2/(\mu M\ s)$	$b_4 = 0.029/s$
$a_5 = 20/(\mu M\ s)$	$b_5 = 1.64/s$

From each state S_{ijk} the subunit can switch into three other states, as can be seen in Fig.1. Given a large ensemble of subunits, rate equations can be written for the fractions x_{ijk} of subunits being in state S_{ijk}. For the state S_{100}, for example, the following rate equation can be found

$$\frac{dx_{100}}{dt} = pa_1x_{000} - b_1x_{100} \\ +b_2x_{101} - a_2cx_{100} \\ +b_5x_{110} - a_5cx_{100} \tag{1}$$

Similar equations can be derived for all the other states. The fraction of open IP_3-receptors, x_{110}^3, together with leakage, determines the Ca^{2+} flux density from the ER into the intracellular space, i.e.

$$\frac{dc}{dt} = \underbrace{v_1c_1x_{110}^3(c_{ER} - c)}_{\text{channel flux}} + \underbrace{v_2c_1(c_{ER} - c)}_{\text{leakage}} - \underbrace{\frac{v_3c^2}{c^2 + k_3^2}}_{\text{pump}} \tag{2}$$

with

v_1	6/s
v_2	0.11/s
v_3	0.9/(μM s)
k_3	0.1μM

and c_1=0.185. The first term denotes the flux density through the receptor channels driven by the gradient of the Ca^{2+} concentration between the ER (c_{ER}) and intracellular space (c). The second term describes a leakage flux density from the ER into the intracellular space and the third term describes the ATP driven calcium pump that removes Ca^{2+} from the intracellular space. Equation 2 and the set of kinetic equations for the x_{ijk} constitute the DeYoung-Keizer model.

Kinetic equations of the type in (1) are suitable when the numbers of channels are large and thus the fractions of channels occupying the states continuous variables. When the number of channels *N* is small, the possible fractions become discrete variables and fluctuations of the numbers of subunits in the states S_{ijk} around their expectation values become large; the standard deviation over mean value is of the order $1/\sqrt{N}$). Under these circumstances, stochastic methods have to be employed to describe the receptor dynamics.

The stochastic models for the IP_3 receptor are based on two main assumptions:

1. The transitions of the receptor states with time can be described by Markov processes
2. The underlying stochastic processes are stationary

The first assumption implies that the transition probabilities *P* of hopping within Δt to any of the three possible states "lmn" when being in state "ijk" at time *t* (see Fig.1) are only determined by the state of the system at time *t* and not by the pre-history, i.e.

$$P = P(ijk,t \mid lmn,t+\Delta t) \tag{3}$$

for all Δt. The second assumption implies that the transition probabilities are symmetric with respect to continuous time translation. Thus by subtracting t from the arguments in (3), one finds that the transition probabilities can only depend on the time difference Δt, i.e.

$$P(ijk,t \mid lmn,t+\Delta t) = P(ijk \mid lmn,\Delta t)\,. \tag{4}$$

Expanding the transition probabilities for small time steps Δt, and only retaining terms of linear order, one finds for the transition probabilities

$$P(ijk \mid lmn,\Delta t) = \Delta t \cdot r_{ijk \to lmn} \tag{5}$$

for transitions into a *different* state where the rates $r_{ijk \to klm}$ are the time derivative of the transition probabilities, and

$$P(ijk \mid ijk,\Delta t) = 1 - \Delta t \sum_{lmn \neq ijk} r_{ijk \to lmn} \tag{6}$$

for the probability remaining in the same state. In the simplest implementation of this scheme, transitions of individual subunits between states can be determined by drawing random numbers from a uniform distribution on the unit interval. Each transition represents a subinterval on the unit-interval [0,1] proportional to the value of its transition probability where the sum of all subintervals representing possible transitions (including the transition into the identical state). The transition that corresponds to the subinterval into which the random number falls is actually performed. Since we use the stochastic DeYoung-Keizer model only to produce reference values for comparison with results from approximate schemes, we do not further elaborate on more efficient simulations schemes. An example of such a simulation for a cluster of 20 IP_3Rs is shown in Fig.2.

12.2.2 The Stochastic Li-Rinzel Model

As can be seen in Fig.2, the transition rates between the states of an IP_3 channel differ significantly in their size for a large range of intracellular Ca^{2+} concentrations. The binding and unbinding of IP_3 is the fastest process, followed by the binding of Ca^{2+} the activation binding-site (approximately a factor of 10 slower) and by the binding of Ca^{2+} to the inactivation binding-site (another factor of 10 slower). Using a type of multiple time-scale expansion, Li and Rinzel (1994) were able to separate the kinetic equations for the IP_3 channel in a fast set, an intermediate set and a slow set of equations. The fast and intermediate sets of equations are solved explicitly, thus eliminating all but the slowest variables. As a result they obtain the following set of differential equations on the slowest time scale with two variables only:

$$\frac{dc}{dt} = -c_1 m_\infty^3 n_\infty^3 h^3 (c - c_{ER}) - c_1 v_2 (c - c_{ER}) - \frac{v_3 c^2}{k_3^2 + c^2}, \tag{8}$$

$$\frac{dh}{dt} = \frac{1}{\tau_h}(h_\infty - h) \tag{9}$$

where the first term on the right hand side of (8) denotes the channel flux density from the ER into the intracellular space, the second term the leak flux density and the third term the pump flux density from the intracellular space into the ER. Equation (9) for h, defined as the sum of the fractions of the subunits in de-inhibited states, i.e. $h=x_{000}+x_{100}+x_{010}+x_{110}$, has the form of a two-state gate. The power h^3 in (8) indicates all three independent subunits have to be de-inhibited in order for the channel to be open. The expressions m_∞^3 and n_∞^3 describe the average fraction of the channels with bound IP_3 and bound activating Ca^{2+}, respectively. The values of the parameters above are

$$n_\infty = \frac{p}{p+d_1} \; ; \; m_\infty = \frac{c}{c+d_5} \; ; \; \tau_h = \frac{1}{a_2(Q_2+c)}$$
$$h_\infty = \frac{Q_2}{Q_2+c} \; ; \; Q_2 = d_2 \frac{p+d_1}{p+d_3} \; ; \; d_i = \frac{b_i}{a_i}. \tag{10}$$

The set of equations (8,9) resemble the form of Hodgkin-Huxley neurons with only sodium channels, where the faster m activation gate-variables are replaced by their steady state value and only the slow in-activation gate is being considered dynamically. The channel current is driven by a calcium gradient instead of a voltage gradient.

The dynamic properties of the Li-Rinzel model in comparison with the DeYoung –Keizer model can be summarized in the bifurcation diagram in Fig.3. At concentrations of IP_3 of less than 0.345μM and larger than 0.644μM, the Li-Rinzel model predicts fixed points, i.e. the Ca^{2+} concentration approaches a constant value after an initial transient. A super-critical Hopf-bifurcation generates Ca^{2+} oscillations for 0.345μM < $[IP_3]$ < 0.644μM. The curves in the regime 0.345μM < $[IP_3]$ < 0.644μM represent the minima and the maxima of $[Ca^{2+}]$ during the oscillations. The Li-Rinzel model (solid line) reproduces the bifurcation diagram of the DeYoung-Keizer model well. The upper Hopf-bifurcation, however, is somewhat shifted. The fixed-points for $[IP_3]$ > 0.644μM and $[IP_3]$ < 0.345μM are reproduced almost exactly.

We consider a cluster of N IP_3Rs, where N can be a small number and thus a stochastic description is necessary. A stochastic version of the Li-Rinzel equations can be obtained by replacing h^3 by the fraction of de-inhibited channels $N_{h-\text{open}}/N$ where N is the total number of IP_3Rs and $N_{h-\text{open}}$ is the number of channels that are de-inhibited, i.e.

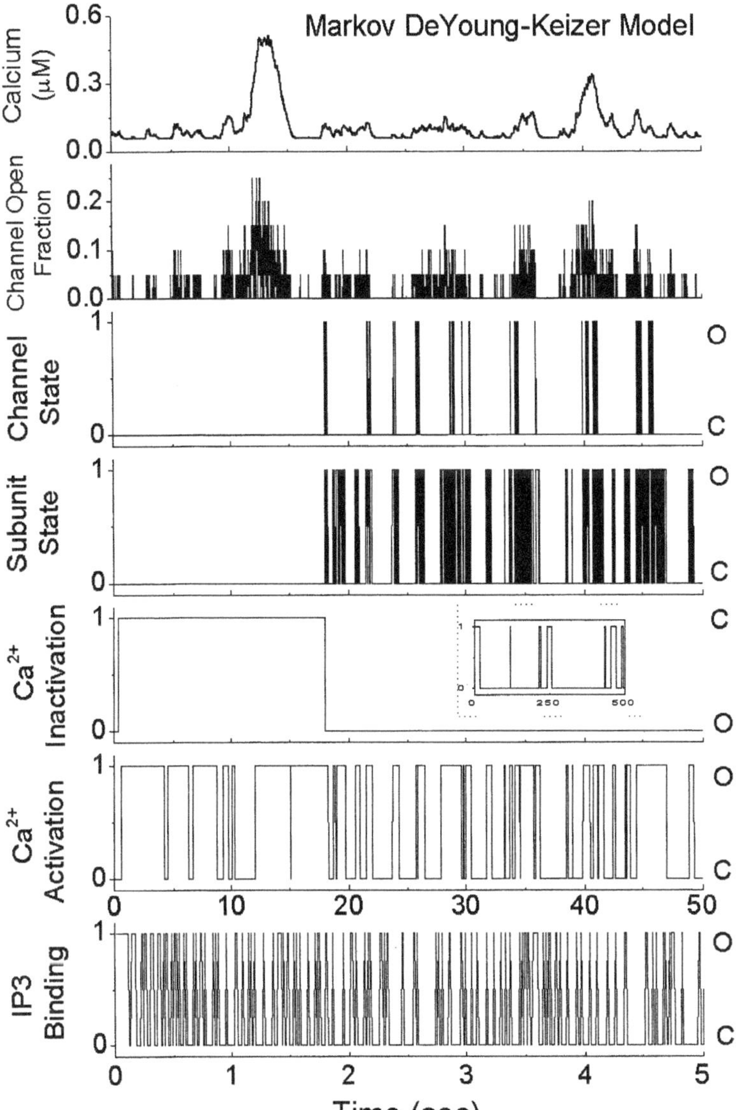

Fig. 2. The time traces of a single IP_3R embedded in a cluster of 20 IP_3Rs at [IP3]=0.3μM, simulated with the stochastic DeYoung-Keizer model are shown. The fastest process is clearly the IP_3 binding and unbinding process (lowest panel). Slower processes are calcium activation and de-activation and the slowest process is calcium inactivation (fifth panel from top). The state of one single subunit is shown in the fourth panel and the single channel state is shown in the third panel. The fraction of open channels is displayed in the second panel. The top panel displays the Ca^{2+} trace released by the cluster.

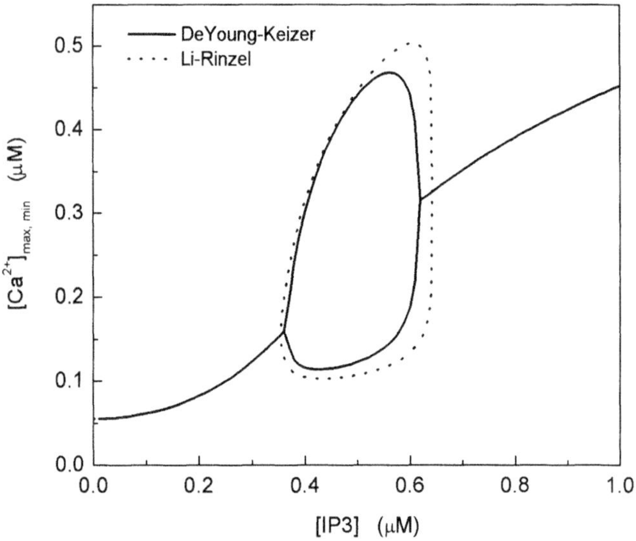

Fig. 3. Bifurcation diagram of the deterministic Li-Rinzel model (dotted line) and the DeYoung-Keizer model (solid line). Oscillations are possible between the two Hopf-bifurcations, i.e. 0.345μM<$[IP_3]$<0.644μM in the Li-Rinzel model. The two curves between the Hopf-bifurcations represent the minima and the maxima of $[Ca^{2+}]$ during their oscillations. For all other values of $[IP_3]$ fixed points are observed.

$$\frac{dc}{dt} = -c_1 m_\infty^3 n_\infty^3 \frac{N_{h-\text{open}}}{N}(c - c_{ER}) - c_1 v_2 (c - c_{ER}) - \frac{v_3 c^2}{k_3^2 + c^2} \tag{11}$$

The opening and closing rates α(p) and β(p), respectively, of this "*h*-gate" can be obtained from (10) as

$$\alpha(p) = a_2 Q_2(p) \quad ; \quad \beta(c) = a_2 c\,. \tag{12}$$

The kinetic scheme for the 4 states of the three *h*-gates of the Li-Rinzel model is given in Fig.4.

Assuming stationary Markov processes, the transition probabilities in linear order of Δt are given by the matrix (see (5))

$$S_0 \underset{\beta}{\overset{3\alpha}{\rightleftharpoons}} S_1 \underset{2\beta}{\overset{2\alpha}{\rightleftharpoons}} S_2 \underset{3\beta}{\overset{\alpha}{\rightleftharpoons}} S_3$$

Fig. 4. The kinetic scheme of the *h*-gates of the Li-Rizel model. The symbol S_n denots an IP_3R with *n* *h*-gates open (not inhibited). The channel is open in the state S_3.

$$\begin{pmatrix} P(0|0,\Delta t) & P(0|1,\Delta t) & P(0|2,\Delta t) & P(0|3,\Delta t) \\ P(1|0,\Delta t) & P(1|1,\Delta t) & P(1|2,\Delta t) & P(1|3,\Delta t) \\ P(2|0,\Delta t) & P(2|1,\Delta t) & P(2|2,\Delta t) & P(2|3,\Delta t) \\ P(3|0,\Delta t) & P(3|1,\Delta t) & P(3|2,\Delta t) & P(3|3,\Delta t) \end{pmatrix} =$$

$$\begin{pmatrix} 1-3\alpha_h\Delta t & 3\alpha_h\Delta t & 0 & 0 \\ \beta_h\Delta t & 1-(\beta_h+2\alpha_h)\Delta t & 2\alpha_h\Delta t & 0 \\ 0 & 2\beta_h\Delta t & 1-(\alpha_h+2\beta_h)\Delta t & \alpha_h\Delta t \\ 0 & 0 & 3\beta_h\Delta t & 1-3\beta_h \end{pmatrix} \tag{13}$$

A simple way to implement the stochastic scheme for N channels is to simulate each channel with the transition probability matrix given by (13) and then count the number of channels in the open-state S_3. If we are mostly interested in a small cluster, such a procedure is sufficiently efficient and easy to implement.

A more efficient scheme can be put together by grouping the IP_3Rs into sets of channels with n open h-gates, where n=0,1,2,3. If there are N_n IP_3Rs in state S_n and the channels are all independent, the probabilities that m IP_3Rs switch into the states S_{n+1} or S_{n-1} (by opening or closing an h-gate) within Δt are given by the binomial distributions (see e.g. (Schneidman et al. 1998))

$$\begin{aligned} P_{n\to n+1}(m,\Delta t) &= \binom{N_n}{m} P(n\,|\,n+1,\Delta t)^m \left(1-P(n\,|\,n+1,\Delta t)\right)^{N_n-m} \\ P_{n\to n-1}(m,\Delta t) &= \binom{N_n}{m} P(n\,|\,n-1,\Delta t)^m \left(1-P(n\,|\,n-1,\Delta t)\right)^{N_n-m} \end{aligned}, \tag{14}$$

with transition probabilities in (13). There are numerous schemes how to update the occupation numbers N_n during one time step. We start with the fastest process (largest transition probability) and then move to the next slower one and so on. We now discuss concrete results obtained with the stochastic Li-Rinzel model.

12.2.3 Calcium release by an isolated cluster of IP_3Rs

We consider a cluster of N IP_3Rs that are spatially close enough so that it can be assumed that the calcium concentration is uniform across the cluster. Given a cluster diameter of about 100nm, the diffusion time

$$\tau_D = \frac{L^2}{D} \approx \frac{(100nm)^2}{50\left(\mu m^2/s\right)} = 200\mu s \tag{15}$$

which is more than three orders of magnitude smaller than the time scale of the Li-Rinzel model (seconds). Trajectories obtained from a cluster of 20 IP_3Rs are shown for three different values of p=[IP_3] in Fig.5. For [IP_3]=0.3μM, i.e. below the Hopf-bifurcation of the deterministic model ([IP_3]=0.345μM), a trace is shown in Fig.5a. While the deterministic equation predicts a fix point (see Fig.3), the stochastic dynamics generates release events where a significant fraction of the

IP_3Rs open up for a short amount of time to release a puff of calcium into the intracellular space. The trajectory for $[IP_3]=0.5\mu M$ is shown in Fig.5b. While the deterministic solution (see Fig.3) predicts periodic calcium oscillations, the stochastic model generates a spike train that looks not very different from the trajectory for sub-threshold $[IP_3]$. The situation is similar for $[IP_3]=0.8\mu M$ in Fig.5c.

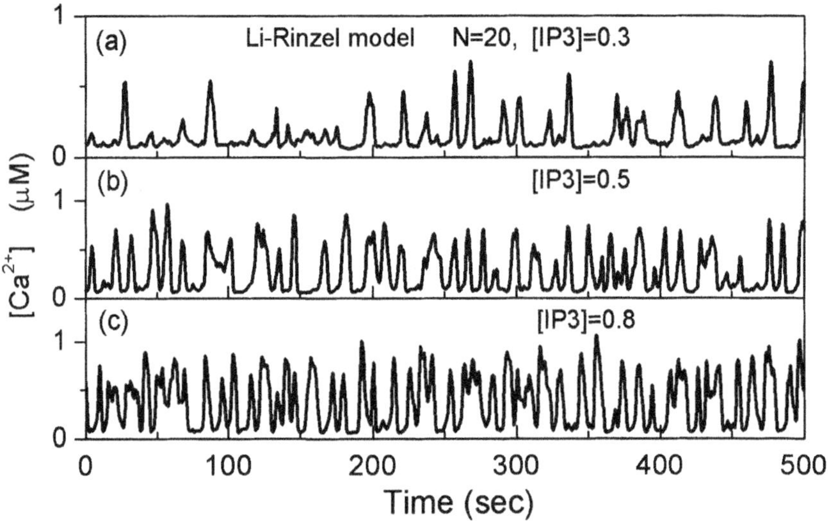

Fig. 5. Calcium traces obtained from the stochastic Li-Rinzel for a cluster of 20 IP_3Rs at three characteristic values of $[IP_3]$. The time (horizontal axis) is measured in seonds.

The most important conclusion from this observation is that a cluster of as few as 20 IP_3Rs, a size which is believed to be realistic, cannot be described by a set of kinetic equations such as the deterministic DeYoung-Keizer model or the Li-Rinzel model. Stochastic methods are necessary.

12.2.4. Langevin Approximation to the Stochastic Li-Rinzel Model

The drawback of the stochastic methods described here is that they become computationally very demanding as one wishes to describe an entire cell with thousands of release clusters or even networks of cells. In the following we describe an alternative scheme resulting in stochastic differential equations (Langevin equations) that are easy to implement and computationally very efficient. The first approximation is that all three *h*-subunits can be replaced by a single two-state subunit with the same opening and closing rates $\alpha(p)$ and $\beta(c)$.

This assumption can be somewhat motivated but justified only by the success of the resulting Langevin equations. The motivation is that all subunits are clamped to the same Ca^{2+} concentration and their opening and closing events are thus somewhat correlated. For a fixed Ca^{2+} concentration, the probability of finding N open single-subunit channels in a cluster of N_t channels obeys the birth-death master equation

$$\frac{\partial P(N,t)}{\partial t} = G^+(N-1)P(N-1,t) - \left(G^+(N)+G^-(N)\right)P(N,t)+G^-(N+1)P(N+1,t) \tag{16}$$

where

$$G^+(N) = (N_t - N)\alpha_h$$
$$G^-(N) = N\beta_h \quad . \tag{17}$$

The following scaling transformation

$$n = N / N_t$$
$$g^+(n) \equiv \frac{G^+(N)}{N_t} = (1-n)\alpha_h$$
$$g^-(n) \equiv \frac{G^-(N)}{N_t} = n\beta_h \tag{18}$$
$$\rho(n) \equiv P(N_t n) \quad ,$$

and the subsequent Taylor-expansion in powers of $\varepsilon = 1/N_t$

$$G^+(N-1) = G^+\left(N_t(n-\varepsilon)\right) = g^+(n-\varepsilon) = g^+(n) + \sum_{i=1}^{\infty}\frac{1}{i!}\varepsilon^i\left(-\frac{\partial}{\partial n}\right)^i g^+(n)$$
$$G^-(N+1) = G^-\left(N_t(n+\varepsilon)\right) = g^-(n+\varepsilon) = g^-(n) + \sum_{i=1}^{\infty}\frac{1}{i!}\varepsilon^i\left(\frac{\partial}{\partial n}\right)^i g^-(n) \quad ,$$
$$P(N+1) = P\left(N_t(n+\varepsilon)\right) = \rho(n+\varepsilon) = \rho(n) + \sum_{i=1}^{\infty}\frac{1}{i!}\varepsilon^i\left(\frac{\partial}{\partial n}\right)^i \rho(n) \tag{19}$$
$$P(N-1) = P\left(N_t(n-\varepsilon)\right) = \rho(n-\varepsilon) = \rho(n) + \sum_{i=1}^{\infty}\frac{1}{i!}\varepsilon^i\left(-\frac{\partial}{\partial n}\right)^i \rho(n)$$

yields after inserting (19) into (16) and comparing coefficients of equal powers in ε the Kramers-Moyal expansion

$$\frac{\partial \rho(n,t)}{\partial t} = \sum_{i=1}^{\infty} \frac{1}{i!} \varepsilon^{i-1} \left(-\frac{\partial}{\partial n} \right)^i M_i(n) \rho(n,t) \quad , \tag{20}$$

with the Kramers–Moyal coefficients

$$M_i(n) = g^+(n) + (-1)^i g^-(n) = (1-n)\alpha_h + (-1)^i n\beta_h \quad . \tag{21}$$

We truncate the Kramers-Moyal expansion after the second term (the diffusion term) and obtain the Fokker-Planck equation

$$\frac{\partial \rho(n,t)}{\partial t} = -\frac{\partial}{\partial n} M_1(n)\rho(n,t) + \frac{1}{2N_t}\frac{\partial^2}{\partial n^2} M_2(n)\rho(n,t) \quad , \tag{22}$$

with the drift and diffusion coefficients

$$\begin{aligned} M_1(n) &= g^+(n) - g^-(n) = (1-n)\alpha_h - n\beta_h \\ M_2(n) &= g^+(n) + g^-(n) = (1-n)\alpha_h + n\beta_h \quad . \end{aligned} \tag{23}$$

Note that this truncation does not guarantee the preservation of the exact stationary probability (Hänggi and Jung 1988). An equivalent Ito-type Langevin equation can be found as

$$\begin{aligned} \frac{dn}{dt} &= \alpha_h(1-n) - \beta_h n + \xi(t) \\ \langle \xi(t)\xi(t') \rangle &= \frac{\alpha_h(1-n) + \beta_h n}{N_t} \delta(t-t') \quad , \\ \langle \xi(t) \rangle &= 0 \end{aligned} \tag{24}$$

with Gaussian, white noise $\xi(t)$. Fox and Lu (1994) and later Schmid et al. (2001) have used such a scheme to describe action potentials generated by clusters of neuronal ion channels. The Langevin approximation of the stochastic Li-Rinzel model thus reads

$$\begin{aligned} \frac{dc}{dt} &= -c_1 m_\infty^3 n_\infty^3 h^3 (c - c_{ER}) - c_1 v_2 (c - c_{ER}) - \frac{v_3 c^2}{k_3^2 + c^2} \\ \frac{dh}{dt} &= \alpha_h(1-h) - \beta_h h + \xi(t) \quad . \\ \langle \xi(t)\xi(t') \rangle &= \frac{\alpha_h(1-h) + \beta_h h}{N_t} \delta(t-t') \\ \langle \xi(t) \rangle &= 0 \end{aligned} \tag{25}$$

It is almost as simple as the deterministic Li-Rinzel model except for the noise term in the second equation.

Integrating these equations with a first order solver, the stochastic term results in one additional term

$$h(t+\Delta t) = \left(\alpha_h(1-h(t)) - \beta_h h(t)\right)\Delta t + \sqrt{2K\Delta t}\,R_t \qquad (26)$$
$$K = \frac{\alpha_h(1-n)+\beta_h n}{N_t},$$

where R_t denotes a Gaussian random number with unit variance. The most significant advantages of the Langevin equations are that they are simple to implement and that their computational demand does not grow with increasing cluster size N_t.

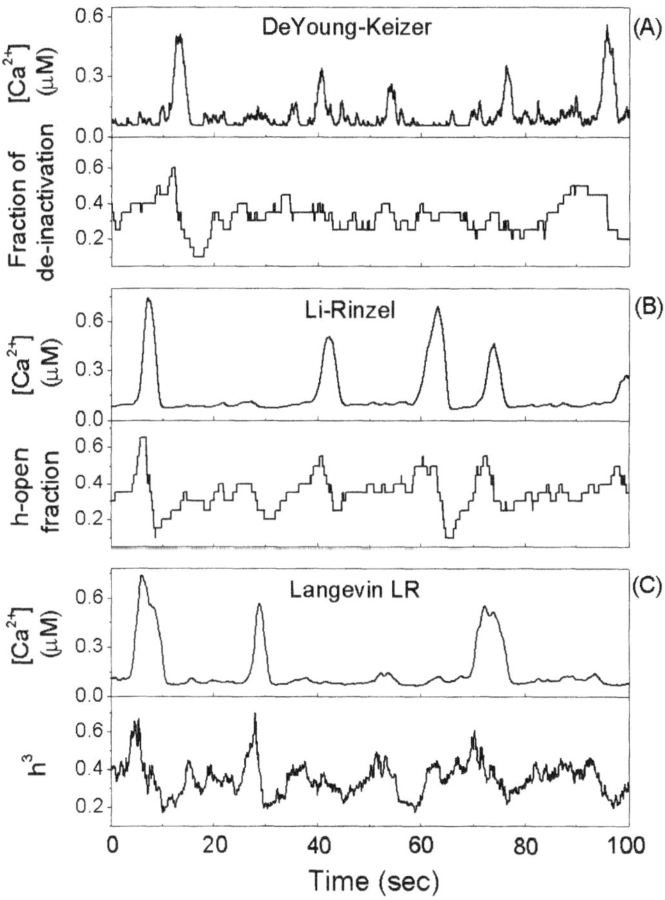

Fig. 6. Simulation results obtained from three models (DeYoung-Keizer (A), Li-Rinzel (B), Langevin (C)) of an isolated cluster of 20 IP_3Rs.

12.3. Results

12.3.1 Calcium Release Dynamics

In this section we compare the time traces of calcium released by isolated, mostly small, clusters of IP_3Rs, obtained from the models described above. In Fig.6, the upper panel (A) shows the intracellular Ca^{2+} concentration obtained with the DeYoung-Keizer model for a time interval of 100s. Below, we show the fraction of channels that are de-inhibited to compare with the corresponding fractions obtained with the other methods. Panel (B) shows a Ca^{2+} trace and the corresponding fraction of *h*-open channels during 100s of simulation obtained with the stochastic Li-Rinzel model. Panel (C) shows the corresponding data obtained with the Langevin approach. The Ca^{2+} trace obtained from the DeYoung-Keizer model has more high-frequency content since the Li-Rinzel and Langevin approach are obtained from the DeYoung-Keizer model after eliminating fast time scales. In all three cases, large fractions of de-inhibited channels correlate with peaks in the calcium release (puffs). For the Langevin approach, the inactivation variable *h* is continuous while it is discrete for the Markov-approaches.

We made more detailed comparisons of the statistical moments of the stochastic Ca^{2+} traces at different cluster sizes and concentrations of IP_3 (Fig.7). The stochastic Li-Rinzel model reproduces the data of the DeYoung-Keizer model remarkably well for cluster sizes larger than 10 channels. More surprising, although the Langevin approach is derived for large cluster sizes, it works very well for clusters larger than 100 channels and reasonably well even for clusters of 20 channels. At 20 channels, the mean value obtained from the Langevin equation is about 7% off the mean value obtained from the DeYoung-Keizer Model. In Fig.8, we compare the mean values and the variances of the Ca^{2+} signal generated by a cluster of 20 IP_3Rs as a function of [IP_3].

The agreement is good throughout. Also the fractions of de-inhibited channels, important for the stochastic Li-Rinzel and the Langevin models, are in good agreement between the models. This is demonstrated in Fig.9, where we compare the fraction of de-inactiavted channels obtained from the DeYoung-Keizer model with the fraction of *h*-open channels of the stochastic Li-Rinzel model and the continuous *h*-open fraction of the Langevin approach for three different of IP_3 concentration.

12.3.2. Puff Amplitude Distributions

Another measure used in experiments to characterize calcium puffs is the puff-amplitude distribution. During each calcium puff, the maximum Ca^{2+} concentration is recorded (Pratusevich and Balke 1996, Bootman et al. 1997; Smith et al. 1998, Izu et al. 1998, Sun et al. 1998, Thomas et al. 1998, Cheng et al.

1999, Jiang et al. 1999, Callamaras and Parker 2000, Marchant and Parker 2001, Rios et al. 2001). We have computed histograms of these puff-amplitudes using the three stochastic methods.

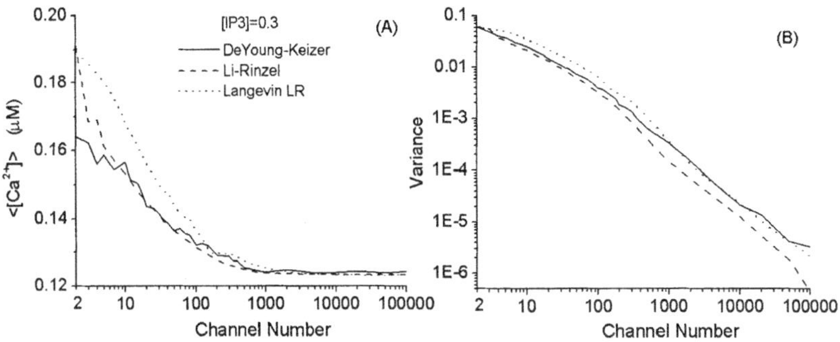

Fig. 7. In panel (A) the mean cytosolic Ca^{2+} concentrations are shown as a function of the number of IP_3Rs per cluster at a fixed sub-threshold concentration of [IP3]=0.3μM for all three different schemes. In panel (B), the variances of the stochastic Ca^{2+} concentration are shown.

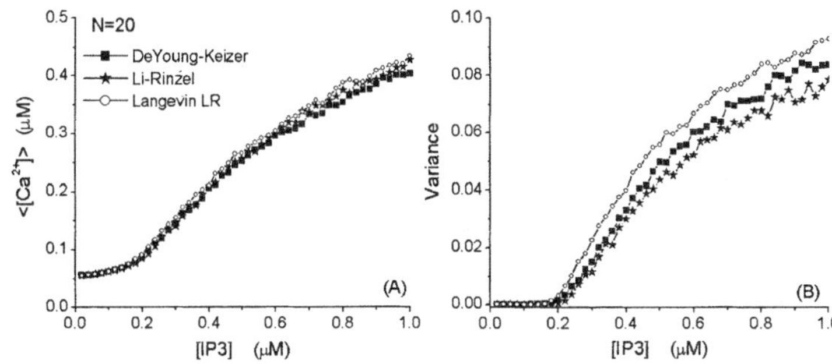

Fig. 8. The mean values (A) and the variances (B) generated by a cluster of 20 IP_3Rs are compared for different concentrations of $[IP_3]$. The three models are compared. Note the different scale in conparison to Fig.7.

It is clear from Fig.6, that the DeYoung-Keizer model generates more small puffs than the other methods since for the other methods, the fast time scales have been eliminated. From the upper panel of Fig.6 it also clear that these fast puffs are smaller puffs. The question therefore is whether the larger and slower puffs produced by the Li-Rinzel and the Langevin model match those of the DeYoung-Keizer model. In order to compare these, we filter the Ca^{2+} time series by eliminating puffs below a cut-off size. The remaining puffs are analyzed with respect to their amplitudes and re-normalized to one. In Figs.10A,B we compare the puff-amplitude distributions of the filtered puffs from the DeYoung-Keizer model with the puff amplitude distributions obtained from the stochastic Li-Rinzel and the Langevin models at $[IP_3]$=0.3μM (Fig.10A) and $[IP_3]$=0.5μM (Fig.10B), using two different cut-off sizes, 0.2μM(upper panels A) and 0.4μM (lower panels B). The distributions are in qualitative agreement at the higher cut- off threshold of 0.4μM.

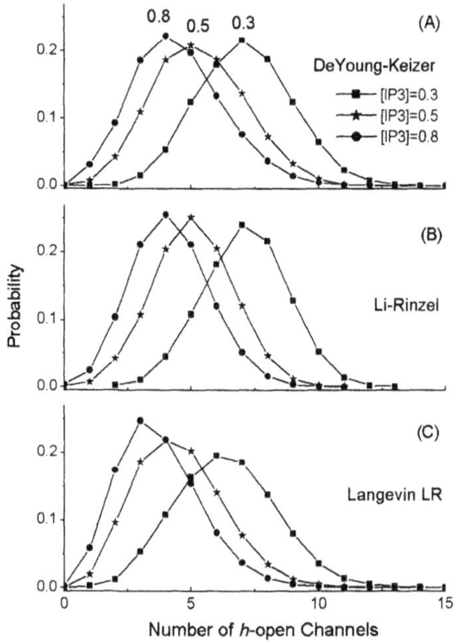

Fig. 9. Distribution of open de-inhibited channels of the DeYoung-Keizer model (A) compared with the distribution of *h*-open channels obtained from the stochastic Li-Rinzel model (B) and the Langevin model (C) for three charecteristic values of $[IP_3]$. The data were obtained from a cluster with 20 IP_3Rs.

Depending on the concentration of IP_3 and the number of channels for the cluster, the puff amplitude distribution can assume quite different shapes. Generally for small concentrations of IP_3, the puff-amplitude distribution is monotonously decaying. When the concentration approaches the critical value at which oscillations set in (first Hopf bifurcation, see Fig.6) in the deterministic system, the puff - amplitude distributions develop additional peaks, indicating characteristic amplitudes. When $[IP_3]$ exceeds the concentration at which the other Hopf bifurcation occurs, a single peak remains at a larger Ca^{2+} concentration. A more detailed sketch of the shapes of the puff-amplitude distributions under various conditions is shown in Fig.11.

12.3.3. Puff-width Distributions

Another quantity that is often measured in the context of calcium puffs is the distribution of the temporal width of a puff released from a single release cluster (Sun et al. 1998, Thomas et al. 1998; Haak et al. 2001). We define the width as the time interval during which the Ca^{2+} amplitude exceeds half its peak value. To compare puff-width distributions obtained from the stochastic DeYoung-Keizer model in a meaningful way with those obtained with the other methods, the same filtering as described in Sect. 3.2 has been applied. The results are shown in Fig.12. It is quite remarkable how the enhanced filtering (B) results in good agreement for all three methods.

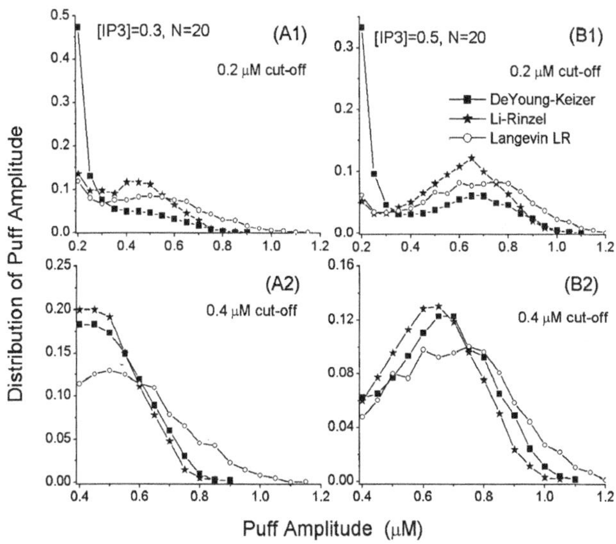

Fig. 10. Puff amplitude distributions of a cluster with 20 IP_3Rs at $[IP_3]$=0.3μM (A) and 0.5μM (B). In the upper panels (A1,B1) we have used a cut-off size of 0.2μM while 0.4μM has been used in the lower panels (A2,B2)

12.3.4. Inter-puff Interval Distributions

Another quantity characterizing Ca^{2+} puffs is the distribution of the time intervals between two consecutive puffs (Marchant et al 1999). To compare inter-puff interval distributions obtained from the stochastic DeYoung-Keizer model in a meaningful way with those obtained with the other methods, the same filtering as described in Sect. 3.2 has been applied. The results are shown in Fig.13. While for a low cut-off value for puffs (panels A1,B1), the DeYoung-Keizer model exhibits an inter-puff interval distribution with more short intervals, additional size filtering with a threshold twice as large (A2,B2) yields good agreement between the inter-puff interval distributions obtained with all three methods.

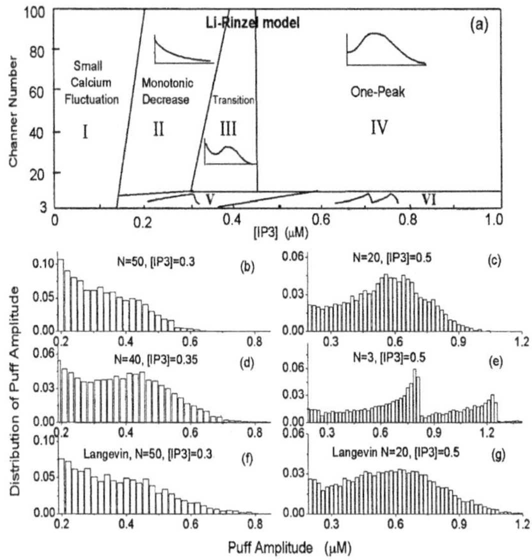

Fig. 11. The *phase diagram* of the puff amplitude distributions is shown in (a) in the $[IP_3]$–N plane. Samples of puff amplitude distributions obtained with the stochastic Li-Rinzel model are shown in b-e in various regimes indicated in (a). Figures f and g show the amplitude distribution obtained with the Langevin method in comparison to those obtained with the full kinetic method in b and c.

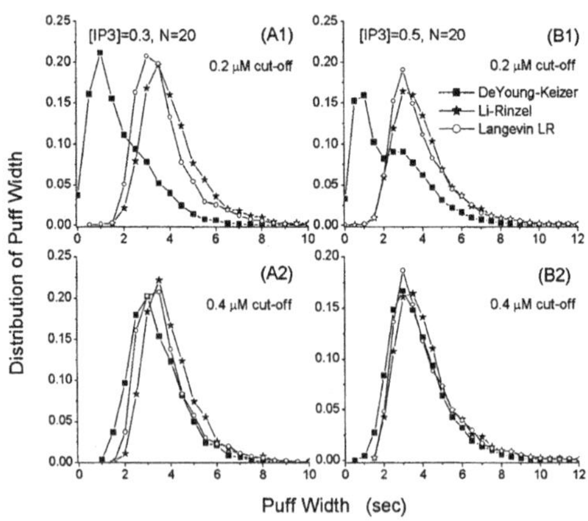

Fig. 12. The puff-width distributions generated from a cluster of 20 IP_3Rs using the three stochastic methods are compared for two different values of $[IP_3]$ and for different cut-off values. The cut-off value in the upper panels (A1,B1) is 0.2μM and in the lower panels (A2,B2) 0.4μM.

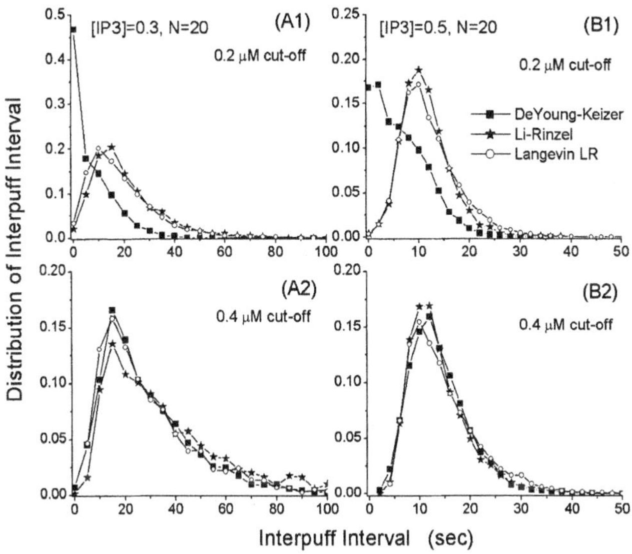

Fig. 13. Distributions of inter-puff intervals generated from a cluster of 20 IP_3Rs using the three stochastic methods are compared for two different values of $[IP_3]$ and for different cut-off values. The cut-off value in the upper panels (A1,B1) is 0.2μM and in the lower panels (A2,B2) 0.4μM.

12.3.5. Large Clusters – The deterministic Limit

In this section, we show that for large channel numbers, the stochastic schemes approach the appropriate deterministic limits. In case of the Langevin approach this is quite obvious since the noise-term in the second equation of (26) vanishes in the limit of large N_t. For the other models, the deterministic limit given by the DeYoung-Keizer model and the Li-Rinzel model will also be approached since the master equations for the multiple subunits can be expanded in completely analogous ways. In Fig.14, we compare the bifurcation diagram for the stochastic Li-Rinzel model and the stochastic DeYoung-Keizer model for a cluster of 10,000 channels with the respective deterministic limits. The minima and the maxima of the stochastic Ca^{2+} traces have been obtained by averaging over local minima and maxima, respectively. While the expectation values agree already very well with the deterministic limits, there are still substantial deviations from the deterministic limit in the bifurcation diagram for large concentrations of IP_3. For larger clusters these differences vanish (not shown here).

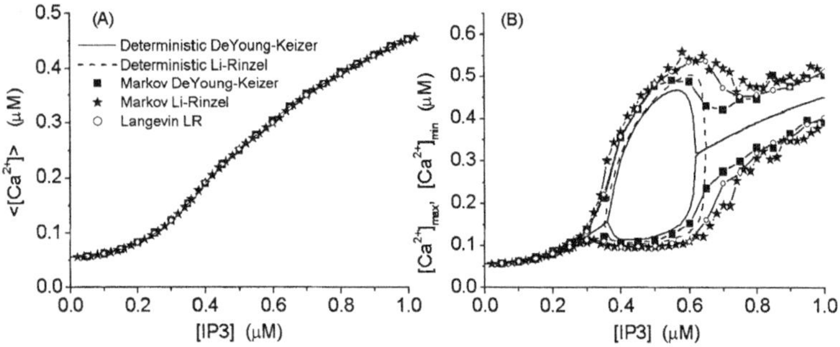

Fig. 14. The average calcium concentration obtained with all three stochastic methods from a cluster of 10,000 IP_3Rs (A). The bifurcation diagrams obtained with all three stochastic methods in comparison with their respective deterministic limits (see also Fig.5) for a cluster of 10,000 IP_3Rs (B).

12.4. Conclusions

Stochastic modeling of IP_3Rs using Markov-methods is computationally very demanding. For whole-cell modeling or even intercellular modeling of Ca^{2+} signaling, computational efficiency becomes an important factor. To meet the need for a fast algorithm, we have put forward a Langevin equation to describe the release of Ca^{2+} by an isolated cluster of IP_3Rs. Although the Langevin equation is derived for large numbers of channels it turns out to produce results that are in good agreement with the DeYoung-Keizer model even for clusters as small as 20 channels. We have compared bifurcation diagrams, puff amplitude distributions, puff-width distributions and inter-puff interval distributions.

This material is based on work supported by National Science Foundation under grant IBN 0078055

References

Bezprovanny I, Watras J, Ehrlich B (1991) Bell-shaped calcium response curves of Ins(1,4,5)P_3- and calcium-gated channels from endoplasmic reticulum of cerebellum. Nature. 351: 751-754.

Bootman, M, Niggli E, Berridge M, Lipp P (1997) Imaging the hierarchical Ca^{2+} signaling system in HeLa cells. The Journal of Physiology 499: 307-314.

Callamaras NJ, Marchant S, Sun X, Parker I (1998) Activation and co-ordination of $InsP_3$ mediated elementary Ca^{2+} events during global Ca^{2+} signals in Xenopus oocytes, Journal of Physiology 509: 81-91.

Callamaras N, Parker I (2000) Phasic characteristic of elementary Ca^{2+} release sites underlies quantal responses to IP_3. The EMBO Journal 19: 3608-3617.

Cheng HW, Lederer J, Cannel MB (1993) Calcium sparks elementary events underlying excitation-contraction coupling in heart muscle. Science 262: 740-744.

Cheng H, Song L, Shirokova N, Gonzalez A, Lakatta EG, Rios E, Stern MD (1999) Amplitude distribution of calcium sparks in confocal images: theory and studies with an automatic detection method. Biophys. J. 76: 606-617.

DeYoung GW, Keizer J (1992) A single-pool inositol 1,4,5-trisphosphate-receptor-based model for agonist-stimulated oscillations in Ca^{2+} concentration. Proceedings of the National Academy of Sciences USA, 89: 9895 - 9899.

Fox RF, Lu Y (1994). Emergent collective behavior in large numbers of globally coupled independently stochastic ion channels. Phys. Rev. E. 62: 2636-2643.

Gonzalez A, Kirsch WG, Shirokova N, Pizarro G, Brum G, Pessah IN, Stern MD, Cheng H, and Rios E (2000) Involvement of multiple intracellular release channels in calcium sparks of skeletal muscle. Proceeding of the National Academy of Sciences USA 97: 4380-4385.

Haak, LL, Song L, Molinski TF, Pessah IN, Cheng H, Russell JT (2001) Sparks and puffs in oligodendrocyte progenitors: cross talk between Ryanodine receptors and inositol trisphosphate receptors. Journal of Neuroscience. 21: 3860-3870.

Hänggi P, Jung P (1988) Bistability in Active Circuits: Application of a novel Fokker-Planck Approach. IBM Jorunal of Research and Development 32:119-125

Izu LG, Wier WG, Balke CW (1998) Theoretical analysis of the Ca^{2+} spark amplitude distribution. Biophysical Journal, 75: 1144-1162.

Jiang Y, Klein MG, Schneider MF. 1999. Numerical simulation of Ca^{2+} sparks in skeletal muscle. Biophys. J. 77: 2333-2357.

Keener J and Sneyd J (2000) Mathematical Physiology. Springer, New York

Li Y, Rinzel J (1994). Equations for IP_3 receptor-mediated Ca^{2+} oscillations derived from a detailed kinetic model: a Hodgkin-Huxley like formalism. Journal of Theoretical Biology 166:461-473.

Lipp P. Niggli E (1998) Fundamental calcium release events revealed by two-photon excitation photolysis of caged calcium in guinea-pig cardiac myocytes. Journal of Physiology 508: 801-809.

Marchant J, Callamaras N, Parker I (1999) Initiation of IP_3 mediated Ca^{2+} waves in Xenopus oocytes. The EMBO Journal 19: 5285-5299.

Marchant JS, Parker I (2001) Role of elementary Ca^{2+} puffs in generating repetitive Ca^{2+} oscillations. The EMBO Journal 20: 65-76.

Mak, DD, Foskett JK (1997) Single-channel kinetics inactivation, and spatial distribution of Inositol Trisphosphate receptors in Xenopus oocyte nucleus. Journal of General Physiology 109: 571-587.

Melamed-Book N, Kachalsky SG, Kaiserman I, Rahamimoff R (1999) Neuronal calcium sparks and intracellular calcium noise. Proceeding of the National Academy of Sciences USA. 26: 15217 - 15221.

Pratusevich VR, Balke CW (1996) Factors shaping the confocal image of the calcium spark in cardiac muscle cells. Biophys. J. 71: 2942-2957.

Rios EN, Shirokova W, Kirsch G, Pizarro MD, Stern H, Cheng H, Gonzalez A (2001) A preferred amplitude of calcium sparks in skeletal muscle. Biophys. J. 80: 169-183.

Schmid G, Goychuk I, P. Hanggi P (2001) Stochastic resonance as a collective property of ion channel assemblies. Europhys. Lett. 56: 22-30

Schneidman E., Freedman B., and Segev I. (1998) Ion channel stochasticity may be critical in determining the reliability and precision of spike timing. Neural Computation 10:1679 - 1703

Smith GD, Keizer JE, Stern MD, Lederer WJ, Cheng H (1998) A simple numerical model of calcium spark formation and detection in cardiac myocytes. Biophys. J. 75: 15-32.

Sun XN, Callamaras NJ, Marchant JS, Parker I (1998) A continuum of $InsP^3$-mediated elementary Ca^{2+} signaling events in Xenopus oocytes. Journal of Physiology 509: 67-80.

Thomas D, Lipp P, Berridge MJ, Bootman MD (1998) Hormone-evoked elementary Ca^{2+} signals are not stereotypic, but reflect activation of different size channel clusters and variable recruitment of channels within a cluster. The Journal of Biological Chemistry 273: 27130-27136.

13. Building a Wave – Models of the Puff-to-Wave Transition

Martin Falcke

Hahn Meitner Institute, Glienicker Str. 100, 14109 Berlin, Germany

Abstract. Ca^{2+} release channels open and close randomly because of random binding and dissociation of Ca^{2+} and IP_3 to their subunits. They are arranged in clusters on the membrane of the endoplasmic reticulum with each cluster containing a small number of channels only. The concentration profile of free Ca^{2+} resulting from open channels in a cluster decays almost not at all within a cluster but considerably on the length of cluster spacing. Hence, the spatial coupling within clusters is strong but weak between clusters. These circumstances set the stage for modelling of intracellular Ca^{2+} as an array of active stochastic elements. This chapter deals with the consequences of random behavior of single channels for global dynamics. First, experimental findings on the relation between elemental and global events are reviewed. Next, a frame work of stochastic modelling on length scales of many cluster spacings is set up. Modelling starts with the stochastic reaction of Ca^{2+} and IP_3 with the receptor subunit. These random events are then coupled by diffusion profiles of single clusters. Global concentration profiles can be obtained as superposition of single cluster profiles.
The random behavior on the cluster level shows up on the global level as nucletion events initiating waves or oscillations with variable period. The consequences for the dependence of the period on experimentally accessible parameters are discussed. We will consider the deterministic limit of the stochastic model. Finally, a more abstract approach is presented allowing to investigate more general principles underlying the stochastic pattern formation.

13.1. Introduction

One of the fascinating aspects of intracellular Ca^{2+} dynamics is the possibility to study how elemental events called sparks, blips and puffs build up global events like waves and oscillations. It can be observed in different cell types (Reber 1996, Heemskerk 2001, Horne 1997, Tovey 2001). Tovey et al. report that puff characteristics are even rather independent of the isoform of the Inositol 1, 4, 5-trisphosphate receptor channel (Tovey 2001). However, there are differences between cells with respect to the details of initiation of global events (Tovey 2001).

In this chapter, we will first look at experimental findings on initiation of global events by elemental events. This is followed by the larger part on modeling of that transition and the global phenomena. Modeling starts from a consideration of the elemental building blocks. The methods used to model a wide range of time and length scales will be explained. They allow for investigation of the stochastic features of global events. We will start with models including many experimental details. In the last part, we will describe a more abstract and simplified approach based on an equation for the propagation of channel states.

The dynamics of cytosolic Ca^{2+} concentration is driven by release from the endoplasmic reticulum (ER) through channels, uptake into the ER by pumps, release and uptake of Ca^{2+} by mitochondria and exchange of Ca^{2+} through the plasma membrane with the exterior of the cell. Ca^{2+} binding proteins buffer Ca^{2+} in all compartments to a high degree. Especially in the cytosol, more than 99% of the Ca^{2+} may be bound to buffers. We will focus on Ca^{2+} exchange between the ER and the cytosol and the role of buffers in this chapter and will consider release through the Inositol 1, 4, 5-trisphosphate receptor channel (IP_3R) only.

Channels open and close randomly. The open probability depends on binding of Inositol 1, 4, 5-trisphosphate (IP_3) to the receptor channel. It is very small, if IP_3 is not bound. Binding of IP_3 is one condition for activation of the channel. However, the crucial feature allowing for communication between the IP_3Rs is that the open probability depends on the cytosolic Ca^{2+} concentration (see Chap.1 and Taylor 1998, Patel 1999). Moderate increase of the concentration favors opening. That causes self amplification of Ca^{2+} release from the ER. If a channel opens, Ca^{2+} diffuses to its neighbor channels and increases the probability so that they open as well and raise the cytosolic Ca^{2+} concentration even more. This self-amplification allows for waves of release propagating through the cell or global oscillations in small cells.

It is again cytosolic Ca^{2+} which terminates release during waves or oscillations. High Ca^{2+} concentration leads to inhibition of the channels. Inhibited channels are closed and can open again only when they have recovered from inhibition. Inhibition is slower than the activating processes by about a factor 1/10. It takes 2-4s in Xenopus oocytes to occur. Recovery from inhibition is again slower and occurs within 10-15s (Ilyin 1994).

The communication between release channels by Ca^{2+} described so far would hardly produce the whole spectrum from small, localized events to waves since release would very likely spread immediately through the whole cell when a channel opens. The continuum of release events is possible because of the spatial organization of channels on the ER membrane. The channels form small closely packed groups, which are called clusters. The clusters are randomly distributed on the ER membrane with distances between 2 and 7 μm in Xenopus oocytes and HeLa cells. That spatial separation allows for isolated events. If the amount of Ca^{2+} released by the channels in a cluster is not sufficient to open channels in a neighboring cluster, release will not spread.

We will go through the sequence from isolated events to a high degree of cooperation in global events in more detail for two cell types. The sequence coincides with going from low to high IP_3 concentration. Photorelease of little amounts of caged IP_3 in Xenopus oocytes causes a small elevations of Ca^{2+} all over the area exposed to increased IP_3 comparable to homogeneous rise of the base level of cytosolic Ca^{2+} (Parker 1996, Parker 1996a). That elevation of Ca^{2+} increases like $[IP_3]^3$ up to the threshold of spontaneous wave creation. In that regime, small discrete events - called blips - can be observed (Parker 1996, Parker

1996a). We will use the term "blip" with the meaning of small puff. A puff reported in detail (Yao 1995) and visualized by 40μM calcium green-1 dye shows a rise time of about 60ms and reaches a fluorescence amplitude of about 20 intensity units. The average decay of 13 puffs in the same system could be fitted to a diffusion process with an effective diffusion coefficient of 25 $\mu m^2 s^{-1}$. Most of the puffs are observed in a layer 5-7μm beneath the plasma membrane. The density of puff sites is about 1 per $30\mu m^2$ (Yao 1995). Puffs originate from spots smaller than 1μm in diameter (Parker 1996a).

The amplitude of single events is typically given as fluorescence F normalized by the background value F_0. It shows a skewed distribution with a maximum F/F_0 at 1.5 and a tail as high as 3.5, if it is measured with Oregon Green 488 BAPTA-1 (Sun 1998). The signal mass distribution of discrete events is exponential (Sun 1998). The smallest events resolved correspond to a signal mass of $2\ 10^{-20}$ mol and large events to $2\ 10^{-18}$ mol. Signal mass continued to increase longer than peak fluorescence in the time course of puffs. That suggests that Ca^{2+} release was still going on even while fluorescence was declining (Sun 1998). A typical current underlying a puff is 2.5pA (Sun 1998). That current can serve to estimate the number of open channels in a cluster during a puff. Sun et al. obtained an estimate of 8 channels (Sun 1998). That is close to the estimate by Mak et al. of 5 (Mak 1997).

The FWHM (full width at half magnitude) of small puffs is about 1.5μm and larger ones reach approximately 4-6μm (Parker 1996a, Sun 1998). The shortest discrete events last for about 50ms. The duration distribution shows a peak at about 150ms and an exponential decay up to 600ms (Sun 1998). Duration is defined here as the time the fluorescence is above 1/2 of the peak value (full duration at half maximum FDHM).

The time interval between puffs recorded with constant IP_3 concentration from a $(3\mu m)^2$ region follows a distribution with a peak at about 10s and an exponential decay with a time constant of 14s for intervals larger than 10s (Parker 1996a). Measurements with higher resolution revealed a larger fraction of short intervals between puffs (Parker 1996) and showed that the peak of the distribution is around 3s just below the wave initiation threshold and around 5s just above (Marchant 1999). The oocyte was loaded with 25fmol 3-F-IP_3, which is poorly metabolized, i.e. its concentration is constant for the duration of the experiment (Parker 1996a). That minimizes the effect of IP_3 changes on the results. Measurements of the variation of puff frequency between different sites confirmed the Poisson distribution for intervals longer than 10s. The average puff frequency increases like $[IP_3]^{3.1}$ (Yao 1995).

Puffs occur with a random latency upon photo release of caged IP_3. The mean latency is proportional to the square of the inverse flash strength ($[IP_3]$ is assumed to be proportional to flash strength (Parker 1992)). The puff frequency after $[IP_3]$ increase is proportional to the third power of flash strength (Yao 1995). The number of sites responding to $[IP_3]$ increases like (flash strength)$^{4.2\text{-}4.9}$. Hence, the cubic dependence of puff frequency on $[IP_3]$ measured in (Yao 1995) does not simply reflect the dependence of open probability of the single channel on $[IP_3]$.

The increase of the number of responding sites like $[IP_3]^{4.2\text{-}4.9}$ hints that there might be cooperation by Ca^{2+} additionally to that one by IP_3 binding. That cooperation may reach across nearby neighbored puff sites. Puff sites spaced less than 2.2μm apart showed high correlation whereas those more than 6μm apart did not show any correlation in activity (Yao 1995).

At even higher $[IP_3]$, some puffs trigger abortive Ca^{2+} waves. A further increase leads to steady and repeated waves (Callamaras 1998). The amount of Ca^{2+} needed to trigger a wave is larger than that liberated by a single puff (Marchant 1999). Marchant et al. estimated the amount necessary to trigger a wave with a certain probability (Marchant 1999). The IP_3 concentration was chosen so that spontaneous waves occurred with a mean latency of 5s upon photo release of IP_3. A Ca^{2+} signal mass of approximately 2 10^{-17} mol was sufficient to trigger a wave with high probability. A 50% probability was observed with approximately 2.4 10^{-18} mol. The largest puffs observed released 1.8 10^{-18} mol, but typical puffs release 0.1-0.6 10^{-18} mol only (0.34 10^{-18} mol average) (Marchant 1999). In summary, waves cannot be initiated by a single puff - i.e. by a single cluster - but the cooperation of several clusters is necessary to set off such a global event (Marchant 1999).

The mechanism by which waves are initiated is not an increase in puff amplitude. The amplitude of puffs preceding wave initiation was constant. The critical amount of Ca^{2+} is raised by increasing puff frequency, if waves are initiated by a step increase of $[IP_3]$ (Marchant 1999). That causes an elevated average Ca^{2+} concentration at the site of wave initiation prior to the start of the wave.

Waves arise preferentially from distinguished sites in the cell, which were christened focal sites (Lechleiter 1991, Marchant 2001). These sites show a higher density of clusters than non-focal sites. Their average spacing was approximately 5μm whereas non-focal sites showed an average spacing of 7.3μm (Marchant 2001). Additionally, focal sites show a higher puff activity suggesting that they are more sensitive to IP_3 than non-focal sites (Marchant 2001).

The necessity for cooperation of several puffs to initiate a wave allows for the occurrence of single puffs in between consecutive waves. Marchant et al. measured the characteristics of these puffs in dependence on wave phase and period (Marchant 2001). The mean rate of rise of fluorescence at a release site during a wave is only about 15% faster than that one of an isolated puff (Callamaras 1998), i.e. approximately the same number of channels open in both situations. The measurements demonstrate typical differences between repetitive wave initiation with short and long periods. The peak of the previous wave was chosen as the reference time. Puffs did not occur in the first 7s after a wave passed. Then, puffs occur with increasing frequency. That increase continues till initiation of the next wave for short period waves (period $<$ 15s). The puff amplitude quickly reaches a certain level, which then stays constant for the last 30-40% of the wave period. Waves with intermediate periods (15-50s) exhibit an increase in puff frequency from 7s after the previous wave till about three quarter through the cycle. The amplitude again soon reaches its steady level, which it then keeps for most (approximately 60%) of the cycle. For these waves, the probability of wave initiation by artificial creation of supercritical Ca^{2+} stimuli was measured too. It reached a steady high constant level after about 70% of the period. Finally, waves with periods longer than 50s did not show any essential variation in puff amplitude or puff frequency during the 60% of the wave cycle preceding the next wave (Marchant 2001).

A noticeable difference between short and long period waves is the behavior of base level Ca^{2+} concentration (Marchant 2001). It declines till emergence of the next wave between short period waves. Base level Ca^{2+} reaches a minimum at about half the period and than slowly rises till initiation of the next wave between long period waves. That increase in base level Ca^{2+} preceding waves was called pacemaker Ca^{2+}.

Wave propagation is saltatorily just above threshold, i.e., release hops from one cluster to the next and single release events can be distinguished. The average wave velocity just above wave threshold is about $10\mu ms^{-1}$ and increases with increasing IP_3 up to about $50\mu ms^{-1}$ (Marchant 1999). Taking wave velocity as an indicator for the excitability of the system set by IP_3, it was found that this is constant for at least 5s upon photo release of IP_3, i.e. the IP_3 concentration does not change at least for that time.

In experiments with large global increase of $[IP_3]$, an initial wave of high Ca^{2+} sweeps across the whole oocyte. The high fluorescence signal then declined over a few minutes giving way to waves of high Ca^{2+}. The wave pattern started irregularly consisting of many wave fragments. As Ca^{2+} declines to base level, waves become more sharply defined and nucleate less frequently from a limited number of sites (Yao 1995). Parker et al. assume that this development of wave patterns is due to adaptation of the receptor to the high level of agonist (Yao 1995).

A wide range of average periods of global events was found in Xenopus oocytes with the shortest ones of approximately 10s and the longest ones up to 2min (Marchant 2001). The time necessary for inhibition of the channels and subsequent recovery is about 10s in Xenopus oocyte (Ilyin 1994) and is in the range of the short periods. Hence, the dynamics of the channel alone cannot account for the long period oscillations of 60s and more. Especially for long periods, the nucleation of a wave by the cooperative action of a few puffs could be demonstrated (Marchant 1999). Activity of a single puff site is typically not sufficient to set off a wave but it needs a supercritical nucleus formed by a few puff sites for the wave to be initiated. Since puffs are stochastic events, the formation of such a supercritical nucleus occurs with a certain probability only. According to Marchant et al., two processes determine the time elapsing between two consecutive waves: The recovery from inhibition caused by the first wave and the creation of a supercritical nucleus for the second wave (Marchant 2001). The probabilistic character of nucleation introduces variability into the wave period. Marchant et al. report a standard deviation up to 40% for long period waves (Marchant 2001). That supports the interpretation of long period repetitive waves as nucleation phenomena.

The interplay between puffs and waves described so far is not unique to Xenopus oocytes. Puffs are observed e.g. in HeLa cells as well (Bootman 1997a).

IP_3 is increased in these cells by stimulation with different agonists activating the phosphoinositide pathway (histamine, ATP, acetylcholine) or by a membrane-permeant IP_3 ester $InsP_3BM$ (Bootman 1997a, Bootman 1997b, Thomas 1999). Increasing stimulation causes higher intracellular IP_3 production. HeLa cells are much smaller than Xenopus oocytes i.e. a typical length of a few 10μm (Bootman 1997a). Puff sites show a spacing of 2-7μm (Thomas 1998). Hence, there are less puff sites than in Xenopus oocytes. Staining of HeLa cells with antibodies against type 1 and 3 IP_3R (representing 90% of total IP_3R mRNA (Thomas 1999)) reveals that IP_3R are most dense around the nucleus and least dense at the plasma membrane. Puff sites are located close to the nucleus. The average density of puff sites responding to stimulation with $InsP_3BM$ and histamine decreases faster than IP_3R density with increasing distance from the nucleus (Thomas 1999). Most puffs were found within 3μm of the nuclear envelop (Lipp 1997). Puffs evoked by thimerosal stimulation showed a broader distribution (Thomas 1999). Waves traveling into regions further away from the nucleus revealed an average cluster spacing of 7.6μm there (Bootman 1997a).

Elementary events in HeLa cells exhibit a broad amplitude and lifetime distribution. The amplitude varies by a factor of 40 and the lifetime from approximately 100ms to 900ms (Thomas 1998). Amplitude and lifetime are weakly correlated only. Amplitude distributions obtained from different cells show a shift to higher amplitudes and larger standard deviation with increasing stimulation (Thomas 1998). The range of amplitudes observed was the same for all histamine concentrations but the average amplitude increased by a factor 2.5 when histamine was changed from 0.1μM to 1μM. The concentration of free Ca^{2+} at the channel mouth of an open channel reaches values up to 100μM (Swillens 1998). Small values of a few hundred nM, like reported for the measurements in HeLa cells, can be obtained as average values in a volume of about 1fl around the channel mouth (Swillens 1998).

Individual puff sites show large or small spread of the amplitude of the puffs they produced. A correlation between amplitude and spread of amplitude was not observed (Thomas 1998). Thomas et al. interpret that as caused by different buffering and resequestration environments of the different puff sites. That is supported by the observation that puff sites close to the nucleus display longer decay times than those far away from it (Lipp 1997) since the nucleus does not resequester Ca^{2+}. According to Thomas et al. (1998), the variation seen in puffs generated by a single site could be due to loose coupling of the channels. The existence of puff sites with small spread in amplitude alongside those with large spread could reflect different isoform composition (Thomas 1998).

The range of puff amplitudes corresponds to a range of current underlying individual puffs from 1 to 25pA (Bootman 1997a). Analysis revealed that the variation is not due to recruitment of neighboring puff sites but to a varying number of open channels within one cluster (Bootman 1997b). That suggests that channel clusters contain between 1 and 25 channels. Comparing signal mass, Thomas et al. conclude that the number of channels in clusters in HeLa cells might be similar to that in Xenopus oocytes (Thomas 1998).

A transition from abortive to fully regenerative waves with increasing strength of stimulation is observed in HeLa cells too (Bootman 1997a). Puffs precede wave initiation. The elementary events measured in the context of wave initiation span a broad range of characteristics. Larger puffs have a $t_{1/2}$ for the rising phase of approximately 42ms and for the decay of approximately 180ms and spatial spreading was 4-7μm. Small isolated events show a $t_{1/2}$ for the rising phase of about 12ms and for the decay of approximately 47ms and spatial spreading was approximately 1.3μm. The estimated current associated with the large puffs is approximately 25pA and with the small ones approximately 1pA (Bootman 1997a). Most of the cells stimulated with histamine showed one wave initiation site only (70%) and the remaining ones two or more initiation sites (Bootman 1997a). Although, some sites might have been missed by the imaging procedure.

Analogously to Xenopus oocytes, an increase in basal Ca^{2+} and puff activity precedes a wave immediately in HeLa cells too (Thomas 1998, Bootman 1997b). In HeLa cells however, the puff amplitude or frequency or both may increase and lead to wave initiation (Bootman 1997b). In difference to the well defined wave initiation threshold in Xenopus oocytes, repeated identical suprathreshold stimulation however not always leads to a fully regenerative wave but sometimes produced abortive signals only or pacemaker activity may stop before a global response was set off (Bootman 1997b). The velocity of abortive waves is about 10μms^{-1} in average reaching 40μms^{-1} maximum. Interestingly, the amplitude of the global signal is proportional to the number of initiation sites.

The consecutive activation of single channels of a cluster could be observed in a HeLa cell. A small initial event set off a sequence of a few step like Ca^{2+} increases most likely representing opening of further channels in the cluster. In the example pictured in (Bootman 1997a) the third channel opened approximately 80ms after the first one. That was followed by a steep concentration rise not resolved into single events anymore. Events of the size of the very first one in this row are observed in isolation as well. Hence, the build up of global events from blips via puffs to waves could be completely resolved. A difference of that hierarchy to wave initiation in Xenopus oocytes is that wave initiation was prepared by activity from a few puff sites but a single puff site sets off the wave in HeLa cells. The sites participating in the liberation of pacemaker Ca^{2+} could be 10μm or further apart, whereas cooperative action of several nearby puff sites (focal sites) was required in oocytes. The ability of a few widely distributed sites to generate a global response is most likely due to the smaller volume of the HeLa cell. One puff can raise the base level of free cytosolic Ca^{2+} globally in a small cell. That generates a global communication between puff sites in difference to the local communication in large cells.

13.2. Cluster Models

Modeling must start from a consideration of the essential features of intracellular Ca^{2+} dynamics allowing for local events as well as the transition to global events.

Local events occur spontaneously, i.e. they are stochastic events. They are characterized by amplitude and lifetime distributions. The probabilistic nature of the local dynamics appears in the global events too. There are abortive waves for parameters close to the onset of wave propagation. Their propagation might be terminated by fluctuations. In the regime of reliably propagating waves, there is variability in the time intervals between nucleation of waves as another manifestation of stochasticity. Hence, the model must be a stochastic model for all length and time scales to cover this behavior. The occurrence of puffs independently of waves demonstrates the existence of two different degrees of coupling. There is the strong coupling of the channels of a single cluster and the weaker coupling between different clusters. The channels belonging to one cluster are very close to each other compared to the distance of clusters. Coupling is due to cytosolic Ca^{2+}. The strong decay of the Ca^{2+} profile around open channels together with the two length scales of channel spacing creates the two levels of coupling. The comparison of the length scales of the Ca^{2+} profile and the channel spacing shows that the profile does not decay within one cluster but the decrease over the distance of cluster spacing may reach two orders of magnitude. Hence, the model must take into account the spatial dependence of the concentration profiles around clusters and the spatially discrete arrangement of clusters.

13.2.1. Stochastic Channel Models

The processes causing random behavior in intracellular Ca^{2+} dynamics are the transitions between the different states of the channel subunits and the channel. Channels open and close randomly. They stay open typically for a few milliseconds in a single opening event. The opening and closing probability depends on the state of the channel subunits. The opening probability is the highest, if a minimum number of subunits are activated and very low otherwise. The findings of Mak et al. suggest that the change of open probability is due to a change of the average opening rate whereas the closing rate is more or less constant (Mak 1998, Mak 2001).

Different models have been developed for the IP_3 receptor channel (Tang 1996, Sneyd 2002). They differ with respect to the set of binding sites on a subunit, assumptions on subunit interaction and binding sequences. The model by Moraru et al. (1994) was set up to fit measurements of open probability. Behavior of a subunit depends on the state of other subunits in that model. The model by DeYoung and Keizer (1992) considers subunits as independent. It assumes 8 states for a subunit. It rather tries to capture the basic mechanisms and hence can be mathematically simpler than a model including many details (DeYoung 1992, Keizer 1994). The model by Sneyd et al. (2002) captures more detailed experimental findings on subunit dynamics like Ca^{2+} dependent binding of IP_3, sequential binding of Ca^{2+} and IP_3 and saturating binding rates. It assumes 11 possible subunit states (Sneyd 2002).

Moraru et al. model the type 1 IP_3 receptor channel. They assume the dissociation constants to depend on the number of ions bound to the channel complex (Moraru

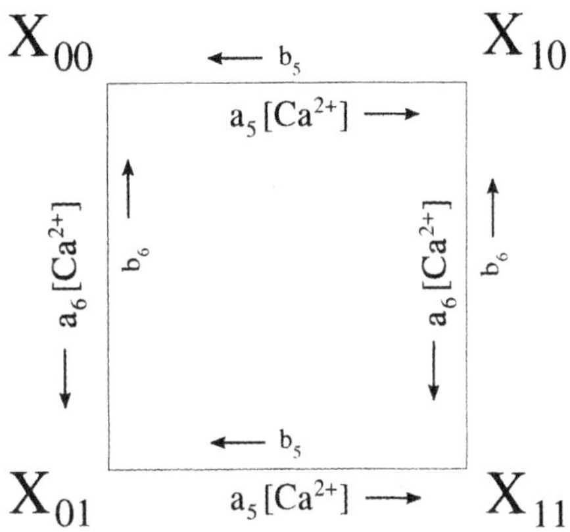

Fig. 1. The four Ca^{2+} binding states X_{ik} of a subunit in the reduced DeYoung Keizer model. The index i corresponds to the activating site and k to the inhibiting site. An index is 1 if the site has an ion bound and 0 otherwise. The a_i are binding rate constants and the b_i unbinding rates.

1999). The subunits are considered to be identical having one Ca^{2+} binding site, a low affinity IP_3 binding site and a high affinity IP_3 binding site. The dissociation constants for Ca^{2+} were chosen as $2.6^{(n-1)}0.7\mu M$, with n being the number of Ca^{2+} ions bound. Binding of IP_3 to the high affinity binding site was modeled with a dissociation constant of 0.3μM if no Ca^{2+} was bound and 1.05μM otherwise. A state-independent dissociation constant of 10μM was appointed to the low affinity IP_3 binding site. The active states of the channel are those with one or two IP_3 molecules to the high affinity site and one or two Ca^{2+} ions bound; additionally the states with at least two IP_3 molecules bound to the low affinity site. Hence, Ca^{2+} dependent inhibition occurs when three or more Ca^{2+} molecules are bound. IP_3 binding to the low affinity site overrides inhibition. Experiments showed that the opening and closing rates depend on the Ca^{2+} concentration (Moraru 1999). Accordingly, the opening and closing rates in the model of the different active states are state dependent and reach from 2 to $40s^{-1}$ (opening) and 150 to $1000s^{-1}$ (closing). With these assumptions, the model reached an excellent fit of the stationary opening probabilities measured experimentally by the authors.

Swillens et al. (Swillens 1998) consider the binding to different subunits of the channel complex as coupled as well. Their model was developed for the type 1 receptor channel, too. The model allows for binding of up to two Ca^{2+} ions for activation, up to two ions for inhibition and for the binding of one IP_3 molecule (Swillens 1998). In a later version, binding of a second IP_3 molecule was introduced (Swillens 1999). The channel is open, if two IP_3 molecules and two activating Ca^{2+} ions are bound and closed in all other states. The open state is identical with that binding configuration rather than a spontaneous opening (and closing) in activated states like in the model by Moraru et al. The binding and unbinding rates of the Ca^{2+} binding sites depend on whether the first or second ion binds. Both binding processes are positive cooperative, i.e., the second ion binds faster than the first one and the dissociation constant for the second ion is smaller than that of the first one. The rate values were obtained from experimentally measured opening times (Bezprozvanny 1994) and a fit to the stationary open probability distribution (Bezprozvanny 1991). The model was used to simulate single channel opening events (Swillens 1998) and puffs (Swillens 1999). An average burst length of single channel openings of 26ms was found. The Ca^{2+} amplitude was defined as the average of a 0.9fl volume surrounding the channel and reached peak values of 300nM and a mean value of 201nM.

The model by DeYoung and Keizer was set up as a deterministic model by the authors (DeYoung 1992) and used later on as a stochastic scheme by Falcke et al. (Falcke 2000b, Falcke 2003a). The model assumes identical, independent subunits. Each subunit has three binding sites: an activating binding site, an inhibiting binding site and a binding site for IP_3. Subunits are activated, if IP_3 is bound, Ca^{2+} is bound to the activating site and is not bound to the inhibiting site. The channel is activated, if at least three subunits are activated. The rapid opening and closing of the channel in the activated state on a millisecond time scale is not modeled as single events in the stochastic interpretation of the DeYoung Keizer model. It was accounted for by choosing a single channel current corresponding to the time average of the channel current. This approach is possible since the channel deactvates or inhibits on a time scale, which is long compared to a single opening event and hence averaging arises naturally as we will see below.

We can further simplify the model by taking advantage of the time scale separation between IP_3 binding and dissociation and Ca^{2+} binding and dissociation. DeYoung and Keizer assumed the IP_3 processes to be two orders of magnitude faster than the other reaction rates. That implies that the binding state of IP_3 will be in a stationary distribution most of the time. Since we know the probability to find a subunit with IP_3 bound or not bound, we can lump pairs of states with identical Ca^{2+} binding configuration into one state. That leaves four lumped states of a subunit corresponding to Ca^{2+} bound or not bound to the activating and inhibiting binding sites. Given one of these four states, the subunit is in one of the substates of IP_3 binding with the probability given by the stationary distribution. Such a four state scheme is shown in Fig.1. The transitions corresponding to binding of Ca^{2+} ions depend on concentration of free Ca^{2+} at the location of the channel. In summary, we reached a scheme with four states. The transitions between the states are the stochastic events.

13.2.2. Modeling Concentrations

The Ca^{2+} in a cell is transported through channels and by pumps, diffuses in the cytosol and the endoplasmic reticulum and reacts with buffers. These processes are described by reaction diffusion equations for the cytosolic concentration of free Ca^{2+} c, the concentration in the ER E and the buffer concentrations (with Ca^{2+} bound) in the cytosol and ER b_i and g_j. The buffer dynamics is binding and dissociation of Ca^{2+} and diffusion. The reaction diffusion equations are simplified by the well justified assumption, that the diffusion coefficient of a buffer molecule does not change upon binding of a Ca^{2+} ion. Then, the total concentration of buffer B_i (free + Ca^{2+} bound) obeys $\partial B_i/\partial t = D_b \nabla^2 B_i$. If the total concentration of buffer B_i is homogeneous in the beginning, it stays homogeneous since $\nabla^2 B_i = 0$. That allows to express the concentration of free buffer as B_i-b_i. The index 'i' of the buffer variable denotes different buffers like endogenous stationary buffer (s), endogenous mobile buffer (m), exogenous mobile buffer (ex) or a buffer with very slow reaction rates (slow). Stationary buffers do not diffuse.

Living cells and their subcompartments are of course spatially three dimensional objects. Nevertheless, the reaction diffusion problem is often reduced to two spatial dimensions. The fluxes between the compartments cytosol and ER are then scaled by the volume ratio $\gamma = V_{cytosol}/V_{ER}$. That leads to the spatially two-dimensional partial differential equations:

$$\begin{aligned}
\frac{\partial c}{\partial t} &= D_C \nabla^2 c + (P_l + P_C(r,t))(E-c) - P_p \frac{c^2}{k_d^2 + c^2} \\
&\quad - \sum_i \left[k_{b,i}^+ (B_i - b_i) c - k_{b,i}^- b_i\right] \\
\frac{\partial E}{\partial t} &= D_E \nabla^2 E - \gamma \left[(P_l + P_C(r,t))(E-c) - P_p \frac{c^2}{k_d^2 + c^2}\right] \\
&\quad - \sum_j \left[k_{g,j}^+ (G_j - g_j) E - k_{g,j}^- g_j\right] \\
\frac{\partial b_i}{\partial t} &= D_{b_i} \nabla^2 b_i + k_{b,i}^+ (B_i - b_i) c - k_{b,i}^- b_i \\
\frac{\partial g_j}{\partial t} &= D_{g_j} \nabla^2 g_j + k_{g,j}^+ (G_j - g_j) E - k_{g,j}^- g_j
\end{aligned} \tag{1}$$

The first term in the equation for the cytosolic Ca^{2+} concentration *c* is the diffusion term. The second and third term model the Ca^{2+} flux through the membrane of the ER. P_l is the coefficient for a leak flux proportional to the concentration difference *E-c*. The function $P_c(r, t)$ describes location and opening state of channel clusters. $P_c(r, t)$ assumes a positive value at the location of a cluster, if channels in the cluster are open. We include the number of open channels by the size of the area where $P_c(r, t)$ is larger than zero. That area equals $N^o \pi R_S^2$. R_S is a radius reflecting the spacing of channels within a cluster and N^o is the number of open channels. The third term of the equation for *c* models the action of the pumps transporting Ca^{2+} from the cytosol into the ER. We describe that as a spatially continuous flux density depending on *c*. The last two terms of the equation describe the reaction of free Ca^{2+} with buffers.

The equation for the dynamics of the Ca^{2+} concentration in the ER *E* includes diffusion, the flux through the membrane and the reaction with buffers inside the ER. The last two equations describe the dynamics of the buffers with Ca^{2+} bound, which is diffusion and reaction with free Ca^{2+}.

When a channel opens, a concentration profile builds up quickly in the cytosol and inside the ER. A stationary state is reached when free Ca^{2+} has spread far enough so that removal of Ca^{2+} from the cytosol back into the ER balances the flux through the channel. The dependence of the stationary profile on buffer and cluster characteristics is shown in Fig. 2, left panel. The amplitude of free Ca^{2+} increases with the dissociation constant of the buffer K_{ex} (Fig.2, A) since less Ca^{2+} is bound when K_{ex} increases. High mobility of buffers leads to low peak values of free cytosolic Ca^{2+} (Fig.2, B). Buffer molecules with Ca^{2+} bound can rapidly diffuse away and a large diffusion flux of unoccupied buffer molecules towards the cluster enhances buffering (facilitated diffusion (Keener 1998)). The rates k_{ex}^{+} and k_{ex}^{-} vary while K_{ex} was kept constant in Fig.2, C. The faster the buffer the lower are the peaks of free Ca^{2+}. The fast buffer is not in equilibrium (i.e. the stationary state of binding and unbinding) around a channel mouth. There is an excess of Ca^{2+}-free buffer because of the directions of the diffusion flux for bound and Ca^{2+}-free buffer. Bound buffer diffuses away from the cluster and Ca^{2+}-free buffer towards the cluster. The excess vanishes for infinitely fast buffers, or in other words more Ca^{2+} is buffered at the channel mouth for faster binding rates. As the last buffer parameter, we vary the total concentration B_{ex}. Increasing the value of B_{ex} decreases the peak of free Ca^{2+}. Furthermore, localization of the concentration profile of free Ca^{2+} becomes stronger with increasing buffer concentration beyond the decrease in amplitude. That can be explained by the sink effect of gradients in buffered systems described in (Wagner 1998).

The dependence of the single cluster profiles on some cluster characteristics is shown in the right panel of Fig.2. The peak concentration at the center of the cluster depends linearly on P_c for the range shown. The amplitude saturates for very large values of P_c (data not shown). Plot (C) deals with the dependence on the total average concentration of Ca^{2+} in the cell C_0. That total amount corresponds to the total amount of free Ca^{2+} and Ca^{2+} bound to fast buffers, if slow variables are involved (see below). The change of the parameter C_0 in panel (C) mimics the behavior of single cluster profiles during reactions of Ca^{2+} with slow variables.

As we will see below in conjunction with the two time scale analysis, the more Ca^{2+} is bound by slow buffers, the lees is available for fast buffers to form stationary profiles on a fast time scale. Finally, we vary the pump strength by a factor 5. The relative impact on the peak values is negligible. Pumping is more important for the profile values at large distances.

The concentration fields and the cluster dynamics are mutually coupled. On the one hand, the transition rates for binding transitions depend on the concentration of free Ca^{2+} as shown in Fig.1, on the other hand, the concentration fields depend on the configuration of open channels.

Before we start to look at simulations of global events, we consider more carefully a few features of the system. We start with an inspection of time scales. The channel has opening and closing transition probabilities $p_o dt$ and $p_c dt$ leading to dwell times of a few milliseconds in the activated state. When the channel opens, the concentration of free Ca^{2+} at the location of the channel rises within microseconds. The spatial profile is strongly localized and reaches peak values of 30-100μM within a few nanometers. The dissipation of the Ca^{2+} peak upon closing of a channel is comparably fast. Hence, the local Ca^{2+} concentration at the channel mouth can follow the opening and closing dynamics. The local Ca^{2+} concentration enters the transition probabilities for the subunit states (see Fig.1). Hence, the transition probabilities for spontaneous opening and closing have to be compared to the time scale of change of the subunit state by ligand binding or dissociation. The probability $p_b(t_1,t)$ that a Ca^{2+} binding site binds an ion the first time in $(t,t+dt)$ when no ion was bound at t_1 is:

$$p_b(t_1,t) = \exp\left[-a_j \int_{t_1}^{t} dt' c(t'))\right] a_j c(t) dt.$$

Here, *c(t)* denotes the concentration of free Ca^{2+} at the location of the channel. The integral in this expression averages *c(t)* over several opening and closing events since p_o, $p_c \updownarrow a_j c(t)$ holds for the IP_3R. We take advantage of that. Instead of simulating single opening and closing events we assume the channel to be open, if at least three subunits are in the activated state. At the same time we reduce the channel flux so as to generate a concentration at the channel corresponding to the time average over many openings and closings on the millisecond time scale. That simplification eliminates the fastest time scale of the channel dynamics from the simulation. However, that is not of any use as long as the fast dynamics of the concentration fields in the microsecond range has to be simulated dynamically. Here, a closer look at the pattern type in dependence on the IP_3 concentration suggests a simplification eliminating that time scale too. Propagating waves are the first global events observed, when the concentration of IP_3 is increased. Propagating waves exist as trigger waves or as phase waves. Propagation is due to a spatial phase gradient in phase waves and occurs in oscillatory systems. They will not be considered here.

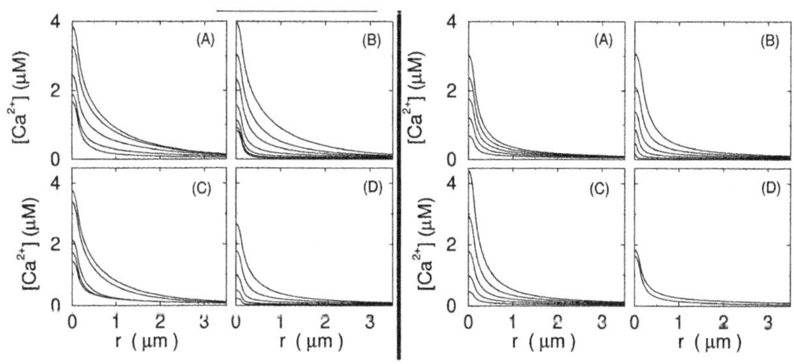

Fig. 2. Left panel: Single cluster profiles of cytosolic Ca^{2+} for different parameters of the exogenous buffer. Parameters not mentioned are like in Table 1. (a) Lines from bottom to top K_{ex}=0.25, 1.25, 6.25, 31.25, 156.25 μM, k_{ex}^+ was kept constant; (b) Lines from bottom to top D_{ex} = 320, 160, 80, 40, 20, 10, 0 $\mu m^2 s^{-1}$; (c) Lines from bottom to top k_{ex}^+ = 312.5, 62.5, 12.5, 2.5, 0.5 $(\mu Ms)^{-1}$. K_{ex} is kept constant; (d) Lines from bottom to top B_{ex} = 320, 160, 80, 40, 20 μM. Right panel: Single cluster profiles of cytosolic Ca^{2+} for different cluster parameters. Parameters not mentioned are like in Table 1. (a) Lines from bottom to top P_c = 160, 260, 360, 460, 560 s^{-1}; (b) Lines from bottom to top R_s = 20, 40, 60, 80, 100, 120 nm; (c) Lines from bottom to top C_0 = 52, 62, 72, 82, 92 μM; (d) Lines from bottom to top P_p = 125, 25 μMs^{-1}. (Falcke 2003b)

The waves at low IP_3 - as we will see below - are trigger waves. The excitation at one site triggers the excitation at a neighboring site i.e. trigger waves require spatial coupling. Therefore, spatial coupling between the clusters is given right from the onset of global events. Global events start when puffs are strong enough to generate profiles reaching the neighboring clusters, since spatial coupling is mediated by the concentration profile of free Ca^{2+}. Moreover, the increase in free Ca^{2+} at the neighboring cluster must be sufficient to increase the opening probability to values a few times that of base level in order to observe propagation.

We want to simulate the behavior of the system for IP_3 concentrations just at the onset of propagation and larger. We can conclude from the above, that puffs build up profiles of free Ca^{2+} in that regime extending far enough to increase the opening probability at their neighbor's site substantially. We now use this observation to simplify the simulations. We assume, that these profiles build up fast compared to the lifetime of a puff or - expressed in technical terms - do an adiabatic elimination of the concentration dynamics around a cluster with open channels. If a channel opens or closes in the simulation, the concentration profile around its cluster is set immediately to the stationary profile belonging to the current number of open channels in the cluster.

Figure 2 shows that these stationary profiles are well localized on the length scale of cluster spacing. The concentration increase due to open channels is about two orders of magnitude smaller at the next neighbor's location than at the cluster with open channels itself. That increase is the difference between the base level concentrations (when all channels are closed) and the concentration profile when channels are open and we call it single cluster profile. It approaches 0 for large distances from the cluster. In taking advantage of the localization of single cluster profiles we represent the complete concentration field as sum of the base level and a superposition of all single cluster profiles.

The single cluster concentration profiles for all possible numbers of open channels can be calculated in advance to the simulation. During the simulation, we appoint a single cluster profile to a cluster according to its number of open channels. The superposition of all profiles is the global concentration field. We introduce the vector $U(r, t)$ the components of which are the concentration fields included in the simulation. The solution for a single cluster with $N_i^o(t)$ open channels in the center of a large circular area be $U^s(r,N_i^o(t))$and the base level be U^b. The complete concentration field expressed in these variables is:

$$U(r,t)=U^b+\sum_i\left[U^s(r-r_i,N_i^o(t))-U^s(\infty,N_i^o(t))\right]$$

with the summation including all clusters.

The adiabatic elimination (a.k.a. quasi steady state approximation) we have used for the concentrations requires all buffer rates to be fast. However, slow buffers are used in experiments. The method to deal with slow time scales in our system is multiple time scale analysis. We will sketch that procedure here for a two dimensional system.

We start with defining the total Ca^{2+} concentration C_T:

$$C_T = c + \sum_i b_i + \frac{1}{\gamma}\left[E + \sum_j g_j\right].$$

The first summation includes all cytosolic buffers the second one all luminal buffers. One of the cytosolic buffers be a buffer with slow dynamics, i.e. its reaction rates are much smaller than all other rates in the system. We write its dynamics as

$$\frac{\partial b_{slow}}{\partial t} = D_{slow}\nabla^2 b_{slow} + \varepsilon\left[k^+_{slow}(B_{slow} - b_{slow})c - k^-_{slow} b_{slow}\right]$$

with $\varepsilon \leftrightarrow 1$ and k^+_{slow}, k^-_{slow} in the range of the rates of the other buffers. Hence, the factor ε indicates the order of magnitude of the ratio of the binding and dissociation rates of slow buffers to fast buffers. Multiple time scale analysis can be applied if fast and slow processes can be distinguished in a system, i.e., if $\varepsilon \leftrightarrow 1$ holds (Falcke 2003b). It is based on the assumption that during the time it takes the fast processes in the system of PDEs to reach their stationary state the slow variables can be considered as constant. The goal is to obtain a partial differential equation for the dynamics on the slow time scale and to keep the adiabatic elimination for the fast variables. The stationary solution of the fast variables will of course depend on the slow variables.

The multiple time scale analysis requires to identify the terms in the dynamics belonging to fast or slow processes. The differential time scales are introduced by assuming that the time derivative can be split into a partial derivative with respect to fast processes – on the time scale t - and one with respect to slow processes – on the time scale τ:

$$\frac{\partial}{\partial t} \Rightarrow \frac{\partial}{\partial t} + \varepsilon\frac{\partial}{\partial \tau} \quad \text{with } \tau = \varepsilon\, t.$$

The diffusion term in the dynamics of b_{slow} will be small too since the slow reaction rates cannot build up steep gradients. Therefore, b_{slow} changes over long distances only, analogously to the change on long time scales We take that into account by introducing the length scale $\zeta = \varepsilon^{1/2} r$. That leads to a modified spatial derivative in the same spirit as above:

$$\nabla \Rightarrow \nabla_r + \varepsilon^{1/2}\nabla_\zeta.$$

Finally, the dynamics for the total Ca^{2+} concentration obeys:

$$\frac{\partial C_T}{\partial t} + \varepsilon\frac{\partial C_T}{\partial \tau} = \left\{\nabla_r^2 + \varepsilon\nabla_\zeta^2\right\}\left\{D_c c + \sum_{l \neq slow} D_{b_i} b_i + \frac{1}{\gamma}\left[D_E E + \sum_j D_{g_j} g_j\right]\right\} + \varepsilon D_{slow}\nabla_\zeta^2 b_{slow}.$$

Here we define the variable A as

$$A = D_c c + \sum_{i \neq slow} D_{b_i} b_i + \frac{1}{\gamma}\left[D_E E + \sum_j D_{g_j} g_j \right]$$

and reach

$$\begin{aligned} \frac{\partial C_T}{\partial t} + \varepsilon \frac{\partial C_T}{\partial \tau} &= \left\{ \nabla_r^2 + \varepsilon \nabla_\zeta^2 \right\} A + \varepsilon D_{slow} \nabla_\zeta^2 b_{slow} \\ \varepsilon \frac{\partial b_{slow}}{\partial \tau} &= \varepsilon D_{slow} \nabla_\zeta^2 b_{slow} + \varepsilon \left[k_{slow}^+ (B_{slow} - b_{slow}) c - k_{slow}^- b_{slow} \right] \end{aligned} \qquad (2)$$

An $\varepsilon^{1/2}$- time scale, as could arise from a $\partial^2/\partial r \partial \zeta$- term, does not appear since A does not depend on r as we will see below.

The total Ca^{2+} concentration still contains b_{slow}. The fast components of C_T are $C_T - b_{slow}$, and with $\partial b_{slow}/\partial t=0$ we obtain:

$$\frac{\partial (C_T - b_{slow})}{\partial t} + \varepsilon \frac{\partial (C_T - b_{slow})}{\partial \tau} = \left\{ \nabla_r^2 + \varepsilon \nabla_\zeta^2 \right\} A + \varepsilon D_{slow} \nabla_\zeta^2 b_{slow} - \varepsilon \frac{\partial b_{slow}}{\partial \tau}. \qquad (3)$$

The terms of order 1 of this equation give the dependence of C_T on the fast time scale:

$$\frac{\partial (C_T - b_{slow})}{\partial t} = \nabla_r^2 A. \qquad (4)$$

Now, we solve for the stationary solution and obtain $\nabla_r^2 A=0$. We require no flux boundary conditions. That leads to the solution $A=A(\tau,\zeta)$ of that equation with A^0 being independent of r and t. The complete solution on the fast time scale involves the (1) as well. It is the stationary solution of (1) with $A = A^0$.

The terms of order ε in (3) are the differential equation determining the dependence of A^0 on the long length scale and the slow time scale ζ and τ:

$$\frac{\partial (C_T - b_{slow})}{\partial A^0} \frac{\partial A^0}{\partial \tau} = \nabla_\zeta^2 A^0 + D_{slow} \nabla_\zeta^2 b_{slow} - \frac{\partial b_{slow}}{\partial \tau}. \qquad (5)$$

The function $A^0(\tau,\zeta)$ is determined up to an arbitrary constant only. That constant can be fixed by the total amount of Ca^{2+} in the cell: $\rightrightarrows_V dV\, C_T = VC_0$. Here, V denotes the volume of the cell and C_O an average total concentration. Note, that A^0 does depend on the configuration of open channels, since channel opening or closing redistributes Ca^{2+} from one buffer to another one or to free Ca^{2+}. The changes of A^0 due to opening and closing of channels is the fast dynamics of A's whereas the slow dynamics obeys 5.

We can again use the single cluster profile superposition even with A^0 depending on ζ and τ, if computational speed is a concern. Note that the base level of concentrations is not spatially constant anymore. Spatial variations in base level concentration may cause some diffusion of free Ca^{2+} and buffers. We assume, that this diffusion is negligible compared to the diffusion of A^0 and the diffusion in the contribution from open channels. That assumption turns the system of coupled PDEs, which had to be solved to obtain the base level, into algebraic equations to be solved at each point in space, thus saving computing efforts. In order to calculate the single cluster profiles, we again need to apply an approximation. The contribution from the open channels decays to zero on length scales shorter than the length scale of variations of A^0 or in other words A^0 is constant where the channel contribution is different from zero. Hence, we can calculate the single cluster profile by assuming A^0 to be spatially constant with the value of A^0 at the location of the cluster. Renaming ζ and τ as r and t again we obtain:

$$U(r,t) = U^b(A^0(r,t)) + \sum_i \left[U^s(r-r_i, N_i^o(t), A^0(r_i,t)) - U^s(\infty, N_i^o(t), A^0(r_i,t))\right].$$

We have set up the channel dynamics and the concentration dynamics by now and can start simulations. Before we do so, I would like to summarize one iteration step of a simulation. Starting from an initial configuration of open channels $\boldsymbol{N}^o(t)$ we compute the concentration of free Ca^{2+} in the cytosol and the transition rates from that. Then, random of the states of the channel subunits transitions corresponding to binding or dissociation of Ca^{2+} are determined and (2) and (5) are integrated one iteration step. The subunit transitions provided a new configuration of open channels and the cycle starts anew.

13.3. Stochastic Wave Patterns

A set of simulations for different IP_3 concentrations is shown in Fig.3. Only puffs are found at low concentrations, i.e. release events are localized and not coordinated on a length scale of several cluster distances. That changes with the onset of global events at a slightly higher $[IP_3]$ (Fig.3, A). These global events are waves emerging from a nucleation area. They are very rare for low IP_3 concentrations and may travel across the whole system (Fig.3, A, first and second peak) or fade away before they reach the system boundary (third peak). That parameter regime of abortive wave propagation is characterized by a distribution of the probability for a wave to travel a certain distance before being destroyed by fluctuations rather than steadily propagating waves (Falcke 2000b). It can be understood more easily in one spatial dimension. We start from the probability p_N, that a cluster, with at least one open channel, excites a neighboring cluster. This probability for certain parameters is shown in Fig.4. From there, we can easily calculate the probability that an excitation travels k clusters as p_N^k.
We return to the two dimensional simulations. Increasing $[IP_3]$ leads to more frequent waves (Fig.3, B) and almost every wave travels across the whole system now. The time interval in between waves is not completely regular like the period

of a deterministic oscillation, but we can determine an average interwave time interval T_{av} and its standard deviation ΔT_{av}, e.g. the simulation shown in (Fig.3, B) has T_{av}=53.22±8.37s. Both T_{av} and ΔT_{av} decrease with increasing [IP_3] (Fig.3, D) leading to almost regular oscillations with a period of about 17s at high [IP_3] (Fig.3, C).

That scenario found while going from low to high [IP_3] agrees with experimental observations (Marchant 2001). Especially, the large range of periods of 8 times the shortest one reported from the experiments is captured. Short periods could be set essentially by the longest time scale of the channel dynamics T_d being the transition to the inhibited state and recovery from it.

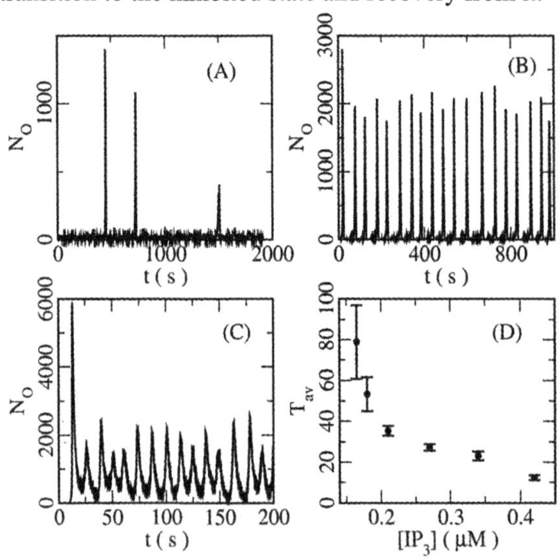

Fig. 3. Simulations and oscillation characteristics for different IP_3 concentrations: N_O is the global number of open channels. (A) I=0.15μM, (B) I=0.18μM, (C) I=0.42μM, (D) T_{av} and its standard deviation in dependence on [IP_3]. See Table 1 for further parameters. (Falcke 2003a)

However, long periods last 3-8 times longer and cannot be explained by the channel dynamics alone. That is illustrated in Fig.5. It shows the number of open channels N_O in panel A and the number of excitable channels N_E in panel B. An excitable channel has at least three subunits in X_{00}. Waves traveling through the whole system appear as spikes in N_O or downward spikes in N_E. N_E reaches again a high level after a wave has passed within T_d=18s, i.e. the difference between T_d and T_{av} is very obvious. That difference demonstrates that the channel dynamics alone cannot account for the long average periods, since the channels are in a state ready to support another wave after time T_d. To explain the long periods, we start from the observation that waves emerge from small areas - a nucleus - and than spread through the whole system. That is called wave nucleation and is observed in systems like e.g. the FitzHugh-Nagumo-model with additive noise (Hempel 1999) or an integrate and fire model with additive noise (Jung 2000) as well.

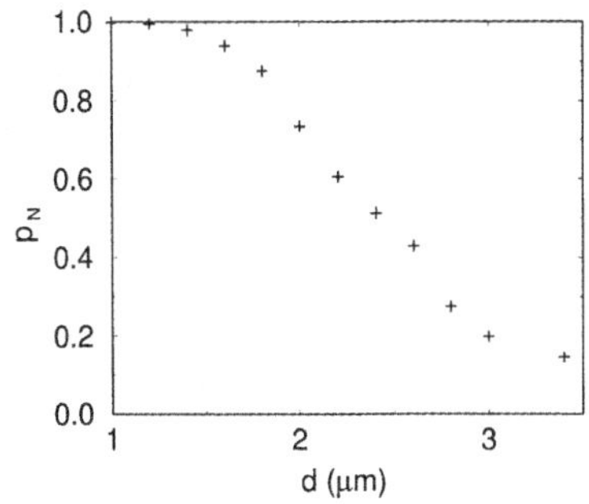

Fig. 4. The probability p_N that an active cluster activates a neighboring cluster in dependence on the cluster spacing d. The flux density ((single channel flux)/d) was kept constant by multiplying P_c for d=1μm with d. For details see (Falcke 2000b)

Nucleation of global events is probabilistic, because a single puff activates a neighboring cluster with a certain probability only - not with certainty. The nucleation probability p_n is small compared to the puff probability because a large supercritical nucleus of a few clusters is needed. The larger a nucleus, the smaller is the curvature of its boundary. The smaller the curvature of the boundary, the larger is the number of active neighbors of an inactive cluster just outside the nucleus and hence the probability that that cluster is activated. That means, the larger the nucleus the larger is the probability that it grows. Deactivation, inhibition and fluctuations hinder the growth of a nucleus. In that way, a critical size of a nucleus arises.

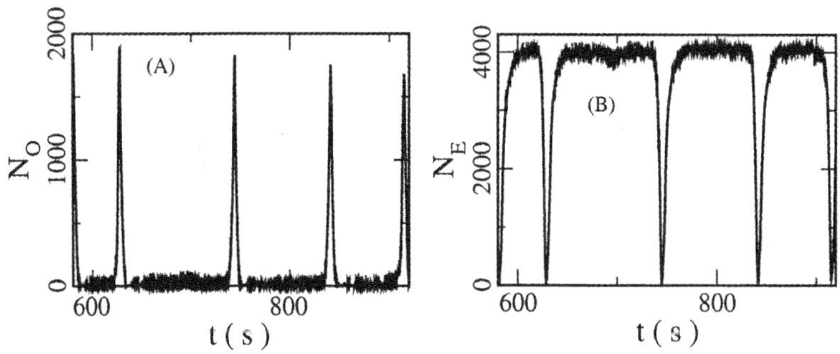

Fig. 5. The number of open channels (A) and the number of excitable channels (at least 3 subunits in X_{00}) (B) during repetitive nucleation of waves. The transition from $X_{01} \rightarrow X_{00}$ was sped up by a factor of 2.5 compared to the standard parameters of Table 1. See Table 1 for parameters except B_{ex}=70μM and N_F=10. (Falcke 2003a)

The nucleation probability is very small just after a wave has traveled across the system because of inhibition of most of the channels by the high Ca^{2+}-concentration during the wave. That causes the deterministic part T_d of the time elapsing between two consecutive waves. T_d is determined essentially by the transition rates from the activated state (X_{10}) to the inhibited state (X_{11}) and recovery from inhibition ($X_{11} \rightarrow X_{01} \rightarrow X_{00}$). However, p_n is still small after recovery from inhibition. Hence, the next wave does not emerge immediately but it takes some time before another global event can be set off. I would like to illustrate that with the example of Fig.7, B. The average period T_{av} of the complete

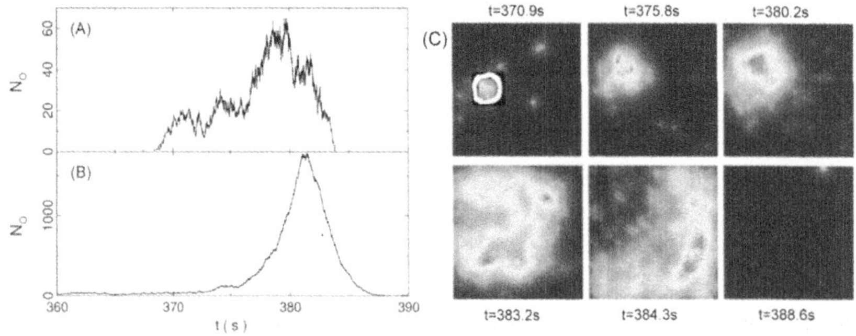

Fig. 6. Nucleation of a wave in a regime with long wave periods. Panel (A) shows the number of open channels in the area where the wave nucleates. The nucleation area is bound by a white line in panel (C) frame t=370.9s and comprises several clusters. Panel (B) shows the global number of open channels. Panel (C) shows 6 snapshots of the simulation area taken at the times indicated. Shown is Ca^{2+} bound to exogenous buffer. The gray level scaling is a minimum-maximum-scaling with black indicating minimal values and white maximal ones. (Falcke 2003a)

run over 1000s is approximately 65s. The nucleation probability is the inverse of the stochastic part T_{st} of T_{av}, i.e. $p_n=T_{st}^{-1}=(T_{av}-T_d)^{-1}=0.0097s^{-1}$. Hence, the small value of p_n accounts for the larger part of T_{av}.

A nucleation is shown in Fig.6. The puff activity leading to the supercritical event starts in the area marked in Fig.6 in the frame t=370.9s. That area is outside but close to a focal site. It comprises a few clusters and the interaction of these clusters allows for stochastic release to go on for a few seconds. The focal site is activated by the ongoing puff activity in its vicinity (Fig.6, frame t=375.8s). It strongly amplifies puff activity and sets off a wave. That wave propagates across the whole simulation area (Fig.6, frames t=380.2s, t=383.2s, t=384.3s). The maximum number of open channels in panel A occurs when the wave travels through the area marked in frame t=370.9s. Ca^{2+} release terminates first in the area where the wave started (frame t=383.2s) due to inhibition. Puff activity is reduced to almost none after the wave ceased (Fig.6, frame t=388.6s). Release events of comparable magnitude to that in panel A of Fig.6 between 370s and 375s can be observed not leading to nucleation of a wave. These failed nucleation events illustrate that it takes a supercritical number of active clusters to initiate a global event.

The idea of nucleation processes setting long periods implies that the stochastic part of the period should depend on system size and the number of focal sites. Simulations with systems only a quarter in size of those used otherwise in this chapter containing one focal site instead of four showed that the stochastic part of the period is indeed proportional to the inverse of system size. Jung et al. suggest for a stochastic integrate and fire model that there is an optimal system size with minimal relative standard deviation (Jung 2000).

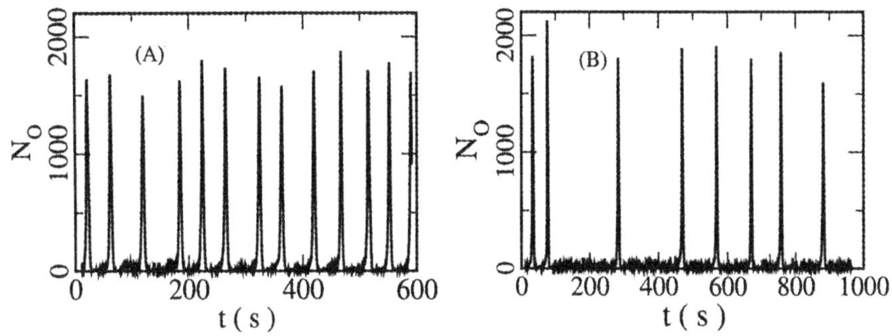

Fig. 7. Repetitive wave nucleation for two concentrations of exogenous buffer: (A) B_{ex}=60μM, (B) B_{ex}=75μM. The number of open channels is shown. The average period is in (A) T_{av}=47.40∓10.00s, (B) T_{av}=121.17∓57.23s. The transition from $X_{01} \rightarrow X_{00}$ was increased by a factor of 2.5 compared to the standard parameters of Table 1. See Table 1 for further parameters except N_F=10. (Falcke 20003a)

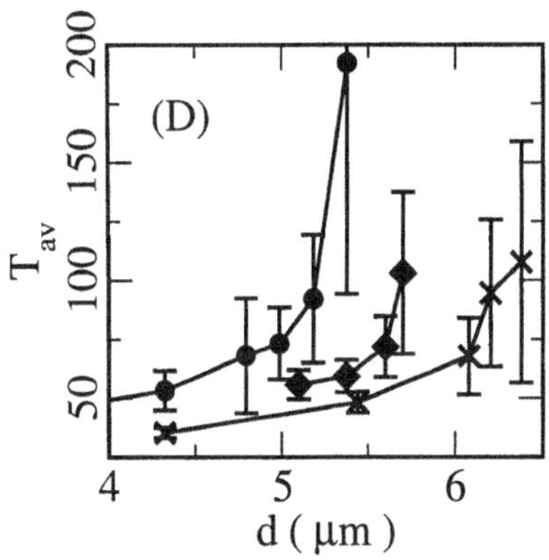

Fig. 8. The average period T_{av} of repetitive wave nucleation and its standard deviation in dependence on the cluster spacing. See Table 1 for further parameters except N_F=10, I=0.18μM (circles), N_F=5, I=0.195μM (diamonds) and N_F=5, I=0.21μM (x). (Falcke 2003a)

Since nucleation of a wave is the cooperative action of a few clusters, the nucleation probability p_n depends very sensitively on parameters determining spatial coupling. Three of these parameters are the buffer concentrations, the cluster spacing and the number of additional clusters in focal sites. Fig.7 compares simulations with two different concentrations of exogenous mobile buffer B_{ex}. An increase of B_{ex} by 25% extends the average period from 47±10s to 121±57s. Such an increase of the interwave interval was observed for wave nucleation with Ryanodine receptor channels in rat ventricular myocytes (Lukyanenko 1999).

Varying the cluster spacing leads to a range of periods again from 17s up to 192±97s (see Fig.8). Abortive waves occur for large cluster spacing. Waves disappear completely, if the spacing is increased even more. These data show that the average period in the nucleation regime does strongly depend on the strength of spatial coupling. Hence, it will as well depend on the number of additional clusters in focal sites or vary for different random realizations of cluster arrays with identical average characteristics. Simulations with different random realizations of a cluster array showed differential T_{av} but a varying relative standard deviation $\Delta T_{av}/T_{av}$ as well. That implies that there are optimal arrays having minimal $\Delta T_{av}/T_{av}$.

Long period oscillations are possible for a variety of parameters of the local dynamics (Fig.9). They can be found for $[IP_3]$=0.15μM and $[IP_3]$=0.42μM or fast and slow recovery from inhibition and a range of inhibition rates. The nucleation characteristics are rather robust with respect to changes of the local dynamics. The weak spatial coupling and the small average number of excitable channels per cluster of about 5 are similar for the different simulations in Fig.9. Oscillation periods were sensitive to changes in these two parameters.

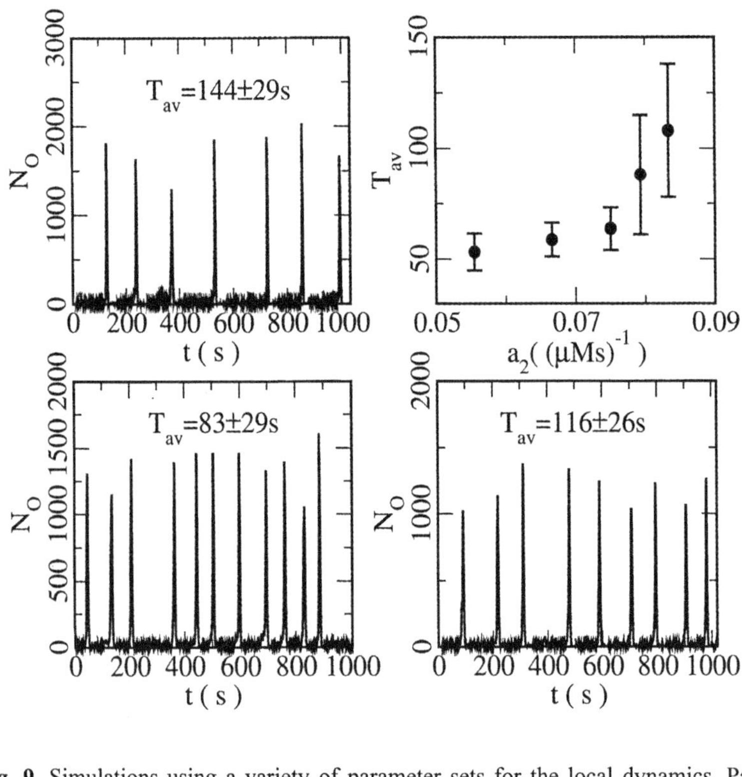

Fig. 9. Simulations using a variety of parameter sets for the local dynamics. Parameters different from those in Table 1 are: top left B_{ex}=60 μM, I=0.18μM, d_2=0.83s^{-1} (increased deactivation), top right T_{av} for different inhibition rate constants a_2 (note a_2=a_4) B_{ex}=60 μM, I=0.18μM, bottom left B_{ex}=60μM, I=0.42μM, N_K^{max}=13, d=4.6μm, bottom right B_{ex}=60μM, I=0.15μM, N_F=7. (Falcke 2003a)

Especially, the sensitivity for the average number of excitable channels per cluster brings up the question for the deterministic limit. The deterministic limit assumes infinitely many channels per cluster while keeping the flux density and $N_K^{max} R_S^2$ constant. While increasing N_K^{max} the single channel radius R_s was rescaled with $(N_K^{max})^{-1/2}$ thus conserving the parameters entering the deterministic limit.

Simulations showed that T_{av} changes by a factor of 5 while going from 25 to 328 channels per cluster at low [IP_3]. Results with even higher values of N_K^{max} strongly suggest that the deterministic limit is not oscillatory but is a stationary state with high activity compared to the activity between waves during long period oscillations (Fig.10).

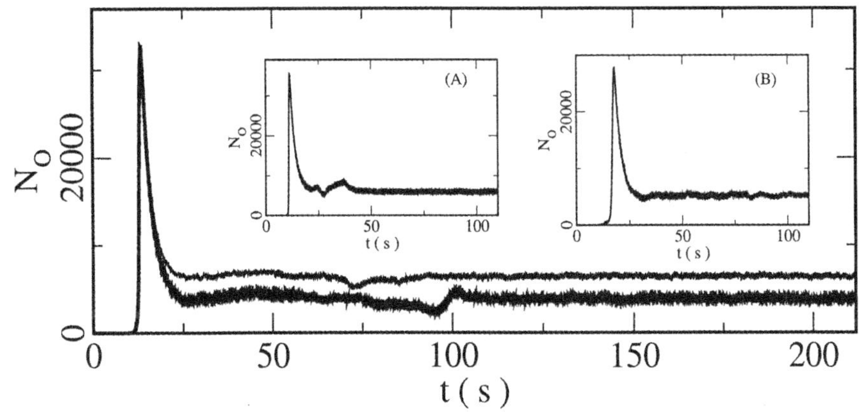

Fig. 10. Simulations with N_K^{max} =1130, I=0.15μM and N_F=5. The single channel radius was chosen proportional to $(N_K^{max})^{-1/2}$ as explained in the text. Shown is the case corresponding to Fig.3, (A) (thick line) and with recovery from inhibition sped up by a factor of 2.5 (thin line). The insets show simulations with (A) I=0.42μM, N_K^{max} =546 and (B) I=0.27μM, N_K^{max} =685. (Falcke 2003a)

That means that the oscillations are completely due to fluctuations i.e. stochastic channel behavior. Obviously, the fluctuations occurring with realistic channel numbers are large enough to push the system out of the attractive region (stable manifold) of the high activity stationary state, which is observed for channel numbers approaching the deterministic limit. Therefore, that stationary state cannot be observed in the dynamics at these parameter values. We find again a high activity stationary state for larger IP_3 and large N_K^{max} (Fig.10, insets). That means that short period oscillations with small N_K^{max} too are due to fluctuations as well.

Finally, we compare simulation results with the data presented in (Marchant 2001). Marchant et al. plot the periods and their standard deviation for a range of IP_3-concentrations. We present these values with simulation data in Fig.11 showing the relative standard deviation ($\Delta T_{av}/T_{av}$) versus T_{av}. The data in (Marchant 2001) were obtained with 33 oocytes from 5 frogs. Accordingly, we include into the comparison periods obtained with different parameters. The agreement especially in the range of 50-80s-periods is good. That confirms Parker's hypothesis that the stochastic part of periods is due to wave nucleation and that this process can be modeled with the simulation concept presented here.

We have used fast buffers only till now. However, not all buffers used in experiments are fast but slow buffers like e.g. EGTA are applied as well. Hence, it

is worthwhile to check if the important result of the dependence of the period on buffer concentrations holds for slow buffers too. We will look at an 'extreme' case with high IP_3 concentration, large numbers of channels in a focal site and rather small spacing. If slow buffers can extend the period for such a system, we can expect that it is possible in more realistic systems with smaller numbers of channels per cluster, less clusters in focal sites and larger cluster spacing too.
When we use slow buffers, we have to apply the equations resulting from the two time scale analysis presented above. Figure 12 shows simulations with increasing concentration of slow buffer. The system exhibits oscillations with an average period of about 20s around an increased level of activity in case of low concentration (Fig.12, A, left panel). Only single waves occur at high concentration of slow buffer (Fig.12). The interwave interval is much longer than the period of the oscillations at low buffer concentrations. Hence, we observe a transition from short period oscillations to a nucleation regime with long average periods with increasing slow buffer concentration, i.e. slow buffer too extends average periods.

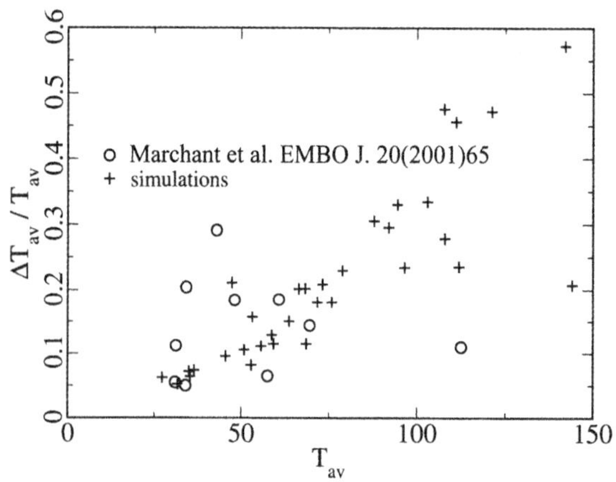

Fig. 11. Comparison of the experimental data with simulations. Shown is the relative standard deviation of $\Delta T_{av}/T_{av}$ versus T_{av}. (Falcke 2003a)

Finally, the question for a unifying picture for the dynamics remains, which explains the nucleation processes at low IP_3 and short period oscillations at high concentrations. If one assumes that the dynamics can be reduced to a plane in phase space - as many deterministic models do (Tang 1996) - a phase space structure like sketched in Fig.13 seems plausible. Each point in phase space corresponds to one state. The motion of the states due to the dynamics of the system generates trajectories. Separatrices are special lines in phase space, which cannot be crossed by trajectories of deterministic systems. Each point in phase space corresponds to one state. The motion of the states due to the dynamics of the system generates trajectories. Separatrices are special lines in phase space, which cannot be crossed by trajectories of deterministic systems. Usually, the Ca^{2+} concentration and the fraction of inhibited channels are used to characterize the state of the system.

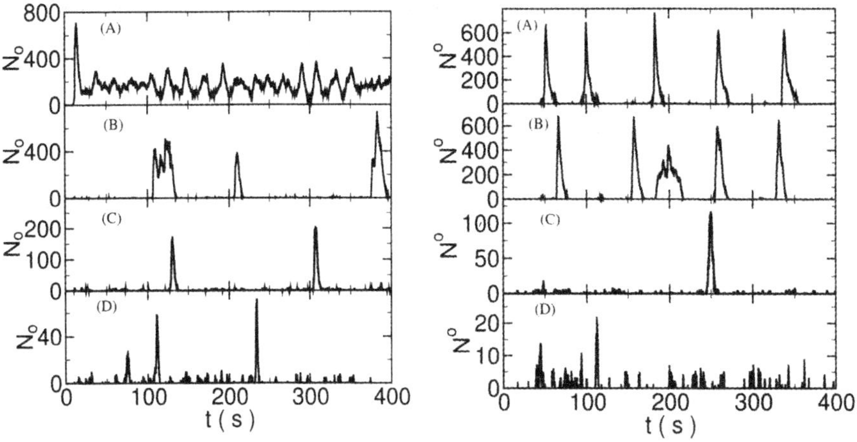

Fig. 12. Number of open channels N_O in dependence on time for different concentrations of slow buffer B_{slow}. Left panel: d=3.968μm, $[IP_3]$=0.24μM, (A) B_{slow}=0μM, (B) B_{slow}=1200μM, (C) B_{slow}=1400μM, (D) B_{slow}=1600μM. Right panel: d=3.443μm, $[IP_3]$=0.36μM, (A) B_{slow}=400μM, (B) B_{slow}=600μM, (C) B_{slow}=1200μM, (D) B_{slow}=1600μM. See Table 1 for further parameters. (Falcke 2003b)

Fig.13 shows a bistable system with three stationary states: two stable ones and a saddle point. The ingoing separatices of the saddle pass close by the stationary states. The separatrix close to the lower stationary state provides the excitability of the low activity stationary state. Perturbations not crossing the separatrix decay to the stationary state, but the (deterministic) system cannot simply return to the stationary state upon perturbations across the separatrix. The upper stationary state is a focus. The trajectory of the system during a wave in the long period regime is shown in Fig.13, A. Nucleation corresponds to crossing the separatrix close to the lower stationary state because of fluctuations. That is followed by motion close to a deterministic trajectory leading into the vicinity of the upper stationary state. The second ingoing separatrix is very close to that state at low IP_3 concentrations. Hence, fluctuations push the system across it whenever it approaches the high activity stationary state. Finally, it relaxes to the lower stationary state again in a deterministic-like way.

The distance between the separatrix and the upper stationary state increases with increasing IP_3. Therefore, fluctuations do not push the system across it when they approach the stationary state under these conditions. The short period oscillations could now be comprehended as a fluctuation driven transition to an outer revolution of the spiral-like separatrix, when the system approaches the stationary state followed again by a relaxation towards the stationary state similar to the deterministic dynamics. That picture would imply that the single channel dynamics is crucial for setting the short periods.

The stationary high activity state reflects the behavior of the DeYoung-Keizer model. Such a state can be reached in the deterministic model by increasing the

Ca^{2+} flux through channels, which causes an increase of the cytosolic Ca^{2+} concentration. That situation occurs at the location of open channels in a discrete model. The high local Ca^{2+} concentration suppresses oscillatory behavior, which needs intermediate Ca^{2+} concentrations. That suggests that the oscillatory regime in models is lost during the transition from spatially continuous channel density to spatial discreteness of the channel clusters while keeping the average flux density constant. Fluctuations compensate for that by restoring the ability for oscillatory

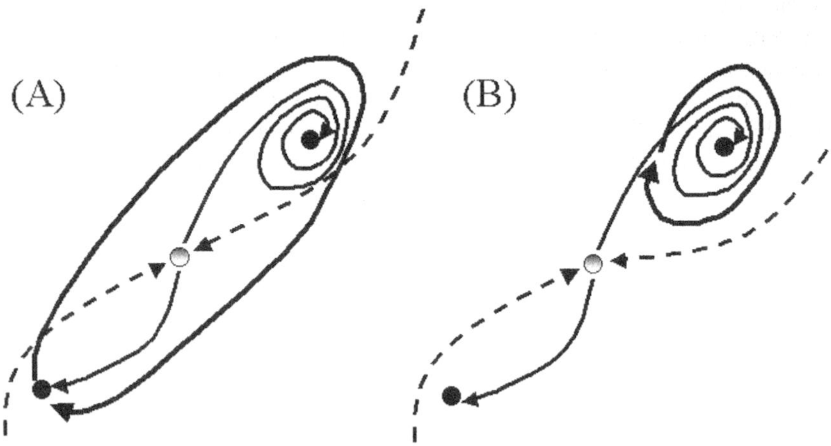

behavior.

Fig. 13. Phase space structure which could be imagined as underlying the observed behavior of the stochastic system. Full dots: stable stationary point, dot with gradient filling: saddle point, thin full lines: outgoing separatrices, dashed lines: ingoing separatrices, thick full lines: trajectories. (A) low $[IP_3]$, (B) high $[IP_3]$. (Falcke 2003a)

A high activity stationary state as one stable state of a bistable dynamics was found in a model for Xenopus oocytes with energized mitochondria too (Falcke 1999, Falcke 2000a). That state was stabilized by the interaction of the Ca^{2+} handling of mitochondria and the ER. The occurrence of such a state in different models of intracellular Ca^{2+} dynamics supports the idea that the cytosol might be a bistable system for high IP_3 (see as well (Wagner 1998)). That stationary state of the local dynamics led to the disappearance of spiral waves without abolishing less curved, long period Each point in phase space corresponds to one state. The motion of the states due to the dynamics of the system generates trajectories. Separatrices are special lines in phase space, which cannot be crossed by trajectories of deterministic systems. waves (Falcke 2000a). It was found that stable and unstable branches of wave train solutions in the period-velocity-phase plane collide creating two saddle node bifurcations. These bifurcations border a gap in the dispersion relation. If spirals would have a frequency in the gap, they cannot exist. Experimentally as well as in the model, it was found that the cytosol might stay for a rather long time in a high activity stationary state and then leave it by a front of low Ca^{2+} release or because of fluctuations (Falcke 1999, Fig.3 E, F). Bistability was assumed as well in a model for the fertilization wave in Xenopus oocyte (Wagner 1998). There, it is suggested by the large width of the excited region observed in experiments.

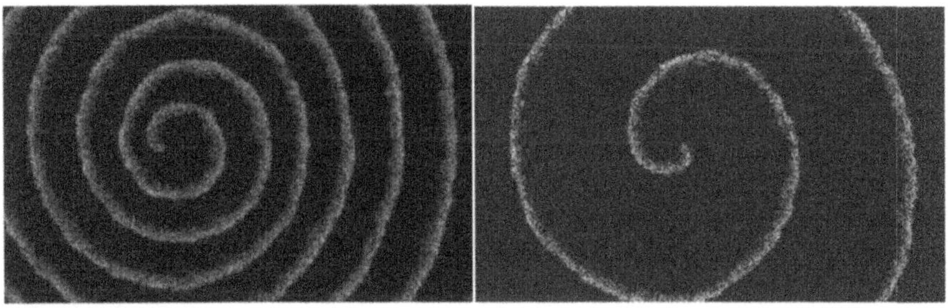

Fig. 14. Simulations with different N_K^{max} but the same deterministic limit. The single channel radius was chosen proportional to $(N_K^{max})^{-1/2}$ as explained in the text. Left panel N_K^{max} =60, right panel N_K^{max} =27. The system size is 1.57x1.31 mm^2. See Table 1 for further parameters

In the simulations presented up to now, we have found a strong dependence of nucleation rates and average periods on the number of channels per cluster. Such dependence exists for other characteristics of patterns too. We have to consider larger length scales in order to note that. Patterns on these large spatial scale are shown in Fig.14, 15 and 16. The spiral in the left panel of Fig.14 was simulated with a larger number of channels per cluster than the one on the right hand side, but again both systems have the same mean field equations. The difference of the wave lengths of the spirals is obvious. The spiral on the left hand side rotates with a period of about 45s and the one on the right hand side with approximately 75s.

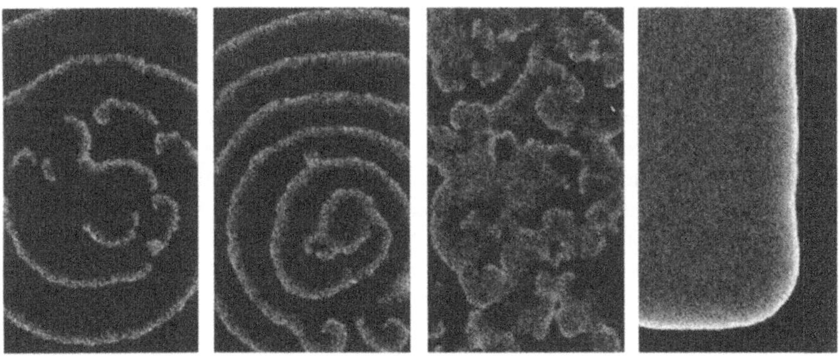

Fig. 15. Simulations with different N_K^{max} but the same deterministic limit. The single channel radius was chosen proportional to $(N_K^{max})^{-1/2}$ as explained in the text. N_K^{max} from left to right 23, 35, 47, 71. For all panels I=0.175μM, E=3.375μM, P=3614s^{-1}, P_p=40μMs^{-1}. The system size is 1.31x1.57 mm^2. See Table 1 for further parameters

The transition to the deterministic limit is demonstrated in Fig.15 on the basis of spatio-temporal patterns. Pulses forming more or less stable spirals are observed at small numbers of channels per clusters. At large numbers, front waves occur. They are moving interfaces switching the medium from a low activity state into a high activity state. They are the trademark pattern for bistable local dynamics. The bistability might disappear in a saddle node bifurcation at very low IP_3 concentrations.

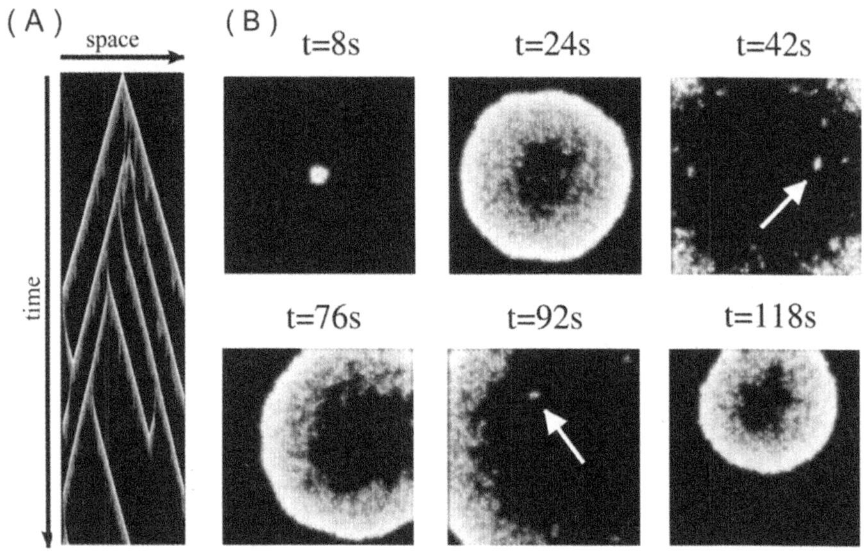

Fig. 16. Stochastic backfiring. Panel (A) shows a typical pattern arising from stochastic backfiring in one spatial dimension. Pulses are emitted by existing pulses in the backward direction and do not emerge spontaneously independently of existing pulses. Panel (B) shows an example for two spatial dimensions. The white arrows point at those active patches, which will become the nucleus for the next circular wave (see Falcke 2000b for details) in the refractory back of a pulse; it is possible that channels were originally activated by the first pulse and have not as yet become inhibited or alternatively they may have already undergone the complete activation-inhibition-deinhibition cycle and hence can be re-opened. We always found the latter scenario to be the case for the sample clusters, which we looked at in more detail during simulations. Note that the Ca^{2+} level in the refractory area is less than in the excited region but higher than the base level. This provides an increased opening-probability as compared to the medium in front of the pulse. If a group of re-opened channels stays open long enough for sufficient decay of the inhibition around it, it can set off another wave. (Falcke 2000b)

Another characteristic pattern – stochastic backfiring – can be observed at intermediate channel numbers for parameters where the system would be in the bistable regime in the deterministic limit (see Fig.16). The emergence of new pulses preferentially from the back of existing pulses is the typical phenomenon of backfiring. Active channels in the refractory tail of the original pulse initiate the emitted pulses. There are in principle two scenarios which can lead to active channels in that area. It is possible that they were originally activated by the first pulse and have not as yet become inhibited or alternatively they may have already undergone the complete activation-inhibition-deinhibition cycle and hence can be re-opened. We always found the latter scenario to be the case for the sample clusters, which we looked at in more detail during simulations. Note that the Ca^{2+} level in the refractory area is less than in the excited region but higher than the base level. This provides an increased opening-probability as compared to the medium in front of the pulse. If a group of re-opened channels stays open long enough for sufficient decay of the inhibition around it, it can set off another wave.

13.4. Ca^{2+} Dynamics Without Ca^{2+}

We have seen in previous paragraphs, that the dynamics of the Ca^{2+} concentration at the channel location is much faster than all other processes. Furthermore, wave nucleation and propagation as well as global oscillations could be described with an adiabatic elimination of the concentration dynamics, too. The final consequence of the adiabatic elimination is - expressed in mathematical terms - that we obtained an algebraic relation between the configuration of open channels and the Ca^{2+} concentration. The differential equation for the Ca^{2+} concentration and buffers dropped out of the system of differential equations describing the dynamics. The only dynamical variables left are the channel states. The algebraic relations between the Ca^{2+} concentration and the channel state enter the differential equations for the channel state via the rate expressions and the form of coupling to neighboring clusters. In the following, we generalize that concept by assuming that the Ca^{2+} concentration of a single cluster profile is proportional to the number of open channels in the cluster. Since the binding rates for Ca^{2+} are proportional to its concentration, we end up with a model in which binding steps are proportional to the number of open channels at a cluster and its neighbors. These assumptions are of course simplifications and approximately valid only. However, they preserve the essential features and lead to a model, which allows for investigation of phenomena independent of experimental details. Furthermore, the model provides a framework and concepts to formulate the findings with more complicated models.

The model is as follows (Bär, 2000). We introduce two stochastic variables for each channel cluster: n_i, the number of activated subunits, and m_i, the number of inhibited subunits. At every time step, the number of activated subunits n_i at site i is changed due to three stochastic processes; activation of additional subunits by binding available Ca^{2+} to their activation domains, de-activation by unbinding Ca^{2+} from active subunits, and inhibition by binding Ca^{2+} to their inhibition domains. Transitions involving binding (but not those involving unbinding) depend on local Ca^{2+}; hence we take these transition rates to depend on the number of open channels at site i, c_i, and on the number of open channels at the nearest neighboring sites $i \pm 1$, $c_{i\pm 1}$. Similarly, there will be binding and unbinding to the inhibitory domain, changing m_i. We neglect direct transitions from the inhibited state back to an activated state. That corresponds to assuming that unbinding of Ca^{2+} from the activating site is very fast once the subunit is inhibited. We will now consider the probabilities for events in a small time interval Δt. That interval has to be chosen small enough to guarantee that all probabilities for elemental events are much smaller than $1/N$, with N being the number of elements participating in the process. We denote by $p^{\pm}_{0(1)}$ the probability to activate (+) / inhibit (-) a subunit per number of open channels at the same site (0) or the neighboring site (1). To compute the actual probabilities, we need to multiply these by the number of open channels. Here, we use the simple expedient of taking this to equal n_i^h/hN_s^{h-1} where the total number of subunits is $N_s=hN$; this is the expected number of open channels for large enough N. Here, h is the number of subunits per channel. This approach allows us to avoid keeping explicit account of each of the independent subunits. Also, we let $p^{\pm}_d$ be the deactivation and deinhibition probabilities which are c independent. Let us define the total probabilities $p^{\pm}=p^{\pm}_0+2p^{\pm}_1$ and the "diffusion constant" $\alpha=p^{\pm}_1/(p^{\pm}_0+2p^{\pm}_1)$. The variable $C_i(t)=(1-2\alpha)c_i(t)+\alpha c_{i-1}(t)+\alpha c_{i+1}(t)$ mimics the amount of calcium at site i due to open channels at sites i, $i \pm 1$.

Our model explicitly consists of the following coupled stochastic processes. n_i is updated according to

$$n_i(t+\Delta t) = n_i(t) + \Delta_n^+ - \Delta_n^+ - \delta_n^+$$

where Δ^+_n is a random integer number drawn from the binomial distribution $B(\Delta^+_n, N_s\text{-}n_i(t)\text{-}m_i(t), p^+C_i(t))$, Δ^-_n is drawn from $B(\Delta_n^-, n_i(t), p^-C_i(t))$, and δ^+_n is drawn from $B(\delta^+_n, n_i(t), p_d^+)$. The equation for m_i reads

$$m_i(t+\Delta t) = m_i(t) + \Delta_n^+ - \delta_m^+$$

where Δ^+_m is drawn from $B(\Delta^+_m, N_s\text{-}m_i(t), p^-C_i(t))$, and δ^+_m is drawn from $B(\delta^+_m, m_i(t), p_d^-)$. In all these formulas,

$$B(x,y,p) = \begin{pmatrix} x \\ y \end{pmatrix} p^x (1-p)^{y-x}$$

is the binomial distribution. IP_3 binding is included by rescaling the number of subunits with the probability that IP_3 is bound

As a first step, we consider a simplified version of the channel dynamics with the inhibition process excluded (all $p^-=0$), i.e. a subunit is activated whenever Ca^{2+} is attached to its activating site. Thus we take $m_i=0$, and arrive at the one-variable model for the number of activated subunits n_i. Let us first focus on fairly small N_s. Examples of the stochastic dynamics for several values of parameters are shown in Fig.17. At small α, an initial seed almost always ultimately dies giving rise to so-called abortive calcium waves. At larger values of α the region of activated channels typically expands at a finite rate. This transition mirrors what has been seen in many experimental systems (Callamaras 1998).

As is well known for statistical models such as the contact process, the critical value of α can be accurately determined by computing the distribution of survival times $\Pi(t)$ for the activation process started from a single active site. For $\alpha<\alpha_c$, the distribution falls exponentially at large t as the wave of activation eventually dies out. On the contrary, at $\alpha>\alpha_c$, $\Pi(t)$ asymptotically reaches a constant value, since a non-zero fraction of runs produce ever-expanding active regions. The data (not shown) indicate that α_c is inversely proportional to the number of subunits per site N_s. At $\alpha=\alpha_c$, the distribution function exhibits a power-law asymptotic behavior with the slope determined by the universality class of directed percolation (DP) (Broadbent 1957). For example, $\Pi(t)$ for a group of open clusters at the critical value of α_c for h=3, N_s=10 and γ=0.1 (see below for a definition of γ) shows a power law dependence like $t^{-0.159}$. That is consistent with directed percolation. This is perhaps not too surprising. According to the Janssen-Grassberger DP conjecture (Janssen 1981, Grassberger 1982), any spatio-temporal stochastic process with short range interactions, fluctuating active phase and unique non-fluctuating (absorbing) state, single order parameter and no additional symmetries, should belong to the DP class.

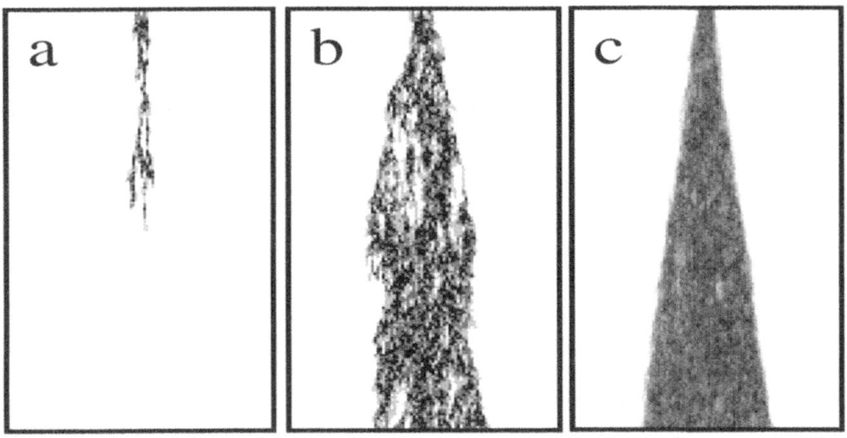

Fig. 17. Evolution of an initial seed of 5 clusters of open channels in the middle of a lattice. Simulation with the simplified model without inhibition for p^+N_s/h=1, p^+_d=0.2, h=3, and N_s=3, α=0.1 (a), N_s=3, α=0.25 (b), and N_s=3, α=0.25 (c). The horizontal axis is spatial coordinate (100 sites), and the vertical axis is time (1000 iterations)

Fig.17, c shows the opposite limit where the dynamics becomes almost deterministic. If we let N_s go to infinity and fix $pN_s/h \rightarrow P$, we can use a mean-field description in terms of the fraction of activated subunits $\rho_i=n_i/N_s$,

$$\dot{\rho}_i = \left[(1-2\alpha)\rho_i^h + \alpha\rho_{i+1}^h + \alpha\rho_{i-1}^h\right](1-\rho_i) - \gamma\rho_i . \qquad (6)$$

We rescaled time $t'=Pt/\Delta t$ and introduced $\gamma=p_d/P$. For all $h\geq2$ and if $\gamma<\gamma_{cr}$ holds, (6) possesses two stable uniform solutions, $\rho=0$ and $\rho=\rho_0$ and one unstable solution ρ_u, where $\rho_{0,\,u}$ are real roots of the algebraic equation $\rho^{h-1}(1-\rho)=\gamma$. The front is a solution connecting these two stable fixed points. This front has a unique propagation velocity.

For small α, the discreteness of our spatial lattice causes the front to become pinned, as the probability of activating subunits at the neighboring site $O(\alpha\rho_0{}^h)$ becomes smaller than the threshold value for excitation probability $O(\rho_u)$. The stationary front solution is described by the recurrence relation

$$(1-2\alpha)\rho_i^h + \alpha\rho_{i+1}^h + \alpha\rho_{i-1}^h = \gamma\frac{\rho_i}{1-\rho_i}.$$

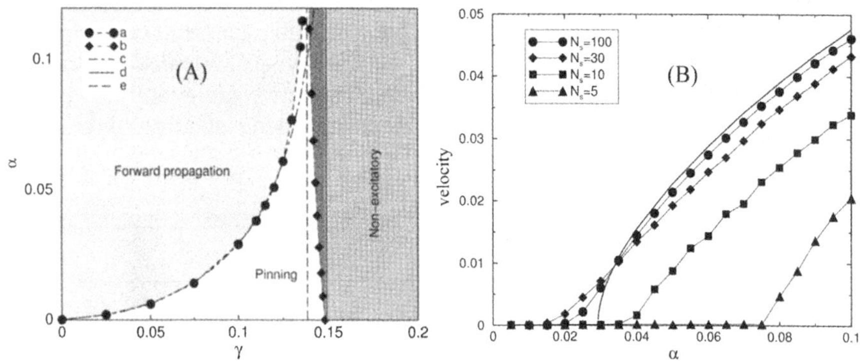

Fig. 18. (A) Phase diagram of the mean field equation of the simplified model (6) without inhibition for h=3: **a** bifurcation line separating forward-propagating fronts from pinning region; **b** same for backward propagating fronts; **c** small-α approximation of pinning line; **d** line $\gamma=4/27$ separating the region of non-existence of the excited state; **e** Maxwell line $\gamma=0.138$...separating forward and backward front propagation in the continuum limit. (B) The average front speed as a function of α for the simplified stochastic model without inhibition at $h=3$, $\gamma=0.1$, $p^+N_s/h=1$, $p^+_d=0.1$, and different values of N_s. Solid line indicates the mean-field limit $N_s\rightarrow\infty$ (Bär 2000)

The bifurcation line which separates pinned and moving fronts, can be found in the limit of small α by using the ideas of Mitkov et al. (Mitkov 1999) that only a few points forming the front are different from 0 or ρ_0. Indeed, in this limit, the values of ρ_i quickly (as α^i) approach 0 and ρ_0 away from the front at $i\rightarrow\pm\infty$ respectively. We can thus replace ρ_i by ρ_0 and 0 everywhere to the left and to the right of the front position except for $\rho_\mp$ at the two sites nearest to the front, i-1 and i+1. Solving the resulting set of two algebraic equations up to α^2, one can obtain the values of $\rho_\mp$. At any γ, there is a critical value of α_m at which the real solution $\rho_\mp$ vanishes. The family of these values α_m forms the bifurcation line for front pinning in (γ, α) plane.

At large α, discreteness of the mean field model becomes insignificant, and the discrete dynamics can be replaced by its continuum limit

$$\partial_t\rho = (1-\rho)(\rho^h - \alpha\partial_x^2\rho) - \gamma\rho,$$

which of course has no front pinning. Instead, α can be scaled out and there is a specific value of γ at which the system goes from forward to backward propagating fronts. Figure18 A shows the phase diagram of the mean field (6) for $h=3$. All the data (except possibly at the non-generic case $\gamma=0$) are consistent with expected $(\alpha-\alpha_m)^{1/2}$ scaling (Mitkov 1999).

How does one get from DP behavior to deterministic pinning/depinning? To investigate this issue, we have performed simulations for the front speed as a function of α at various finite values of N_s; with the results given in Fig.18, B. At large N_s, the velocity approaches the mean field prediction as long as $\alpha>\alpha_m$. Close to the critical value α_m, the velocity deviates from the mean-field dependence proportional to $(\alpha-\alpha_m)^{1/2}$ because of thermally activated "creep"; fluctuations let the front to overcome potential barriers associated with finite site separation, and lead to exponentially slow front propagation (see, e.g., Aranson 2000). Directed percolation regime is not observed at large N_s since the DP critical value α_c is less than α_m. At smaller N_s, the relative magnitude of the fluctuations grows, and the DP threshold value α_c exceeds α_m. Now, the front propagation is determined by fluctuations rather than discreteness, and the critical state exhibits the properties of directed percolation.

13.5. Concluding Remarks

The stochastic modeling of intracellular Ca^{2+} is a rather new branch of theoretical work. The results of research till now reported in this chapter here and the chapter "Langevin Modeling of Intracellular Calcium Dynamics" suggest, that intracellular Ca^{2+} may be a medium in which spatial and temporal structures are completely caused by random fluctuations. That is of course an exciting possibility, but poses questions too. Oscillations and waves, behaving very much like deterministic systems do, are observed experimentally. We have seen, that they can be described by a stochastic simulation with the deterministic limit of a bistable local dynamics. However, is there a deterministic description for these oscillations ? How far can the concept of dynamic regimes be applied ?

Beside these fundamental considerations reaching beyond the specific stochastic approach, more specific questions remain too. How can the nucleation rate be calculated ? Or, can a criterion for the occurrence of wave nucleation be found ? What is the role of pacemaker Ca^{2+} ? With respect to spatio-temporal patterns, wave characteristics – well investigated for deterministic models - need to be determined anew, taking channel numbers into account.

I think, that intracellular Ca^{2+} dynamics will provide much insight into the Physics of random media and spatio-temporal pattern formation.

Table 1. Parameters for simulations. The subunit kinetic is the DeYoung-Keizer-Model as described in (Falcke 2000b). The first figure in column "Value" refers to Figs. 2, 3, 5-11, the second to Fig.12, and the third to Figs.14, 15 The concentration E of free Ca^{2+} in the ER was kept constant for the simulations of Figs. 14, 15. Furthermore, the fast buffer approximation was used in these simulations so that on- and off-rates do not appear in the equations.

Parameter	Value	Unit
leak flux coefficient P_l	2.5 10^{-3}; 2 10^{-4}; 10^{-4}	s^{-1}
channel flux coefficient P	600; 650; 3555	s^{-1}
single channel radius R_s	0.0354; 0.092; 0.01	μm
cluster spacing outside focal sites	4.33; 3.968; 3.5	μm
max. number of channels per cluster N_K^{max}	25; 15; 27	
number of additional clusters in focal sites N_F	5; 25; 0	
pump flux coefficient P_p	40; 45; 80	μMs^{-1}
pump dissociation coefficient K_d	0.2; 0.2; 0.1	μM
volume ratio V_{cyt}/V_{ER}	10.0; 5.4; 5.4	
total Ca^{2+} concentration C_0	5.385; 5.385; -	mM
concentration of free Ca^{2+} in the ER E	-; -; 6.751	μM
diffusion coefficient D of free cytosolic Ca^{2+}	223; 223; 223	$\mu m^2 s^{-1}$
diffusion coefficient D of free luminal Ca^{2+}	40; 223; -	$\mu m^2 s^{-1}$
diffusion coefficient D_m of endogenous mobile buffer	11.26; 11.26; 11.26	$\mu m^2 s^{-1}$
diffusion coefficient D_{ex} of exogenous buffer	32.0; 32.0; 32.0	$\mu m^2 s^{-1}$
diffusion coefficient D_{Em} of luminal mobile buffer	1.0; 10.0; -	$\mu m^2 s^{-1}$
on-rates of fast buffers: k_s^+, k_{ex}^+, k_m^+, k_{Es}^+, k_{Em}^+	500; 500; -	$(\mu Ms)^{-1}$
dissociation constants of buffers k_i^-/k_i^+:		
K_s	2; 2; 9	μM
K_m	2; 2; 6.72	μM
K_{ex}	0.247; 0.247; 0.247	μM
K_{Es}	500; 500; -	μM
K_{Em}	5; 5; -	μM
K_{slow}	0; 0.15; -	μM
IP_3 concentration	0.15; 0.36; 0.1	μM
Total concentrations of buffers:		
B_s	200; 200; 95.31	μM
B_m	50; 50; 19.1	μM
B_{ex}	60; 60;40	μM
B_{Es}	100; 100; -	mM
B_{Em}	5; 5; -	mM
B_{slow}	0; 40; -	μM

Subunit kinetics, $b_i=a_i d_i$:		
a_2, a_4	0.0555; 0.2; 0.2	$(\mu Ms)^{-1}$
a_5	2.222; 2.222; 20	$(\mu Ms)^{-1}$
d_1	0.13; 0.13; 0.13	μM
d_2	3.776; 1.049; 1.049	μM
d_3	0.943; 0.943; 0.943	μM
d_4	0.520; 0.144; 0.144	μM
d_5	0.72; 0.72; 0.0823	μM
$a_6=(a_2I+a_4d_1)/(d_1+I)$		
$b_6=(b_2I+b_4d_3)/(d_3+I)$		

References

Aranson, I., Malomed, B.A., Pismen, L.M. and Tsimring, L.S. (2000). Crystallization kinetics and self-induced pinning in cellular patterns, Phys.Rev.E 62(1): R5-R8

Bär, M., Falcke, M. Levine, H. Tsimring, L.S. (2000). Discrete stochastic modeling of calcium channel dynamics, Ohys.Rev.Lett. 84: 5664-5667

Berridge, M., Bootman, M. and Lipp, P. (1998). Calcium - a life and death signal, Nature 395: 645-648.

Bezprozvanny, I., Watras, J. and Ehrlich, B. (1991). Bell-shaped calcium-response curves of Ins(1,4,5)P_3- and calcium-gated channels from endoplasmatic reticulum of cerebellum, Nature 351: 751-754.

Bezprozvanny, I. and Ehrlich, B. (1994). Inositol(1,4,5)-trisphosphate Ins(1,4,5)P_3-gated channels from cerebellum: conduction properties for divalent cations and regulation by intraluminal calcium, J.Gen.Physiol. 104: 821-856.

Bootman, M., Berridge, M. and Lipp, P. (1997a). Cooking with calcium: The recipes for composing global signals from elementary events, Cell 91: 367-373.

Bootman, M., Niggli, E., Berridge, M. and Lipp, P. (1997b). Imaging the hierarchical Ca^{2+} signalling in HeLa cells, J.Physiol 499 (2): 307-314.

Broadbent, S.R. and Hammersley, J. M. (1957). Proc. Camb. Phil. Soc. 53: 629

Callamaras, N., Marchant, J., Sun, X.-P. and Parker, I. (1998). Activation and co-ordination of $InsP_3$-mediated elementary Ca^{2+} events during global Ca^{2+} signals in Xenopus oocytes, Journal of Physiology 509 (1): 81-91.

Callamaras, N. and Parker, I. (2000). Phasic characteristics of elementary Ca^{2+} release sites underlies quantal responses to IP_3, The EMBO J. 19 (14): 3608-3617.

Callamaras, N., Marchant, J., Sun, X.-P. and Parker, I. (1998). Activation and co-ordination of $insp_3$-mediated elementary Ca^{2+} events during global Ca^{2+} signals in Xenopus oocytes, Journal of Physiology 509(1): 81-91.

DeYoung, G. and Keizer, J. (1992). A single-pool inositol 1,4,5-trisphosphate-receptor-based model for agonist-stimulated oscillations in Ca^{2+} concentration, Proc.Natl.Acad.Sci USA 89: 9895-9899.

Dupont, G. and Swillens, S. (1995). Quantal release, incremental detection, and long-period Ca^{2+} oscillations in a model based on regulatory Ca^{2+}-binding sites along the permeation pathway, Biophys.J. 71: 1714-1722.

Falcke, M., Hudson, J., Camacho, P. and Lechleiter, J. (1999). Impact of mitochondrial Ca^{2+} cycling on pattern formation and stability, Biophys.J. 77: 37-44.

Falcke, M., Or-Guil, M. and Bär, M. (2000a). Dispersion gap and localized spiral waves in a model for intracellular Ca^{2+} dynamics, Phys.Rev.Lett. 84 (20): 4753-4756.

Falcke, M., Tsimring, L. and Levine, H. (2000b). Stochastic spreading of intracellular Ca^{2+} release, Phys.Rev.E 62: 2636-2643.

Falcke, M. (2003a). On the role of stochastic channel behavior in intracellular Ca^{2+} dynamics, Biophys.J. 84(1): 42-56

Falcke, M. (2003b). Buffers and oscillations in intracellular Ca^{2+} dynamics, Biophys.J. 84(1): 28-41

García-Ojalvo, J. and Schimansky-Geier, L. (2000). Excitable structures in stochastic bistable media, J.Stat.Phys. 101 (1/2): 473-481.

Grassberger, P. (1982), On Phase Transitions in Schlögl's Second Model, Z.Phys. B 47: 365

Hempel, H., Schimansky-Geier, L. and García-Ojalvo, J. (1999). Noise-sustained pulsating patterns and global oscillations in subexcitable media, Phys.Rev.Lett. 82: 3713-3716.

Heemskerk, J., Willems, G., Rook, M. and Sage, S. (2001). Ragged spiking of free calcium in ADP -stimulated human platelets: regulation of puff-like calcium signals in vitro and ex vivo, J.Physiol. 535 (3): 625-635.

Horne, J. and Meyer, T. (1997). Elementary calcium-release units induced by inositol trisphosphate, Science 276: 1690-1693.

Ilyin, V. and Parker, I. (1994). Role of cytosolic Ca^{2+} in inhibition of $InsP_3$-evoked Ca^{2+} release in Xenopus oocytes, J.Physiol. 477 (3): 503-509.

Jafri, M. S. and J. Keizer (1995). On the roles of Ca^{2+} diffusion, Ca^{2+} buffers and the endoplasmatic reticulum in IP_3-induced Ca^{2+} waves, Biophys.J. 69: 2139-2153

Janssen, H. K. (1981) On the Nonequilibrium Phase Transitions in Reaction-Diffusion Systems with an absorbing state, Z. Phys. B 42: 151

Jung, P. and Galley, P. (2000). The heartbeat of extended clocks, Ann.Phys. 9 (9-10): 697-704.

Keener, J.P. and Sneyd, J. (1998). Mathematical Physiology, Springer Publ., New York

Keizer, J. and DeYoung, G. (1994). Simplification of a realistic model of IP_3-induced Ca^{2+} oscillations, J.theor.Biol. 166: 431-442.

Lechleiter, J., S.Girard, E.Peralta and D.Clapham (1991). Spiral calcium wave propagation and annihilation in Xenopus laevis oocytes, 252: 123-129.

Lipp, P., Thomas, D., Berridge, M. and Bootman, M. (1997). Nuclear calcium signalling by individual cytopasmic calcium puffs, The EMBO Journal 16 (23): 7166-7173.

Lukyanenko, V. and Györke, S. (1999). Ca^{2+} sparks and Ca^{2+} waves in saponin-permeabilized rat ventricular myocytes, J.Physiol. 521 (3): 575-585.

Mak, D. and Foskett, J. (1997). Single-channel kinetics, inactivation, and spatial distribution of inositol trisphophate (IP_3) receptor in Xenopus oocyte nucleus, J.Gen.Physiol. 109: 571-587.

Mak, D. and Foskett, J. (1998). Effects of divalent cations on single-channel conduction properties of Xenopus IP_3 receptor, Am.J.Physiol. 275 (Cell.Physiol.44): C179-C188.

Mak, D., McBride, S. and Foskett, J. (2001). Regulation by Ca^{2+} and Inositol 1,4,5-trisphosphate ($InsP_3$ of single recombinant type 3 $InsP_3$ receptor channels: Activation uniquely distinguishes type 1 and type 3 $InsP_3$ receptors, J.Gen.Physiol. 117: 435-446.

Marchant, J., Callamaras, N. and Parker, I. (1999). Initiation of IP_3-mediated Ca^{2+} waves in Xenopus oocytes, The EMBO J. 18 (19): 5285-5299.

Marchant, J. and Parker, I. (2001). Role of elementary Ca^{2+} puffs in generating repetitive Ca^{2+} oscillations, The EMBO Journal 20 (1 & 2): 65-76.

Mitkov, I. Kladko, K. and Pearson, J.E. (1999). Tunable pinning of burst waves in extended systems with discrete sources, Phys.Rev.Lett. 81: 5453-5456

Moraru, I., Kaftan, E., Ehrlich, B. and Watras, J. (1999). Regulation of type 1 Inositol 1,4,5-trisphosphate-gated calcium channels by InsP $_3$ and calcium, J.Gen.Physiol. 113: 837-849.

Parker, I. and Ivorra, I. (1992). Characteristics of membrane currents evoked by photorelease of inositol trisphosphate in Xenopus oocytes, Am.J.Physiol. 263: C154-C165.

Parker, I. and Yao, Y. (1996). Ca^{2+} transients associated with openings of inositol trisphosphate gated channels in Xenopus oocytes, J.Physiol.(Lond.) 491: 663-668.

Parker, I., Choi, J. and Yao, Y. (1996a). Elementary events of $InsP_3$-induced Ca^{2+} liberation in Xenopus oocytes: hot spots, puffs and blips, Cell Calcium 20 (2): 105-121.

Patel, S., Joseph, S. and Thomas, A. (1999). Molecular properties of inositol 1,4,5-trisphosphate receptors, Cell Calcium 25 (3): 247-264.

Reber, B. and Schindelholz, B. (1996). Detection of a trigger zone of bradykinin-induced fast calcium waves in PC 12 neurites, Pflügers Arch.-Eur.J.Physiol. 432: 893-903.

Sneyd, J. and Dufour, J.-F. (2002). A dynamic model of the type-2 inositol trisphosphate receptor, Proc.Nat.Ac.Sci. USA 99 (4): 2398-2403.

Sun, X.-P., Callamaras, N., Marchant, J. and Parker, I. (1998). A continuum of $insp_3$-mediated elementary Ca^{2+} signalling events in Xenopus oocytes, Journal of Physiology 509 (1): 67-80.

Swillens, S., Champeil, P., Combettes, L. and Dupont, G. (1998). Stochastic simulation of a single inositol 1,4,5-trisphoasphate-sensitive Ca^{2+} channel reveals repetitive openings during blip-like Ca^{2+} transients, Cell Calcium 23 (5): 291-302.

Swillens, S., Dupont, G. and Champeil, P. (1999). From calcium blips to calcium puffs: Theoretical analysis of the requirements for interchannel communication, PNAS 96 (24): 13750-13755.

Tang, Y. and Othmer, H. (1996). Simplification and analysis of models of calcium dynamics based on IP_3-sensitive calcium channel kinetics, Biophys.J. 70: 246-263.

Taylor, C. (1998). Inositol trisphosphate receptors: Ca^{2+}-modulatd intracellular Ca^{2+} channels, Biochimica and Biophysica Acta 1436: 19-33.

Thomas, D., Lipp, P., Berridge, M. and Bootman, M. (1998). Hormone-stimulated elementary Ca^{2+} signals are not stereotypic, but reflect activation of different size channle clusters and variable recruitment of channels within a cluster, J.Biol.Chem. 273 (42): 27130-27136.

Thomas, D., Lipp, P., Tovey, S., Berridge, M., Li, W., Tsien, R. and Bootman, M. (1999). Microscopic properties of elementary Ca^{2+} release sites in non-excitable cells, Current Biology 10: 8-15.

Tovey, S., de Smet, P., Lipp, P., Young, K. W., Missiaen, L., Smedt, H. D., Parys, J., Berridge, M., Thuring, J., Holmes, A. and Bootman, M. (2001). Calcium puffs are generic $InsP_3$-activated elementary calcium signals and are downregulated by prolonged hormonal stimulation to inhibit cellular calcium responses, J.Cell Science 114 (22): 3979-3989.

Wagner, J., Li, Y.-X., Pearson, J. and Keizer, J. (1998). Simulation of the fertilization Ca^{2+} wave in Xenopus laevis eggs, Biophys.J. 75: 2088-2097.

Yao, Y. and Parker, I. (1995). Quantal puffs of intracellular Ca^{2+} evoked by inositol trisphosphate in Xenopus oocytes, J.Physiol.(Cambridge) 482 (3): 533-553.

Glossary

acetylcholine neurotransmitter that functions at cholinergic chemical synapses, found both in the brain and in the peripheral nervous system. It is the neurotransmitter at vertebrate neuromuscular junctions.

ADP A ribonucleoside 5'-diphosphate serving as phosphate group acceptor in the cell energy cycle.

allosterism is a property of an enzyme or any protein, of which the reactivity is modified by binding of an effector at a site other, (allo), than the site where the substrate or the reactant binds. This leads to a conformational change, altering positively or negatively the biologically-relevant activity of the protein.

ATP A ribonucleoside 5'-triphosphate functioning as a phosphate group donor in the cell energy cycle; carries chemical energy between metabolic pathways by serving as a shared intermediate coupling endergonic and exergonic reactions.

buffer usually any substance which binds protons, used here also for binding Ca^{2+}.

BAPTA Ca^{2+} chelator, buffers Ca^{2+}.

Calcineurin is the serine/threonine protein phosphatase 2B. It binds calmodulin and its activity is regulated by Ca^{2+}.

calmodulin ubiquitous Ca^{2+} binding protein whose binding to other proteins is governed by changes in intracellular Ca^{2+} concentration. Its binding modifies the activity of many target enzymes and membrane transport proteins.

CaM is an abbrevaiation of calmodulin

cyclic ADP ribose Cyclic ADP-ribose is a cyclic nucleotide, that (unlike cAMP) directly modulates calcium-induced calcium release, a major mechanism of calcium mobilization.

cholecystokinin A peptide hormone secreted by cells in the small intestine that stimulates contraction of the gallbladder (release of bile) and secretion of digestive enzymes from the pancreas. In addition, cholecystokinin is produced by neurons in the enteric nervous system, and is widely and abundantly distributed in the brain.

cytoplasm The portion of a cell's contents outside the nucleus but within the plasma membrane; includes organelles

cytosol The continuous aqueous phase of the cytoplasm, with its dissolved solutes; excludes the organelles

dehydrogenase Enzyme that catalyses the removal of pairs of hydrogen atoms from its substrate.

diacyl-glycerol (DAG) lipid produced for instance by the cleavage of inositol phospholipids in response to extracellular signals. Composed of two fatty acid chains linked to glycerol, it serves as signaling molecule to help to activate protein kinase C.

dissociation constant An equilibrium constant (K_d) for the dissociation of a complex of two or more biomolecules into its components; e.g. dissociation of a substrate from an enzyme.

differential equation An algebraic relation between a function and its derivatives or derivatives of a function. Partial differential equations involve partial derivatives (→). Most laws of Physics are differential equations, e.g. Fick's law of diffusion links the first time derivative describing the change of a concentration *u* to the spatial derivative of its flux density. The flux density is itself the spatial derivative of the concentration: $\partial u/\partial t = \partial/\partial r \; D \; \partial u/\partial r$. D is the diffusion coefficient. The solution of differential equations are the functions in dependence on the independent variables (e.g. $u(r,t)$).

derivative, derived function The derived function of a function $f(t)$ is defined as $(f(t+h)-f(t))/h$ in the limit as h tends to zero. If the independent variable t is identified with time, the derived function is the rate of change of f.

dye any coloured or fluorescent indicator, for instance to indicate membrane potential or H^+ concentration, in our context to measure Ca^{2+}.

Fura2 used as dye to measure Ca^{2+}

GABA abbrev. for Gamma-Amino Butyric Acid; inhibitory neurotransmitter

gap junction Communicating cell-cell junction that allows ions and small molecules to pass from the cytoplasm of one cell to the cytoplasm of the next.

glycogen Polysaccharide composed exclusively of glucose units used to store energy in animal cells. Large granules of glycogen are especially abundant in liver and muscle cells.

glycogen phosphorylase → phosphorylase

glycoprotein Any protein with one or more covalently linked oligosaccharide chains. Includes most secreted proteins and most proteins exposed on the outer surface of the plasma memrane.

G-protein GTP-binding protein. There are monomeric and heterotrimeric α, β, γ - G-proteins. Heterotrimeric G-proteins couple to heptahelical receptors. Both forms exist in an active GTP bound state and an inactive GDP bound state. On binding to a ligand-activated receptor, heterotrimeric α, β, γ - G-proteins dissociate into a GTP-bound subunit and a β, γ - complex. Both transmit signals to targets.

EGTA Ca^{2+} chelator, binds Ca^{2+}

endoplasmic reticulum An extensive system of double membranes in the cytoplasm of eukaryotic cells; it encloses secretory channels and is often studded with ribosomes (rough endoplasmic reticulum)

enzyme A protein with catalytic activity, which is restricted to a limited set of reactions, defining the specificity of the enzyme. They are designated by the suffix *–ase,* frequently attached to the type of reaction catalysed. Thus enzymes catalyzing hydrolytic reactions are termed *hydrolases*. Virtually all metabolic reactions are dependent on and controlled by enzymes.

FK506 is the propriety name given to an immunosuppressant by a Japanese company. FK506 is a macrolide antibiotic, isolated from streptomyces tsukubaensis. It is a potent inhibitor of T-cell activation, preventing allograft rejection. FK506 binds to FK-binding protein FKBP

FKBP → FK506

hormone A substance released by an endocrine gland into the bloodstream, which carries it to remote sites in the body.

hepatocyte The major cell type of liver tissue.

Hill coefficient is a measure of cooperativity of ligand binding sites in a protein. The Hill coefficient is estimated from a plot of the Hill equation where log[1/(1-y)] is plotted against log[x], and where y is the fractional saturation of the protein with ligand and [x] gives the concentration of free ligand. The slope is the Hill coefficient. A value greater than 1 gives an estimate of the extent of cooperation among the ligand binding sites on the molecule.

histamine Small molecule derived from the amino acid histidine, released from mast cells and basophils in allergic reactions. Causes irritation, dilation of blood vessels, and contraction of smooth muscle.

hydrolysis The decomposition of organic compounds by interaction with water.

inositol Cyclic molecule with six hydroxyl groups that forms the hydrophilic group of inositol phospholipids.

integrins are members of a large family of transmembrane proteins that act as receptors for cell adhesion molecules.

interleukins belong to a heterogeneous group of cytokines. Up to now, there are about 15 interleukins. They have an essential role in the regulation of proliferation of B and T cells in the immune system.

kinase An enzyme which catalyses the phosphorylation of its substrate by ATP. Thus protein kinases phosphorylate proteins and hexose kinases phosphorylate hexoses.

ligand → receptor

M (molar): Unit of solute concentration corresponding to 1 mole solute per liter of solution.

microsomes membranous vesicles formed by fragmentation of the endoplasmic reticulum of eukaryotic cells; recovered by differential centrifugation.

mutagenesis is achieved by altering the composition of proteins

Na^+/Ca^{2+} and H^+/Ca^{2+} exchangers located in membranes, exchange the ions as indicated according to the concentration gradients on both sides of the membrane.

NAD Nicotinamide Adenine Dinucleotide

NADP Nicotinamide Adenine Dinucleotide Phosphate

partial derivative The derivative of a function f depending on several independent variables (e.g. time t and spatial coordinate r) with respect to one of the independent variables (e.g. partial derivative with respect to time t $\partial/\partial t$, with respect to space $\partial/\partial r$).

peptide A sequence of amino acids held together by peptide bonds. With rare exceptions peptides are unbranched chains joined by peptide bonds between their α-amino and α-carboxyl groups. Peptide can vary in length from 2 (dipeptide) to several hundred (polypeptide) amino acids.

phosphatase An enzyme which dephosphorylates its substrate by hydrolysis of the phosphate ester bond.

phospholipase An enzyme which degrades a phospholipid. There are three main types, phospholipase A, C and D; A removes a hydrocarbon chain from the phospholipid, C cleaves the glycerol/phosphate ester bond and D the phosphate/base linkage.

phospholipase C (PLC) → phospholipase

phospholipid The major category of lipid molecules used to construct biological membranes. Generally composed of two fatty acids linked through glycerol phosphate to one of a variety of polar groups.

phosphorylase An enzyme which catalyses the cleavage of a bond by addition of orthophosphate. Thus glycogen phosphorylase liberates the terminal glucose of glycogen as glucose-1-phosphate.

phosphorylase kinase phosphorylates and activates glycogen phosphorylase The activity of the kinase is regulated by phosphorylation by the cAMP-dependent protein kinase or by dephosphorylation of the serine/threonine protein phosphatase 1 and by Ca^{2+}. Ca^{2+}-regulation is mediated by the δ-subunit of the kinase, which is identical with calmodulin.

phosphorylation Reaction in which a phosphate group becomes covalently coupled to another molecule.

PMCAs (plasma membrane Ca ATPases): Ca^{2+} pumps located in the membrane of the sarcoplasmic and endoplasmic reticulum whose function is to accumulate Ca^{2+} into these organelles, using ATP as the energy source.

polar Hydrophilic molecules or groups that are soluble in water.

protein A polymer of amino acids which may consist of one or more polypeptide chains.

protein kinases → kinase

protein kinase C (PKC) → kinase

pyruvate dehydrogenase → dehydrogenase

receptor Protein that binds a specific signaling molecule (ligand) and initiates a response. Cell surface receptors, such as the acetyl choline receptor and the insulin receptor, are located in the plasma membrane, with their ligand binding site exposed to the external. Intracellular receptors bind ligands that diffuse into or inside the cell.

residue General term for the unit of a polymer.

ryanodine receptor→ receptor, binds ryanodine

sarcoplasmic reticulum Network of internal membranes in the cytoplasm of a muscle cell that contains high concentrations of sequestered Ca^{2+} that is released into the cytosol during muscle contraction.

SERCAs (sarco- and endoplasmic reticulum Calcium ATPases): Ca^{2+} pumps located in the membrane of the sarcoplasmic and endoplasmic reticulum whose function is to accumulate Ca^{2+} into these organelles, using ATP as the energy source.

Substrate Reactant in a reaction catalysed by an enzyme.

transcription (DNA transcription) copying of one strand of DNA into a complementary RNA sequence by the enzyme RNA polymerase

transcription factor Term loosely applied to any protein required to initiate or regulate transcription in eucaryotes. Includes both gene regulatory proteins as well as the general transcription factors.

uniport A transport system that carries only one solute, as distinct from cotransport

vasopressin A nonopeptide secreted by the posterior pituary gland. It elevates blood pressure by the contraction of small blood vessels and aids water resorption by the kidney.

Index

GPSR Compliance

The European Union's (EU) General Product Safety Regulation (GPSR) is a set of rules that requires consumer products to be safe and our obligations to ensure this.

If you have any concerns about our products, you can contact us on ProductSafety@springernature.com

In case Publisher is established outside the EU, the EU authorized representative is:

Springer Nature Customer Service Center GmbH
Europaplatz 3
69115 Heidelberg, Germany

Batch number: 10584666

Printed by Printforce, the Netherlands